Internet of Things

Technology, Communications and Computing

Series Editors

Giancarlo Fortino, Rende (CS), Italy
Antonio Liotta, Edinburgh Napier University, School of Computing, Edinburgh, UK

The series Internet of Things - Technologies, Communications and Computing publishes new developments and advances in the various areas of the different facets of the Internet of Things.

The intent is to cover technology (smart devices, wireless sensors, systems), communications (networks and protocols) and computing (theory, middleware and applications) of the Internet of Things, as embedded in the fields of engineering, computer science, life sciences, as well as the methodologies behind them. The series contains monographs, lecture notes and edited volumes in the Internet of Things research and development area, spanning the areas of wireless sensor networks, autonomic networking, network protocol, agent-based computing, artificial intelligence, self organizing systems, multi-sensor data fusion, smart objects, and hybrid intelligent systems.

** Indexing: *Internet of Things* is covered by Scopus and Ei-Compendex **

More information about this series at http://www.springer.com/series/11636

Gagangeet Singh Aujla • Sahil Garg
Kuljeet Kaur • Biplab Sikdar
Editors

Software Defined Internet of Everything

Springer

Editors
Gagangeet Singh Aujla
Department of Computer Science
Durham University
Durham, UK

Sahil Garg
École de Technologie Supérieure
(Université du Québec)
Montreal, QC, Canada

Kuljeet Kaur
Département de génie électrique
École de Technologie Supérieure
(Université du Québec)
Montreal, QC, Canada

Biplab Sikdar
Department of Electrical and Computer Engineering
National University of Singapore
Singapore, Singapore

ISSN 2199-1073 ISSN 2199-1081 (electronic)
Internet of Things
ISBN 978-3-030-89327-9 ISBN 978-3-030-89328-6 (eBook)
https://doi.org/10.1007/978-3-030-89328-6

This Springer imprint is published by the registered company Springer Nature Switzerland AG
The registered company address is: Gewerbestrasse 11, 6330 Cham, Switzerland

Foreword

Research and development efforts in Software Defined Networks (SDNs) have continued to increase in popularity over recent years, providing the ability to improve network management, support fine-grained control of network elements, enable high speed data exchange and provide the basis for network scalability. Initially, SDN was expected to replace the static architecture of "traditional" decentralized networks by the use of a centralized approach, enabling support for more complex network usage and management. This separation and decoupling of control logic and data path/forwarding logic has also been facilitated with a parallel increase in interest in machine learning and AI, with availability of specialist hardware platforms that can be used to execute ML/AI algorithms. An SDN controller remains a major component in such a decoupled network environment, with significant number of implementations now available, ranging from C++ (NOX) to Java (Floodlight, Beacon, OpenDayLight, ONOS) and Python (POX, Ryu). Although the OpenFlow protocol was a key standard and implementation initially, a number of other vendors have adapted the approach within their own products—e.g. Cisco Systems' Open Network Environment, VMWare NSX, Juniper Network's OpenContrail, VortiQa from NSX semiconductors and Nicira's network virtualization platform. Compatibility of these with OpenFlow varies.

The integration of programmable networks within wider smart cities and Industrial Internet of things environments also opens up a number of possibilities—from more active management of assets within such environments to more effective mechanisms to provide protection against cyberattacks (from DDoS to ransomware). Combining SDN with virtual network functions also offers possibilities to support user-supported functions and combining these with services used to manage the network core.

This book brings together a number of chapters that address both core concepts and emerging themes in SDN technologies, systems and applications. It provides a useful reference for those new to the area to get a better insight into this rapidly developing area, but also provides more in-depth material for those who already make use of SDN approaches in their work. The "Internet of Everything" perspective adopted in this book enables a wider consideration of SDN and programmable network technologies, supporting aspects such as backup/recovery policies, load balancing and traffic filtering. It is also useful to see that coverage includes both data centre-based adoption of SDN along with use of these in edge computing.

It is also good to see the editors of this book originating from three different regions in the world: Canada, Singapore and the UK. I am certain that this geographic spread of research context and background will also provide a more holistic perspective on this emerging area.

Professor of Performance Engineering
School of Computer Science and Informatics
Cardiff University, UK
July 2021

Omer F. Rana

Preface

Nowadays, the global landscape is shifting towards a digitized and autonomous cyber-physical world. *Internet of Things* (IoT) came up as a revolution and overtook the entire global landscape with its presence in almost every sector like smart cities, smart grids, and intelligent transportation. Even more, the technological revolution moved to the machines and systems also, converting them into intelligent commutes that can take real-time decisions and communicate with each other forming an Internet of Systems/Machines. Now, in the above global revolution, we have come up with a new paradigm called the *Internet of Everything* wherein anything that becomes a part has sufficient computing and communication resources. In such an environment where everything can communicate with each other based on the application requirement, the data generated, and data transmitted, is so huge that it cannot be handled by traditional network infrastructure with the same efficiency. The traditional networking based on the TCP/IP model has limitations such as tightly coupled planes, distributed architecture, manual configuration, inconsistent network policies, fallibility, security, and inability to scale. In traditional networks, the control and the packet forwarding parts are integrated at the same place and embedded on the hardware devices. The strongly coupled nature makes it difficult to modify the network policies. Moreover, the distributed architecture is vulnerable to various types of security threats. So, it becomes extremely complicated to troubleshoot the network. Next comes the difficulty to maintain consistency while modifying the network policies. With a substantial increase in the workloads in smart applications, the demand for the increased bandwidth is also expected. The traditional networking domain is statically arranged in such a manner that the growing bandwidth demands of the IT infrastructure end up in the redesigning of the complete topology.

To overcome these issues of traditional networks, a prominent technology named *Software Defined Networking* (SDN), which works on the ideas of Open Flow architecture, is being widely deployed in different network domains. It is logically centralized software capable of controlling the entire network. It advocates the concept of software-based networking and can be termed as a "softwarization" solution that provides network programmability. It solves the issues of the traditional network architecture to accommodate the increased workloads of modern networking. SDN architecture provides better services to manage enterprise, wide area, and data center networks. The control (or network brain) and the data forwarding plane are decoupled in SDN architecture. So, the forwarding nodes become control-free and perform only packet forwarding functionality. Rather than manual configurations of all the network devices by an administrator, SDN allows program-based software control to automate the tasks and increases the flexibility in modifying the network policies dynamically. Another feature of SDN architecture is its centralized architecture which helps to assign, modify, and audit policies with a global network view. These centralized policies are used for routing, maintaining consistency among physical and virtual workloads, load balancing, and global monitoring of the entire network.

The current advances in the industry require an adaptable and dynamic network architecture that provides global visibility and does not require manual reconfiguration. As Industry 4.0 empowers the connected factories and everything, so a restricted policy or limited functionality-oriented network architecture will create a bottleneck in front of the modern era vision of connected things. Thus, SDN can be a suitable solution for handling massive data generated in the Internet of Everything applications. These applications cover almost every domain where networking is partially or fully required, so the future of this area is bright. The current level of research and literature in this area of focus is limited to specific and smaller segments, so this book provides an understanding of the applicability of SDN in a wide range of applications that drive their life's routine. This book provides ample knowledge and a roadmap for academia, researchers, and industry functioning in this area. This book provides a comprehensive discussion on some key topics related to the usage or deployment of SDN in the Internet of Everything applications (like data centers, edge-cloud computing, vehicular networks, healthcare, smart cities). It discusses diverse solutions to overcome the challenges of conventional network binding in various Internet of Everything applications where there is a strong need for an adaptive, agile, and flexible network backbone. This book showcases different deployment models, algorithms, and implementations related to the usage of SDN in the Internet of Everything applications along with the pros and cons of the same. Even more, this book provides deep insights into the architecture of SDN specifically about the layered architecture and different network planes, logical interfaces, and programmable operations. The need for network virtualization and the deployment models for network function virtualization is also a part of this book with an aim towards the design of interoperable network architectures by researchers in the future.

This book is divided into five parts. The first part provides the background about the Internet of Everything and the smart city's ecosystem. The second part focuses on SDN, including the challenges of traditional networks and the development of programmable networks. This part also covers the SDN deployment models, protocols, APIs, layers, network policies, load balancing techniques, and energy optimization approaches. This part ends up with a brief discussion on network function virtualization. The third part discusses the applications of SDN in cloud computing, specifically in data centers and the edge-cloud ecosystem. The associated technologies, design challenges, underlying architectures, and future challenges are discussed. The fourth part entails the security and trust solutions for SDN. The last part provides the application use cases of SDN concerning various problems like high-speed road networks and image processing in industrial IoT.

Durham, UK Gagangeet Singh Aujla

Montreal, QC, Canada Sahil Garg

Montreal, QC, Canada Kuljeet Kaur

Singapore, Singapore Biplab Sikdar
June 2021

Contents

Contributors

Rashid Amin University of Engineering and Technology, Taxila, Pakistan

Gagangeet Singh Aujla Department of Computer Science, Durham University, Durham, UK

Rohit Bajaj Department of Computer Science and Engineering, Chandigarh University, Mohali, India

Rasmeet Singh Bali Department of Computer Science and Engineering, Chandigarh University, Mohali, India

Muhammad Bilal Department of Computer Engineering, Hankuk University of Foreign Studies, Yongin-si, Gyeonggi-do, South Korea

Haotong Cao Jiangsu Key Laboratory of Wireless Communications, Nanjing University of Posts and Telecommunications Nanjing, Nanjing, China
Department of Computing, The Hong Kong Polytechnic University, Hong Kong SAR, China

Amanpreet Singh Dhanoa Department of Computer Science and Engineering, Chandigarh University, Mohali, India

Rajan Kumar Dudeja Department of Computer Science and Engineering, Chandigarh University, Mohali, India

Deepanshu Garg Department of Computer Science and Engineering, Chandigarh University, Mohali, India

Neeraj Garg Department of Computer Science and Engineering, Chandigarh University, Mohali, India

Anwar Ghani Department of Computer Science & Software Engineering, International Islamic University Islamabad, Islamabad, Pakistan

Mudassar Hussain University of Wah, Wah Cantt, Pakistan

Priyanka Kamboj Department of Computer Science and Engineering, Indian Institute of Technology Ropar, Punjab, India

Godfrey Kibalya Department of Network Engineering, Universitat, Politecnica de Catalunya, Barcelona, Spain

Bakht Zamin Khan Department of Computer Science & Software Engineering, International Islamic University Islamabad, Islamabad, Pakistan

Imran Khan Department of Computer Science & Software Engineering, International Islamic University Islamabad, Islamabad, Pakistan

Muazzam Ali Khan Department of Computer Science, Quaid-e-Azam University, Islamabad, Pakistan

S. V. N. Santhosh Kumar School of Information Technology and Engineering, Vellore Institute of Technology, Vellore, India

Fanglin Liu College of Computer Science and Technology, China University of Petroleum (East China), Qingdao, PR China

Haibin Lv North China Sea Offshore Engineering Survey Institute, Ministry of Natural Resources NorthSea Bureau, Qingdao, China

Zhihan Lv College of Computer Science and Technology, Qingdao University, Qingdao, China

Sudip Misra Department of Computer Science and Engineering, Indian Institute of Technology Kharagpur, Kharagpur, India

Ayan Mondal Department of Computer Science, University of Rennes 1, INRIA, CNRS, IRISA, Rennes, France

Sujata Pal Department of Computer Science and Engineering, Indian Institute of Technology Ropar, Punjab, India

Liang Qiao College of Computer Science and Technology, Qingdao University, Qingdao, China

Geetanjali Rathee Department of Computer Science and Engineering, Netaji Subhas University of Technology, New Delhi, India

Shubham Rawat Department of Computer Science and Engineering, Chandigarh University, Mohali, India

Paulo F. Ribeiro Institute of Electrical and Energy Systems, Federal University of Itajuba, Itajuba, Brazil

Bhawana Rudra Department of Information Technology, National Institute of Technology, Mangalore, Karnataka, India

Thanmayee S. Department of Information Technology, National Institute of Technology, Mangalore, Karnataka, India

Rafael S. Salles Institute of Electrical and Energy Systems, Federal University of Itajuba, Itajuba, Brazil

Amritpal Singh Department of Computer Science and Engineering, Chandigarh University, Mohali, India

Gurpinder Singh Department of Computer Science and Engineering, Chandigarh University, Mohali, India

Jingyi Wu College of Computer Science and Technology, Qingdao University, Qingdao, China

Peiying Zhang College of Computer Science and Technology, China University of Petroleum (East China), Qingdao, PR China

Part I
Internet of Everything and Smart City

Chapter 1
Internet of Everything: Background and Challenges

Rajan Kumar Dudeja, Rasmeet Singh Bali, and Gagangeet Singh Aujla

1.1 Introduction

The exponential growth in fields of embedded systems and their computing and communication power leads to the generation of a new era of Internet Technologies. It leads to the generation of the field named the Internet of Things (IoT). The term was first given by Kevin Aston in 1999. IoT is a collection of objects called Things that have sensing capability. These objects also have limited computational power. They can communicate the sensed data using standard protocols. The data on network could be used for further processing and analysis purposes. In simple terms, it is a network of smart objects having sensing and computational capacity. With the advancement in embedded technologies, the production capacity of these smart objects has increased exponentially. Along with its production, the digital transformation era has brought about huge demand for IoT based application. As per an estimate [1], there were 5.8 billion IoT endpoints were there in the market and the demand continues to surge in coming years.

1.1.1 Working of IoT

IoT works on a principle of *Connect, Communicate, Compute, and Action*. It consists of various IoT sensors that connect through wired or wireless manner to the Internet through a gateway node. Sensors measure the state of the environment and

R. K. Dudeja · R. S. Bali
Department of Computer Science and Engineering, Chandigarh University, Mohali, India

G. S. Aujla (✉)
Department of Computer Science, Durham University, Durham, UK

G. S. Aujla et al. (eds.), *Software Defined Internet of Everything*, Technology, Communications and Computing, https://doi.org/10.1007/978-3-030-89328-6_1

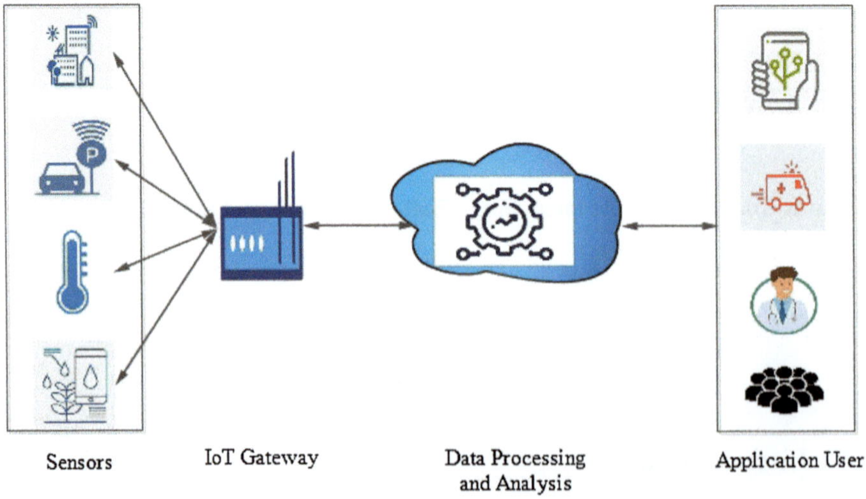

Fig. 1.1 Working of Internet of Things

other measurements and communicate data to the Internet cloud. The Internet uses its huge computational power to perform the required processing on the received data and converts it into useful information. The end-user through the interface application analyzes the information and performs action accordingly [2].

For example, smart homes may have a temperature sensor and air conditioner relay switch connected to IoT controller using standard IoT procedure. It is further connected to the Internet that receives the sensor data. Homeowner acting as end-user continuously monitors the temperature of his home and gives a remote command to control air conditioner as and when required. Figure 1.1 describes the complete working of IoT. It consists of four basic components for its complete working procedure. Following are the description of the components.

1.1.1.1 Sensors

The sensors are electronic components having sense and digital measurement capabilities. These are fitted into the devices called smart objects. A smart object is an electronic component that consists of one or more sensors to monitor the surrounding environmental conditions such as mobile phone. A smart object consists of multiple sensors like an accelerometer, camera, and location tracker. These sensors are majorly playing the role of *collecting* the data. These objects are working in large number of fields such as healthcare, agriculture, medicine, manufacturing, logistics delivery, smart home, smart cities, etc. Figure 1.2 shows some prominent application areas that employ the IoT for more effective operating. These objects are continuously measuring the surrounding environmental conditions and generate data regularly. This leads to the generation of a huge volume of

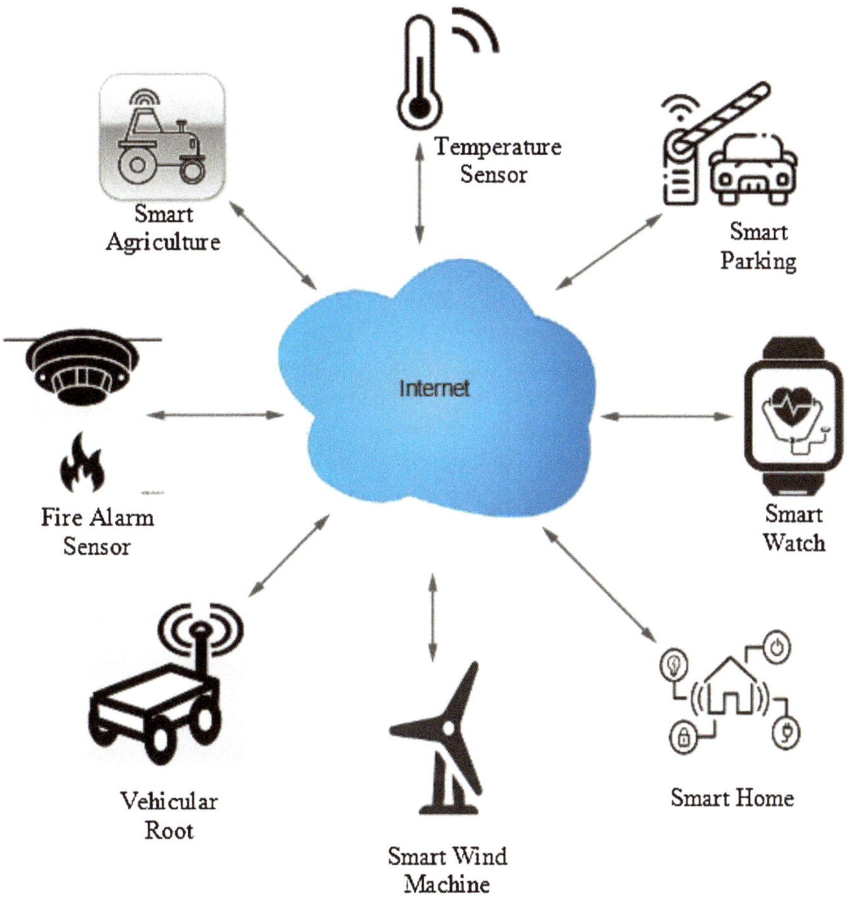

Fig. 1.2 Smart objects used in Internet of Things

data. As the smart objects work in different application areas, the data generated by them are multidisciplinary. These objects are further connected to a controller device called the IoT Gateway node. So the huge volume of multidisciplinary data is also communicated to cloud or edge through gateway node in certain IoT architectures [2].

1.1.1.2 Gateway Node

The gateway node acts as an entry/exit point for IoT devices with the rest of the network. The devices are connected to gateway nodes using standard IoT protocols. It plays a major role of communicating the data. It uses standard protocols like Bluetooth, Wi-Fi, 6LoWPAN, Zigbee, RF Link, Z wave, etc. These protocols work on the different capacity for parameters like topology, range, bandwidth, power consumption, bit rate, etc. The other techniques like RFID and NFC are used for connecting IoT objects. Table 1.1 depicts these parameters for different protocols. Based on a comparison of protocols, different techniques are used to connect different IoT objects. The gateway node has to support multiple ports having different mechanisms if they have to connect multiple functional devices. Other than connecting devices, it connects to the Internet on the other side using standard IP-based protocols.

In an ideal scenario, gateway node transfers the multifunctional data received from connected devices. The huge volume of data generated by the sensors is transferred to the Internet cloud through these gateways. But with the evolution of new paradigms like edge and fog computing, there are gateway nodes that provide additional computational power other than connectivity of smart objects [3]. Handling of this huge volume of data by sensors at source end has been receiving a lot of attention from the research community. This has also led to development of intelligent system that can be integrated with traditional gateway nodes to achieve efficient information dissemination. In such integrated scenarios *compute* along with *communication* of data is performed [4].

1.1.1.3 Data Processing and Analysis

The Data transmitted from sensors through gateways should be received by the devices having enough computational power. Since the amount of data is very large and also multidimensional, it requires intelligent systems that can process the data. The application running on IoT based systems is basically processing the data and converting it into useful information. The data processing techniques have to perform a number of basic functions like denoising of data, feature extraction from data, data fusion, and data aggregation. These processing techniques should be lightweight to implement so that these functions are performed efficiently. The requirement for the type of processing algorithm depends upon the nature of data generated by sensors. The resulted processed data turns out to be information for further course of actions [5].

The resulted information can be analyzed using various Artificial Intelligence (AI) based techniques. It also employs the other techniques like machine learning and deep learning to analyze the data without human involvement. The output produced by above intelligent techniques should be able to produce the correct and useful information required for decision-making and knowledge generation [6]. The decision-making process then results in certain values that are then converted into some physical actions performed by actuators. This knowledge is also stored for further behavioral study of the sensor. The above-discussed process could also get elaborated by Fig. 1.3

The data processing and analysis are generally performed on cloud-centric applications. This leads to generation of huge amount of data on networks that needs to be transferred from the gateway to the cloud. With the advancement in

Table 1.1 Internet of Things protocols

Standard	Bluetooth	Bluetooth 4.0 IE	Zigbee	Wi-fi	6LoWPAN	RF link	Zwave
IEEE Spec.	IEEE 802.15.1	IEEE 802.15.4	IEEE 802.15.4	IEEE 802.11 a/b/g/n	IEEE 802.15.4-2006	IEEE C95.1-2005	Z-wave alliance
Topology	Star	Star	Mesh, Star, Tree	Star	Mesh, Star	–	Mesh
Bandwidth	1 Mbps	1 Mbps	250 Kbps	Up to 54 Mbps	250 Kbps	18 MHz	900 MHz
Power consumption	Very low	Very low	Very low	Low	Very low	Very low	Very low
Max. data rate (Mbits/s)	0.72	5–10	0.25	54	800(Sub-GHz)	1	9600 bits or 40 kbits
Range	≤30m	5–10 m	10–300 m	4–20 m	800 m	≤ 3m	30m
Spectrum	2.4 GHz	2.4 GHz	2.4 GHz	2.4–5 GHz	2.4 GHz	2.4 GHz	2.4 GHz
Channel bandwidth	1 MHz	2400–2480 MHz	0.3/0.6 MHz, 2 MHz	22 MHz	868–868.6 MHz (EU), 902–928 MHz (NA), 2400–2483.5 MHz (WW)	–	868 MHz

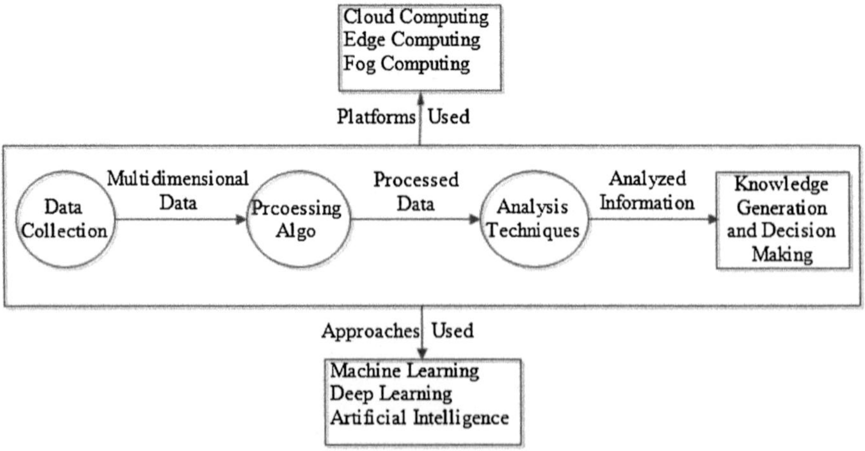

Fig. 1.3 Data processing and analysis

the computing field, edge/fog computing has also been integrated with traditional IoT. They support lightweight data processing and analysis at the source end. This results in an efficient architecture where huge processing is performed on devices connected with sensors results in the transfer of only the required data to the cloud. There is lots of research going in fields on cloud/edge/fog interplay that results in the selection of data processing and analysis at the required end [7, 8].

1.1.1.4 End-User

IoT users are receiving the information through the application interface. There are several different categories for each user. These categories include expert users like healthcare experts, doctors, weather forecasters, engineers, and data scientists. These experts get the analyzed data and perform an action accordingly thereby helping to improve their productivity. The other categories of users are generic persons such as family members, friends, and community services. They have access to data generated by the sensor that they can monitor regularly like a person keeping his home under camera surveillance. The last category includes emergency services like ambulance, fire fighters, and police. These users will get activated by emergency events relevant to their respective domains [9, 10].

1.1.2 Internet of Everything

The key components of IoT are the objects that can form the communication channel. The evolution of distributed computing helps in grater processing capabilities

Fig. 1.4 Evolution of Internet of Everything

at the distributed end. The gaining trend of distributed computing has also resulted in emergence of a new field. This leads to the generation of a new Internet era called the Internet of Everything (IoE). The term was first used by Cisco in 2013 [11]. The field is also a result of further advancements in dynamic information handling at the source end. The integration of Fog and Edge computing results in decentralized processing capacity, providing increased computing power for these kinds of hybrid IoE networks. The primary aim of IoE is to connect anything with the Internet with enough information provided at right time in this digitally driven modern era [12]. Figure 1.4 shows the evolution of IoE from machine-to-machine communication through IoT.

As an intelligent network of people and objects, the scope of IoE is exponentially increasing in multidisciplinary domains across the globe. It is the fusion of advancements in technologies in multiple fields like Information technology, environmental, and biotechnology. It has emerged as the main player for future market growth. The overall growth in IoE will lead to millions of internetworked devices that could result in having increased processing capacity, intelligent decision-making, and improved sensing capacity. This has resulted in development of specialized products based on IoE for organizations that are helping them in improving their operations. It also impacts the interaction methods with the physical environment [13].

IoE can be considered as much broader system than IoT. It is considered to be a superset of IoT along with its variants like Internet of Drone, Internet of Healthcare. It evolves from machine-to-machine communication that is prevalent in IoT to people and people to machine communication formats. The evolvement of IoE results in a better network having better capacity to turn information into actions. It gives rise to new and exciting opportunities with richer experiences for individuals and organizations. It evolves from one pillar called things in IoT to four pillars called People, Process, Data, and Things. IoE is primarily concerned with bringing these pillars in an efficient manner. In simple terms, IoE can be thought as an intelligent network of People, Process, Data, and Things. Figure 1.5 shows the all four components and communication among them in IoE [14].

- *People:* IoE is responsible for making connections among the peoples in a more effective and relevant manner. The people are the main concern to interact with the Internet.
- *Process:* It consists of an efficient collection of processes that converts the collected information into the appropriate actions. It delivers accurate information to the concerned person or object at the appropriate time.

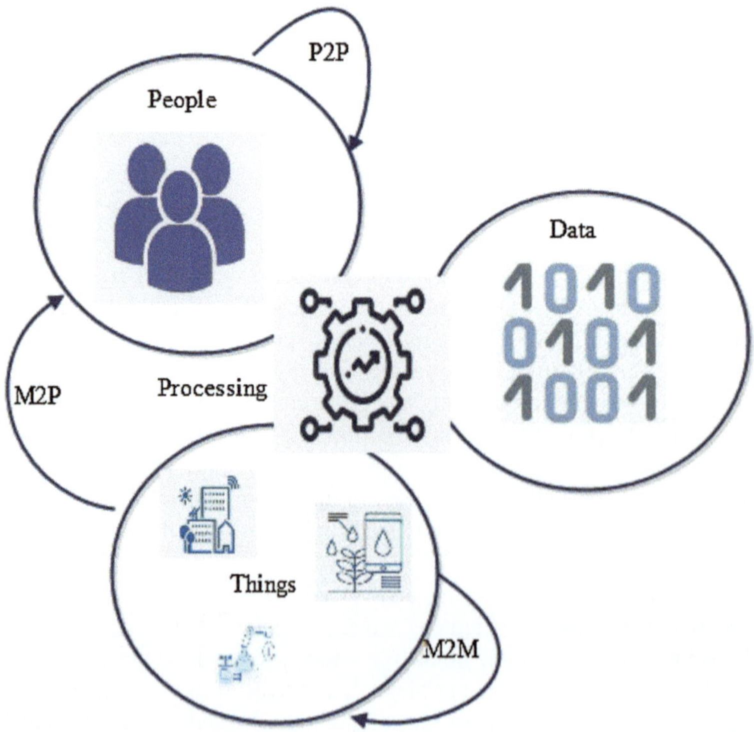

Fig. 1.5 Internet of Everything components

- *Data:* The data is generated during interaction with objects and peoples in multiple forms. There is a requirement of efficient processing and handling of data that could further support the decision-making process and knowledge generation.
- *Things:* These are physical objects having sensing and processing capabilities connected to the Internet. These could also be termed as smart objects.

Recently a broad spectrum of application domains has undergone revolutionary changes globally and produced exciting opportunities due to IoE. IoE has seen the evolution from a smart application in IoT to connecting applications with the user in a much more efficient manner. For example, in smart vehicle management and smart healthcare based IoE systems are helping to connect roads with hospitals for real-time monitoring to save lives. It integrates the people with objects and intelligent processes more efficiently and efficiently such as connecting homes for more comfortable living, connecting food and peoples in supply chain management, and connecting elderly population and their monitoring with healthcare experts. In general, applications of IoE have touched a number of different domains such as healthcare, digital transformation, home automation, energy conservation, security, information exchange and communication, and environmental monitoring[15]. Some of its major applications domains have been discussed in later sections.

1.2 Applications of Internet of Everything

The IoE is touching all aspects of life and it has numerous applications in all of domains . Some of the applications are discussed below.

- *Smart Healthcare:* In the healthcare support, IoE has significant application ranging from in-hospital support to smart wearable devices. There are IoE based techniques available for diagnosis purpose for various diseases. It also has application to assist patients for their regular activities like assisting Alzheimer patients. There are also trackers like smart watch to keep track of routine health related parameters like blood pressure, distance covered [16].
- *Smart Home:* The IoE based number of equipment are into the market that are related to household things. These devices are making a home connected with Internet. These devices can operate autonomously according to environment like air conditioner will get automatically on/off based on set temperature. Other than this there are basic electronic things in house like fan, fridge, washing machine, TV, lights that have to operate autonomously [17].
- *Smart City:* The IoE has huge potential to develop the infrastructure in more organized and efficient manner. There are applications available to upgrade a city to smart city. These applications could help in smart garbage collection, smart parking system, and smart street lights [18].
- *Smart Vehicular Technology:* There are vehicles equipped with sensors and connected with internet are making the drive more safer and comfortable. Even there are vehicles in development process that can run on roads without driver support. The vehicles can use numerous sensor that can help them in judgment of road conditions and traffic [19].
- *Smart Industry:* IOE has brings a revolutionary change in Industry like manufacturing, food and logistics, and packaging. It brings a new era of sensor fitted robots that can replace the humans in the factories. They can work more efficiently and accurately than the human beings. IoE based systems are also used for delivery of logistics in lesser time like Drones [8].
- *Smart Agriculture:* This is one of the major applications of IoE for agriculture based countries. It helps the farmers to check their soil moisture and other parameters digitally on regular basis. It also helps in developing latest equipment that can help the farmer to grow and sell their crops [20].

1.3 Challenges of Internet of Everything

With the growth in the field of IoE and to meet expectations of people and organizations, a number of opportunities have opened up in domains like research and business. However, this also brings some huge challenges along with it in its implementation at a large scale. These challenges are discussed below.

- *Device Security:* It is one of the major concerns for the IoE field. As it encourages the decentralization of each process, we need security at the device level. There are still some design challenges available with many embedded devices. These make them more vulnerable to many security threats. This needs to be incorporated by manufacturers. Other than this, the increase in the computational capacity of devices makes them available for many processes and data for computational purposes. This makes the devices more vulnerable. It leads to attracting several cyberattacks. This points to the need for the generation of a mechanism to provide the security at device level [21].
- *Data Security:* As the data generated by the devices are multidimensional, the requirement for encryptions or any other data security techniques is also different for the type of data streams. The continuous data generation by the objects leads to the generation of a huge volume of multidimensional data. It gives rise to the need for computational mechanisms that are capable enough to provide data security when they are executed them on source end devices. These mechanisms should also be based on lightweight techniques as well as apply data security at the fog level along with other hierarchical levels of computing [21].
- *Scalability:* The data generated by the objects are continuously increasing with time frame. The cloud or edge devices that are going to process and analyze the data should be scalable. There is enough buffer support available with these computing paradigms to support the large volume of data. There should be flexibility to add a new device that generates data with new parameters that does not hamper the computing and analysis process for that data [22].
- *Privacy Issues:* As IoE is touching each domain of personal as well as professional life of a humans, maintaining privacy of data is a major concern. This is especially true for the fields such as healthcare and other domains that manage personalized data. The data generated by the sensors attached to the patient like blood pressure, heartbeat, etc. are critical. So there is the requirement of an efficient privacy policy for such type of healthcare data [22].
- *Need of Standards:* The goal of IoE is to connect anything with the Internet which leads to connecting different types of devices with the Internet. There is also the decentralization of processes to control the overall mechanism. This brings a new challenge in creating the standards to govern the full mechanisms. These standards should be developed by some well-established organizations or open communities. The standards should find the solution to streamline the various protocols and techniques used at the central level as well as distributed level to control the various process involved in implementation as well communication of all objects [23].

- *Device Heterogeneity:* The aim of IoE to connect anything to the Internet will bring a challenge to connect devices having different operating methods to connect. The working method to generate the data by different types of devices like electrical, electrochemical, and electromagnetic is different [23].
- *Compatibility Issues:* There are many heterogeneous devices available in market and the basic connecting mechanism of each of the devices is different. For example, creating an intelligent home system requires different types of IoE devices, based on disparate technologies like Bluetooth, Zigbee, Z-wave, etc. To incorporate these devices with each other on common platforms lead to the generation of huge compatibility issues [24].
- *Bandwidth Issues :* As the size of the market going to increase exponentially, the number of devices is also going to increase. The data transfer by these devices is also huge. Like there is the option to continuous live streaming of video camera of smart home on the mobile phone. These kinds of applications lead to huge bandwidth requirements at the network level to transfer this volume of data.
- *Intelligent Analysis:* The aim of this digital transformation is to ease life with an intelligent decision-making system without human involvement. For the same, there is the requirement of intelligent data analysis techniques deployed on any of the computing platforms. The accuracy of the decision-making depends upon the intelligence of the data analysis techniques. these techniques should be enough capable to handle the unpredictable behaviors of objects that are generating the data [25].
- *Cloud-Edge-Fog Interplay:* As the decentralization of work is increasing in fields like IoE, still many mechanisms need to be handled at a centralized level. Other than this, the availability of computational power at different platforms is not homogeneous. So there should be a smart selection of platforms like cloud, edge, or fog for specific data for its analysis and process to its execution [26].
- *Authentication Mechanism:* As the number of devices is going to increase exponentially, the authentication of such a huge number of devices from different manufacturers is a challenging task. Other than its number, the devices are also heterogeneous. This leads to a requirement of a standard authentication mechanism to tackle the above-discussed issues [24].

1.4 Conclusion

The advancement in embedded technologies has brought about a new era of smart devices that communicate and connect with Internet. Internet of Things has been one of the main technologies that has emerged as a result of this advancement. The integration of distributed computing and decentralization of processing capabilities has resulted in further development of Internet of Things field. However, heterogeneity of devices and application has been a major bottleneck in adaptability of Internet of Things. This leads to development of a new field called Internet of Things that connects people, process, data, and things in effective and efficient manner. With its evolvement and applicability in number of vital domains also brings number of challenges. Efficient handling of huge amount of multidimensional data generated by heterogeneous devices as well as need of standard process and protocols to define Internet of Everything are the major challenges. There is also a need to tackle security at device level as well as data level. So this chapter has discussed the above challenges and highlighted their importance for future models of IoE.

References

1. Ande, R., Adebisi, B., Hammoudeh, M., & Saleem, J. (2020). Internet of things: Evolution and technologies from a security perspective. *Sustainable Cities and Society, 54*, 101728.
2. Xia, F., Yang, L. T., Wang, L., & Vinel, A. (2012). Internet of things. *International Journal of Communication Systems, 25*(9), 1101.
3. Aujla, G. S., & Jindal, A. (2020). A decoupled blockchain approach for edge-envisioned IoT-based healthcare monitoring. *IEEE Journal on Selected Areas in Communications, 39*, 491–499.
4. Zhu, Q., Wang, R., Chen, Q., Liu, Y., & Qin, W. (2010). IoT gateway: Bridgingwireless sensor networks into internet of things, in *2010 IEEE/IFIP International Conference on Embedded and Ubiquitous Computing* (pp. 347–352). Piscataway: IEEE.
5. Krishnamurthi, R., Kumar, A., Gopinathan, D., Nayyar, A., & Qureshi, B. (2020). An overview of IoT sensor data processing, fusion, and analysis techniques. *Sensors, 20*(21), 6076.
6. Kaur, K., Garg, S., Aujla, G. S., Kumar, N., Rodrigues, J. J., & Guizani, M. (2018). Edge computing in the industrial internet of things environment: Software-defined-networks-based edge-cloud interplay. *IEEE Communications Magazine, 56*(2), 44–51.
7. Chaudhary, R., Aujla, G. S., Kumar, N., & Zeadally, S. (2018). Lattice-based public key cryptosystem for internet of things environment: Challenges and solutions. *IEEE Internet of Things Journal, 6*(3), 4897–4909.
8. Singh, A., Aujla, G. S., Garg, S., Kaddoum, G., & Singh, G. (2019). Deep-learning-based SDN model for internet of things: An incremental tensor train approach. *IEEE Internet of Things Journal, 7*(7), 6302–6311.
9. Singh, P., Bali, R. S., Kumar, N., Das, A. K., Vinel, A., & Yang, L. T. (2018). Secure healthcare data dissemination using vehicle relay networks. *IEEE Internet of Things Journal, 5*(5), 3733–3746.
10. Chhabra, S., Bali, R. S., & Kumar, N. (2015). Dynamic vehicle ontology based routing for VANETs. *Procedia Computer Science, 57*, 789–797.
11. Balfour, R. E. (2015). Building the "internet of everything" (IoE) for first responders, in *2015 long island systems, applications and technology* (pp. 1–6). Piscataway: IEEE.
12. Miraz, M. H., Ali, M., Excell, P. S., & Picking, R. (2015). A review on internet of things (IoT), internet of everything (IoE) and internet of nano things (IoNT), in *2015 internet technologies and applications (ITA)* (pp. 219–224). Piscataway: IEEE.
13. Jara, A. J., Ladid, L., & Gómez-Skarmeta, A. F. (2013). The internet of everything through ipv6: An analysis of challenges, solutions and opportunities. *Journal of Wireless Mobile Networks, Ubiquitous Computing, and Dependable Applications, 4*(3), 97–118.
14. Hussain, F. (2017). Internet of everything, in *Internet of things* (pp. 1–11). Berlin: Springer.
15. Fan, X., Liu, X., Hu, W., Zhong, C., & Lu, J. (2019). Advances in the development of power supplies for the internet of everything. *InfoMat, 1*(2), 130–139.
16. Sharma, S., Dudeja, R. K., Aujla, G. S., Bali, R. S., & Kumar, N. (2020). DeTrAs: Deep learning-based healthcare framework for IoT-based assistance of Alzheimer patients, in *Neural Computing and Applications* (pp. 1–13). Berlin: Springer.

17. Kaur, N., & Kumar, R. (2016). Hybrid topology control based on clock synchronization in wireless sensor network. *Indian Journal of Science and Technology, 9*, 31.
18. Singh, I., & Kumar, R. (2018). Mutual authentication technique for detection of malicious nodes in wireless sensor networks. *International Journal of Engineering & Technology, 7*(2), 118–121.
19. Bali, R. S., Kumar, N., & Rodrigues, J. J. (2014). An intelligent clustering algorithm for VANETs, in *2014 International Conference on Connected Vehicles and Expo (ICCVE)* (pp. 974–979). Piscataway: IEEE.
20. Walter, A., Finger, R., Huber, R., & Buchmann, N. (2017). Opinion: Smart farming is key to developing sustainable agriculture, in *Proceedings of the National Academy of Sciences, 114*(24), 6148–6150.
21. Zhang, Z.-K., Cho, M. C. Y., Wang, C.-W., Hsu, C.-W., Chen, C.-K., & Shieh, S. (2014). IoT security: Ongoing challenges and research opportunities, in *2014 IEEE 7th International Conference on Service-Oriented Computing and Applications* (pp. 230–234). Piscataway: IEEE.
22. Chen, S., Xu, H., Liu, D., Hu, B., & Wang, H. (2014). A vision of IoT: Applications, challenges, and opportunities with China perspective. *IEEE Internet of Things Journal, 1*(4), 349–359.
23. Lee, I., & Lee, K. (2015). The internet of things (IoT): Applications, investments, and challenges for enterprises. *Business Horizons, 58*(4), 431–440.
24. Van Kranenburg, R., & Bassi, A. (2012). IoT challenges. *Communications in Mobile Computing, 1*(1), 1–5.
25. Shekhar, Y., Dagur, E., Mishra, S., & Sankaranarayanan, S. (2017). Intelligent IoT based automated irrigation system. *International Journal of Applied Engineering Research, 12*(18), 7306–7320.
26. Wang, W., Wang, Q., & Sohraby, K. (2016). Multimedia sensing as a service (MSAAS): Exploring resource saving potentials of at cloud-edge IoT and fogs. *IEEE Internet of Things Journal, 4*(2), 487–495.

Chapter 2
Smart Cities, Connected World, and Internet of Things

Rafael S. Salles and Paulo F. Ribeiro

2.1 Introduction

The world has undergone constant transformations in several society areas, characterized by a wide digitalization and modernization. Urban environments play a vital role in this process, as these spaces contain concentrations of population, activities, services, and technological advances. More efforts are being made to create models of cities that can adequately absorb cutting-edge technologies and solutions combined with well-being and sustainability [6]. This concern with the quality of life and the environment is a consequence of the problems created so far, such as emission of greenhouse gases, global warming, energy poverty, social inequality in urban centers, among other issues. Those are being faced by new concepts that are linked to a more modern and connected world.

Smart cities are based on this attempt to incorporate different aspects through a technological and holistic perspective [20]. The development of smart cities is supported by the revolution and massive penetration of information and communication technology (ICT), computerization and automation of urban infrastructures. These technologies play a critical factor in this city model's success. With the increase in applications aimed at networked devices, increased use of sensing, applications demand greater importance of information within this giant data flow in urban centers. In this way, it is possible to observe that initiatives and applications based on the internet and ICTs are increasingly emerging in a connected city, focused on modern solutions based on the latest technology and cutting-edge software [25].

The organizational structure that incorporates this concept of smart brings a direction in technological advances, mainly the network structure formed in urban

R. S. Salles (✉) · P. F. Ribeiro
Institute of Electrical and Energy Systems, Federal University of Itajuba, Itajuba, Brazil
e-mail: sallesrds@gmail.com

G. S. Aujla et al. (eds.), *Software Defined Internet of Everything*, Technology, Communications and Computing, https://doi.org/10.1007/978-3-030-89328-6_2

spaces, with innovation in the use of the internet in different infrastructures. It is an effect that culminates in the various components of a city inserted in this context. That includes applications aimed at smart grids, intelligent mobility, universal public health, transparent and people-centered governance, effective and sustainable management of resources and waste, in addition to crucial points such as education, safety, and economy. Such applications focused on a connected world approach are characterized by the protagonism of information, data management, and communication between smart devices through ICTs. That is why the Internet of Things (IoT) emerges as a solution that encompasses all these concepts and allows the establishment of these advances to strengthen smart cities [3]. Below are some established definitions that describe the main concept behind the smart cities.

- "A city well performing in a forward-looking way in economy, people, governance, mobility, environment, and living, built on the smart combination of endowments and activities of self-decisive, independent, and aware citizens. Smart city refers to implementation of intelligent solutions that allow modern cities to enhance the quality of citizens' services" [17];
- "Smart cities connect the physical infrastructure, the IT infrastructure, the social infrastructure, and the business infrastructure to leverage the city's collective intelligence" [18];
- "Smart Cities initiatives try to improve urban performance by using data, information, and ICTs to provide more efficient services, monitor and optimize existing infrastructure, increase collaboration among different economic actors, and encourage innovative business models in private and public sectors" [27];
- "A city is smart when investments in human and social capital, traditional infrastructure, and disruptive technologies fuel sustainable economic growth and high quality of life, with a wise management of natural resources, through participatory governance" [12];
- "Smart cities can be defined as a technologically advanced and modernized territory with a specific intellectual ability that deals with different social, technical, economic aspects of growth based on smart computing methods to develop robust infrastructures and services" [31].

The wide-scale use of sensor technology creates huge volumes of data that smart systems can use to optimize the use of infrastructure and resources. The IoT refers to the enormous implementation of advanced sensors and wireless communication in all kinds of physical objects [12]. In this era of transformations, the connected world allows connections between people, things, devices, applications, and processes. IoT tracking is a modern evolution of the internet that allows massive numbers of devices to connect and interact with people and machines in a networked environment [23]. In addition to supporting these important aspects, it also brings greater flexibility and reliability to the systems, providing adequate solutions to face the complexity that the advances of intelligent cities bring. The IoT is one of the critical elements of smart sustainable cities' ICT infrastructure as an emerging urban development approach due to its great potential to advance environmental sustainability, which is associated with big data analytics and becoming more important in many urban domains [8].

This technological sphere is completed with diverse applications and disruptive technologies such as renewable energy sources, electric vehicles, artificial intelligence (AI) applied to different segments, green technologies, cybersecurity innovations, and many others. These are the results of this evolution applied to smart cities' infrastructures, mainly the so-called critical ones. With the increase in complexity with these various interconnected systems, these technologies help overcome technical barriers and promote the possibility of creating an atmosphere geared towards innovation, sustainability, cooperation, education, and research.

Therefore, this chapter will explore the main concepts of smart cities. It will also address the technological aspects, the holistic and social side, the infrastructures, and how IoT provides these urban spaces' evolution. Lastly, some initiatives that are already being put in place to improve life quality in some cities will be pointed out.

2.2 Smart City Integrated Perspective

The integrated perspective about the smart cities with respect to different factors is comprehensively discussed in this section.

2.2.1 Smart City Overview

Given the starting point of contextualization and definitions raised about smart cities, it is possible to delve into important aspects for understanding the phenomenon as a whole. Smart cities' inherent complexity can be divided into three main aspects: technology, governance, and community. These are interlaced to promote a dynamic system that supports an increase in the performance of infrastructure and well-being through disruptive technologies. Philosophical considerations are also crucial for a complete vision with a sustainable and balanced development [32, 40]. From this set, it is possible to establish a development where the population and all stakeholders appropriate a modern city's benefits. Figure 2.1 illustrates the complexity division of smart cities.

These complexities go back to the objectives, challenges, and the role of each attribute of this system. It is complex because it is necessary to adapt several subsystems to provide a profitable integration that promotes a modern and sustainable city's basic requirement. The literature reports several key components that constitute a smart city. Some of these components, such as smart people, smart economy, smart governance, smart mobility, smart environment, and smart living, are great to describe the features as a role.

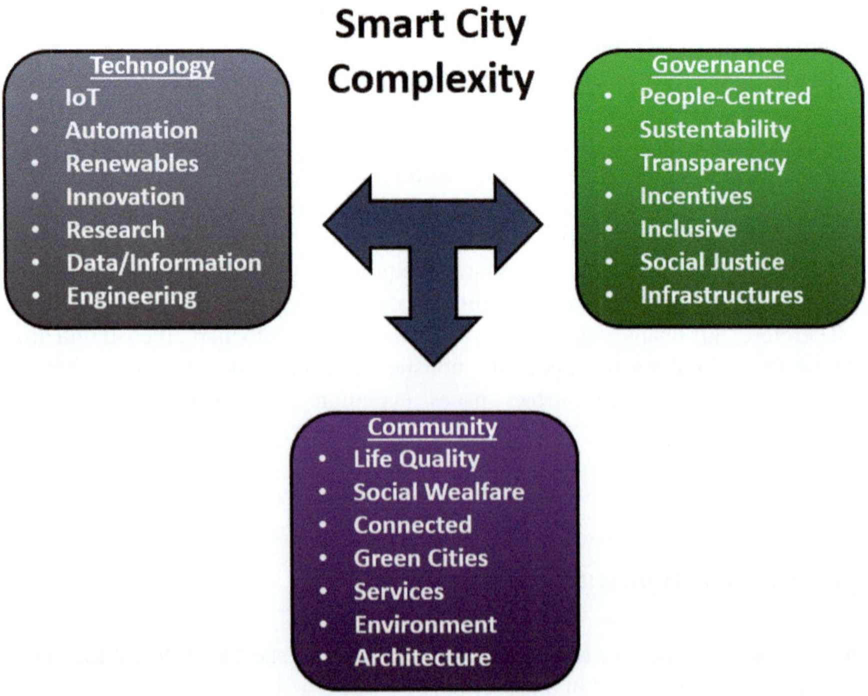

Fig. 2.1 Smart city complexity aspects

The concept of "Smart People" addresses an attempt to modify how citizens interact with services, with the public and private sectors, and with the community, for a more connected way through information and technologies. For this to happen, there must be a framework that promotes individuals' social and digital inclusion so that there is effective participation of the population, which is the biggest beneficiary of the whole process. People in smart cities should connect and communicate with each other to interact online and share physical space with users [38, 39]. Smart people should not only interact with each other via services, but they should also provide data for these services [22]. This data could be critical for decision making and planning by smart city agents, like traffic information, infrastructure expansion, programmable events, services improvements, etc.

Another important point regarding this component is the promotion of intelligent methods in the educational process, which should generate more opportunities, training, professional qualification, and a solid education plan at different levels. It benefits all sectors of society, as qualified and adequately educated citizens make the city develop intelligently. Cities cannot achieve smartness without creativity, education, knowledge, and learning [2]. Within this context, as already mentioned, inclusion plays a significant role. Combined with technological development and advances in infrastructure, it is also necessary that urban centers promote social inclusion, as there is no intelligence in a city with social inequality [35].

"Smart Economy" refers to initiatives and measures to improve the city's local economy, making it strong, making it attractive for innovation and start-ups, making it attractive for investors and new businesses, and creating a business ecosystem and competitiveness. The use of technology in this field must permeate through alternatives that reduce costs and increase productivity, feed the universe of discourse in the economic sector with valuable information with ICTs, promote sustainability for the commercial and industrial segments, and create more jobs.

"Smart Governance" is about changing how the government interacts and connects with stakeholders in a city context. Government actions should reconsider the quality, impact, and scope of services for citizens and civil society organizations. Traditionally governance is "as regimes of laws, administrative rules, judicial rulings, and practices that constrain, prescribe, and enable government activity, where such activity is broadly defined as the production and delivery of publicly supported goods and services" [26]. With the widespread of ICTs, all governance activities which are based on technology are also "Smart Governance" [2]. It improves innovative policies that use information, technology, and business model to serve the city better.

Another point that must be considered is transparency. Within the context of smart cities, public institutions in government must promote data transparency and access to information. It encourages greater participation and accountability concerning society, in addition to inhibiting corruption and political irresponsibility. With data and information available, it is possible to study different ways to apply methods and research to promote better services.

The "Smart Mobility" corresponds to the effort to improve urban transport's performance and quality by adopting innovative mobility solutions, intelligent mobility management, and investment in infrastructure. Managing traffic networks and congestion has been one of the major challenges facing agents in large urban centers. The use of technology and IoT appears as a viable solution for implementation by the monitoring centers. An urban transport network that contains several options and modes, both public and private, will present a more excellent resource to mitigate traffic and mobility problems. Therefore, the smart city must also promote new forms of transport, especially sustainable ones. Some examples are electric vehicles, self-driving cars, shared vehicles (bicycles, scooters, car-sharing, etc.), and infrastructure expansion measures, such as special roads, cycle paths, among others. Every investment and program must contain a people-centered directive to improve people's flow and community mobility.

Focused on the environment, the "Smart Environment" is concerned with better use of resources and sustainability. The city must act on environmental infrastructures like waterways, sewers, and green spaces, and it should also be based on using natural and green energy resources to increase sustainability [10]. For example, electric power systems also go through a transformation phenomenon called smart grids, which seeks to make the system more reliable, sustainable, and robust through ICTs, automation, and renewable energy sources. Another essential point is basic sanitation and waste treatment, which increasingly needs to be managed and used correctly through services that use the technological potential to improve these

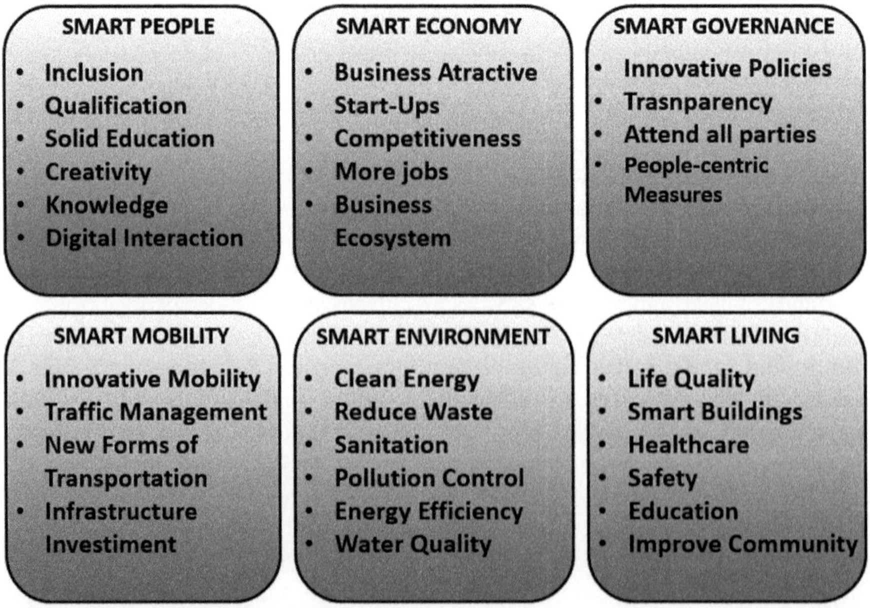

Fig. 2.2 Smart city key components

critical infrastructures. The smart city's objectives are to reduce waste production, control pollution, reduce greenhouse gas emissions, energy efficiency, and ensure water quality and availability.

Lastly, "Smart Living" is focused on improving citizens' quality of life, whether residents or visitors. Smart buildings for education, tourism, healthcare, and public safety comprise the framework. The promotion of civic engagement and the use of IoT-based technologies drive the improvement of individuals' practice in different areas [22]. Public safety is a big concern, for example, affecting the growing urbanization in developing countries [1]. In addition to ensuring that these essential services are served, smart cities must also promote solutions and initiatives that improve the community's daily life. Figure 2.2 illustrates the smart city components, highlighting the key points.

2.2.2 *Smart Cities Goals and Barriers*

The approach of smart cities is shaped by the concepts mentioned above, which allows to detail and highlight some objectives of this proposal to transform urban centers. The focus developed in smart cities is to familiarize the union between the technological environment and the citizen's interfaces with the city so that innovative solutions are generated. Among the objectives of this development is:

- **Focus on Citizen:** improvement on inclusiveness, services, and life quality for people;
- **Resilience and Robustness of Infrastructure:** Modernization of critical infrastructures through IoT, ICT, automation, data processing, etc. Thus increasing performance, efficiency, and also resilient to catastrophes and crises;
- **Huge data processing:** The potential to process and direct data and information in real-time to feed the universe of services discourse, support in city operations, within a context of immersion of data with huge volumes of information.
- **Optimization of Energy and Resources:** Increased energy efficiency, development of smart electrical networks, use of renewable energy sources, measures for decarbonization, waste management, sustainability as priority, and a guarantee of a recycling level;
- **Promote development:** Create a favorable environment for research and development, innovation, economic competitiveness, and modernization of different sectors.
- **Connected Applications:** Use of smart applications in the urban environment, greater user participation in services and processes, AI use to improve experiences, and transparency.

The objectives alone refer to the benefits generated by this intelligent approach. However, a series of challenges arises regarding the complexity of this set of concepts and practices described. It is also important to highlight that every evolution, social, and technological transformation, with disruptive structures, is accompanied by barriers and roadmaps. The barriers placed in the face of advancement are associated with a concern with changes and losses in jobs, security and data privacy, rampant innovation, social reality in different regions, scarcity of resources, etc. Some reports [12, 24] describe these challenges well, which are illustrated in Fig. 2.3.

It is important to affirm that several municipalities in the world have already reached or are searching for the status of smart cities. Despite the difficulties, these cities serve as an example of leadership in the subject for emerging cities. Also, several solutions appear and are used to mitigate barriers and increase cities' performance, mainly based on technologies.

2.2.3 Digitalization and Connected World

The connected, digitized, and computerized world is a powerful feature of smart cities as has been highlighted so far. It is noticeable that this is the character that underlies the smart city concept and even brings technical feasibility to this proposed approach. In this context of connection, some technologies are driving development and have a significant role in guaranteeing the objectives and aspects mentioned in the previous sections.

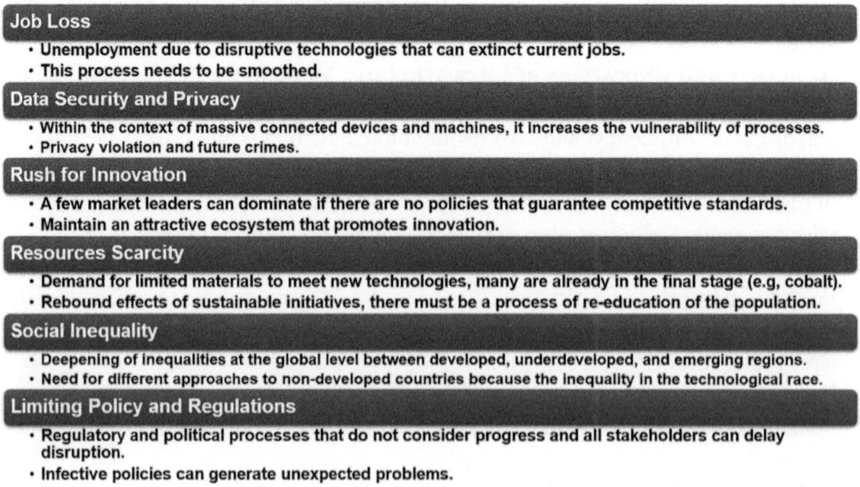

Fig. 2.3 Smart city challenges and barriers

In smart cities, IoT is presented as a tool that provides many particular services that give low-level support to citizens' different applications [28]. This already points out something important within these systems, precisely the variability of devices, equipment, or computers connecting to the same network. These networks are composed of heterogeneous connection loops, communicating through common technologies and protocols, with specific standards. IoT could be explained as a paradigm that allows things to communicate in people's environment through the Internet as if they were computers [34]. Or as objects such as devices, sensors, actuators, and smartphones capable of interacting with each other and cooperating with smart elements to achieve common goals [44]. It is part of this concept that physical objects along with cutting-edge sensors and connectivity transform into smart components and generate a massive amount of data/information. The IoT basis technology implemented at the physical layer is Wireless Sensor Networks. But other technologies, such as Power Line Communications, Bluetooth Low Energy, Radio Frequency Identification (RFID), Digital Enhanced Cordless Telecommunications, Ultra-Wide Bandwidth, and Near Field Communication, are also substantial for IoT applications [41].

In this panorama of increasing the amount of data circulating in the urban environment, Big Data emerges to address solutions precisely to deal with this massive flow. Big data are large and complex datasets that new technologies are necessary before we can use to their full potential [19]. In the document at [14, 45], some key features to define Big Data are listed. The first aspect that can highlight is that machine data production is larger than traditional formats. A smart meter or industrial equipment can produce terabytes in minutes. The second feature is speed, which means that the analysis and treatment transfer solutions must be significantly fast to ensure consistent performance with the smart model. The variety of data types

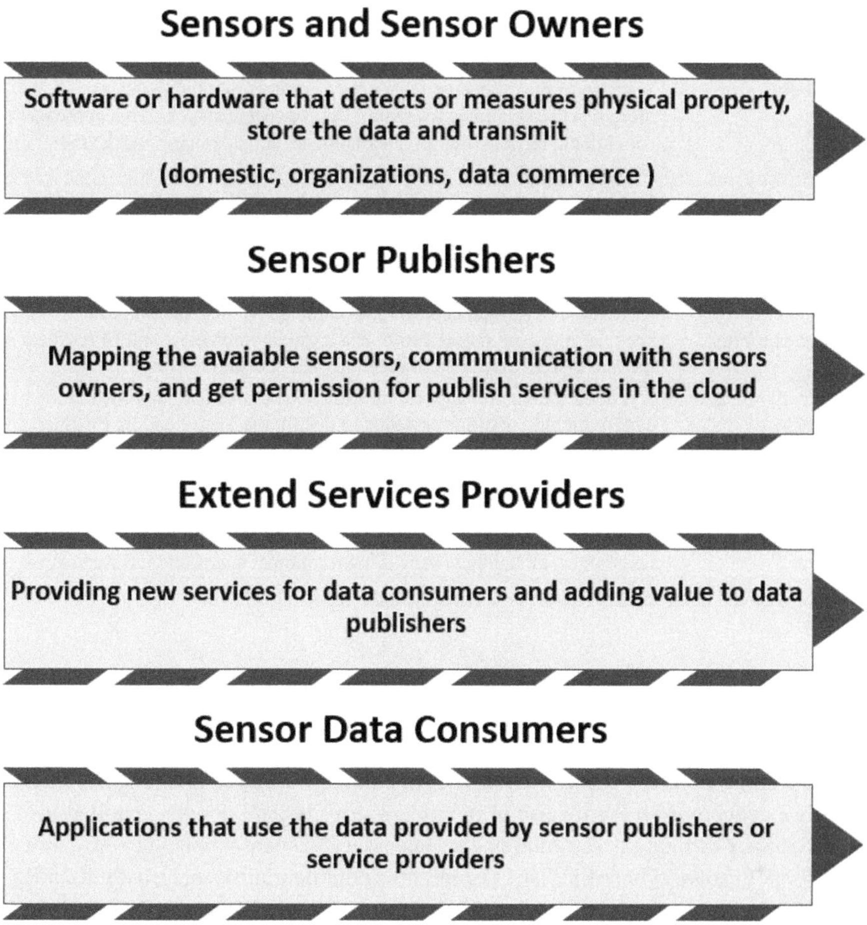

Fig. 2.4 Sensor as services model

is another crucial point, and platforms or applications must be prepared to meet the different types of data possible in intelligent and computerized systems. Finally, the value attribute is presented to differentiate the information in the datasets, that is, in a large universe of data, which are actually representative for a given purpose.

The use of sensors as services also complements these technological biases of the connected city, which may say that the sensor mesh is the glue that unites these different points, parallel with ICTs. Each device connected in a smart city can contain several sensors measuring different variables simultaneously. The sensor as services model is described in some literature [11, 29], and it consists of four conceptual layers: sensors and sensor owners, sensor publishers, extended service providers, and sensor data consumers. Figure 2.4 illustrates this model arrangement.

Table 2.1 Core aspects for Cybersecurity Framework[a]

Core	Description
Digital trust platform	A platform that allows reliable untied connections and manages identities and relationships within the connected system. This approach should help cities to identify, authenticate and authorize people and devices through security mechanisms.
Privacy-by-design	It is a concept that aims to protect people's privacy through the insertion of these concepts in the design process of technologies, processes, and infrastructure.
Cyberthreat intelligence and analysis platform	A platform that identifies the aspects that improve the treatment of data to increase performance. Using AI techniques, this platform can provide high performance from less explicit parameters and better feed the universe of discourse in planning and decision making.
Cyber response and resiliency	It is about being prepared for cyberattacks. It also includes developing a cybernetic investigation resource to inform the identification, treatment, containment, and prevention of organized attacks on different smart city systems.
Cyber competencies and awareness program	This core is about the need for skilled cybersecurity labor. The organizations should maintain concern with training, programs, and task forces, including restructuring traditional teams and infrastructures.

[a]Source: Adapted from "Making smart cities cybersecure" report, 2019 [13]

In this connected world of smart cities, with such complexity and massive information exchange, the cybersecurity concern is growing. The number of devices connected in a network or cloud, with internet access, means that there is more access point, increasing the vulnerability. As pointed out in the challenges section, there is a concern with the misuse of private or security data or even criminal invasions to attack critical infrastructures. It is necessary to spare no effort to structure a framework to solve cybernetic risks to guarantee confidentiality, integrity, reliability, safety, and robustness. It also includes the use of standards, protocols, regulatory policies aimed at greater cybernetic security. In the technical report prepared by Deloitte [13], the primary nuclei for a cybersecurity approach are detailed. Table 2.1 shows the detail described on Deloitte for each vacancy.

These commented aspects form the critical points of the digitized and connected world of smart cities. In this way, it is possible to understand better how the IoT provides technological and social development in urban spaces.

2.3 IoT Enabling Smart Cities

When observing the entire panorama of smart cities, it is noticeable that technological development and the connected framework are enablers of the smart proposal's objectives and benefits. With IoT as a backbone, smart cities can use information resources at a high level of integration and interoperability to provide key urban development elements [23]. IoT's influence leads to a new era of applications and services, and the key components are sensors, mobile phones, RFID tags, etc. [7].

Combining different types of devices with smart devices and ICs, mainly wireless communication, the possibility of new applications and solutions based on adaptive systems, cloud intelligent systems, and innovative products and services in different areas is created. One of the disruptive effects of the IoT is the fusion of information technology with other technologies, like consumer technology, medical devices, or vehicles [12]. It allows the people and things to be interconnected anytime, anywhere, anyplace, anything, and anytime using any pathway or any service [33].

Among the various areas with applications for IoT, the following can be highlighted in the context of smart cities:

1. **Smart Grid:** Electrical Power grids have also undergone transitions and modernization to a more sustainable and robust format. The smart grid uses innovative technologies such as intelligent and autonomous controllers, advanced software for data management, and two-way communications between power utilities and consumers to create an automated and distributed advanced energy delivery network [15, 21]. IoT has shown to be a key technology and with outstanding suitability for applications to face the challenges. In addition to supporting the expansion of distributed generation, mainly with the increase in renewable sources penetration, smart meters and advanced monitoring and communication technologies bring many benefits. Such as demand response programs, high network performance management, greater participation of the consumer in operation, increase in energy efficiency, and increase in power systems' reliability in general.
2. **Environment Monitoring:** Concern about climate change means that organizations and governments are continually monitoring these environment factors. Smart cities use a range of sensors to collect climatic data and various parts of the urban space to assess pollution, air quality and observe climatic disturbances. Through IoT, these sensors bring a range of information that feeds the universe of research discourse and applications associated with entities that promote sustainability and combat problems related to the environment.
3. **Smart Water Supply:** The IoT is applied to the network of sensors and smart meters to monitor the network of water supply within cities. With this, it is possible to increase the performance of the supply operation, more leak detection, reduce water waste, and help the correct and efficient use of this resource together with better consumption management.
4. **Waste Management System:** One important feature in waste management is environmental sustainability [9]. The use of IoT is in the implantation of sensors in garbage bins. This way, the collection centers, and recycling entities can have information on the waste. It can assist in directing recycling, selective and efficient collectors, better management of the collection and the staff involved, and re-education of the population about waste disposal.
5. **Smart Traffic:** Strategy widely used in large urban centers and smart cities consists of adaptive traffic management, provided with real-time data and

information through traffic sensors and cameras [5]. The objective is to avoid congestion and also improve urban mobility [4]. These devices can also help manage public transport and bring forecasts or aid in decision making for planning. The advances lead to communication between vehicles and information derived from them, which can also increase the range of applications and increase urban traffic performance.

6. **Smart Parking:** The implementation of IoT in this area consists of monitoring parking spaces, so that city users have access to information about parking location and available spots, in addition to benefits such as greater ease for online payments. It is a significant bottleneck in municipalities, especially in large centers. I also offer re-education and an intelligent solution to avoid traditional measures such as fines-only control.
7. **Smart Healthcare:** Intelligent medical treatment involves the use of devices that provide information on patients in real-time and monitor hospital beds and organize the care structure. A Wireless Body Area Network, which is based on a low-cost wireless sensor network technology, could benefit patient monitoring systems in hospitals, residential, and work environments [30]. The sensors have the capability of measuring blood flow, respiratory rate, blood pressure, blood PH, body temperature, among others, which are collected and analyzed by remote servers [43]. Citizens of smart cities can benefit from personalized treatment, better medical care, disease prevention, improved prescription of medicines, etc.
8. **Smart Factory:** This concept is widespread as Industria 4.0, and it is the information revolution within the industrial environment of several segments. Through IoT, the phenomenon involves a de-verticalization of the means of production. The information is not contained only in the field of management but extends to the factory floor. It allows better management of resources, allocation of employees, reliability, and security. Advanced ICT solutions in all industry layers allow a more excellent and better exchange of information, characterized by equipment and intelligent devices. For example, a flow control valve in the field traditionally is only a final control element. However, in Industry 4.0, it may also have another role based on the information generated by intelligent valves that have smart control embedded, health monitoring, data management for providing valuable data for the company.

Figure 2.5 illustrates the roles of the IoT in enabling smart cities. These concepts are of paramount importance to understand how Software-Defined Networks is a solution in IoT and the connected world due to its complexity. Software-Defined Networking (SDN) emerges as one of the noteworthy forms of networking concepts, which helps in lubricating a convenient and efficient network control flow that facilitates the cost of investment, which avails a considerable number of users [36]. SDN can be used as the overlay for the implementation of IoT into the real world [7]. It will be discussed intensely further in this book.

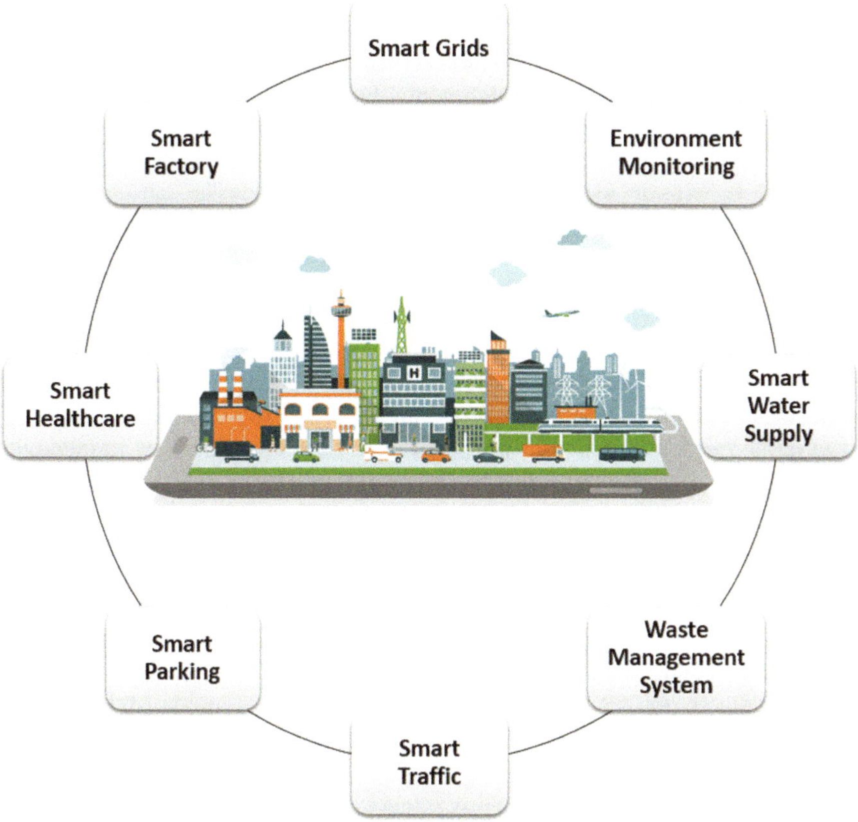

Fig. 2.5 IoT applications for smart cities

2.4 Case Studies

Some case studies will exemplify how smart cities, the IoT, and the connected world are developing in real municipalities to consolidate the concepts discussed in this chapter. This visualization is important to indicate that certain places in the world are reaching important status and with applications that impact society.

- **Harnessing City Data—Singapore [42]:** Singapore has used data and information to improve services and create economic value. Several applications are benefiting from this approach, with emphasis on the health area. The city-state has an elderly population that needs assistance to have an independent life and quality of care. ConnectedLife is an example of an application in the healthcare industry. By combining medical-grade sensor technology, and cutting-edge AI, the ConnectedLife solution facilitates early detection and intervention of various common chronic conditions. It enables continuous monitoring and personalized treatment to improve the quality of life and clinical outcomes of people living with chronic conditions. Real-time information and insurance companies with better data and allows families to protect their elderly members.

- **Water-Conscious Urban Development—Fukuoka, Japan [16, 42]:** Use of IoT and ICT for water supply programs. The city has developed a system that can monitor and control the water flow and supply pressure employing specific sensors. It helps in identifying intervention of water leaks. The information generated by the sensors makes it possible, based on more faithful models, to reconnect forecasts of water demand and effective distribution in the supply. It is also possible to observe a re-education process to not spare efforts for efficient water management.
- **Cubes for Smart Recycling—Seoul, South Korea [37]:** In one of the world's largest urban areas, there was a problem with the frequency of garbage collection and insufficient public waste bins. Presenting overflow of waste bins and also low recycling rates. In partnership with the Ecube Labs company, they installed 85 Clean Cubes for common and recyclable waste in the city center to resolve the issues. With this, they were able to monitor the status and level of filling of the Clean Cubes in real-time and observe and report the efficiency of the collection in the capital. As a result, there was also an improvement in the performance of route management for collection. The associated benefits are eliminating overflow of waste bins, 66% reduction in the frequency of collection, cleaner public areas, and reduction of collection costs by 83%.
- **Using data to predict and mitigate floods—Calgary, Canada [37]:** The city of Calgary is experiencing seasonal flood problems, and these events are regular but unpredictable. Seasonal floods usually bring water quality issues and can also cause costly damage, around millions of dollars. The first players in this type of service to contain this problem did not directly access the data, which slowed the response time and increased costs. The solution was to use and expand Plant Information (PI) systems that collect upstream river flows and rainfall data. The PI system also allows for a high-performance report of water quality. And the front line for responses to these types of events began to access data and information directly. The problems were mitigated, thus presenting high performance in forecasting, response, and monitoring.
- **Envision Charlotte's Smart Energy Now—Charlotte, USA [37]:** The city of Charlotte, approximately in 2016, was looking for an expansion and a boost in urban growth and development. As a result, the challenges of large cities' evolution arise, which even come up against energy consumption. With that, Duke Energy invited Verizon Wireless to run an innovative project focused on educating office workers about changes in habits and ways of working that can collectively impact energy use in the city. Envision's smart grid captured information from 61 buildings using smart meters. This information fed into the program's central office and also provided payment for the individual buildings. Thus were raised tools and directions in the use of energy, such as: turning off unused lights, adjusting thermostats for different situations, and reducing hours with the light-on. Envision Charlotte has reduced energy consumption by 8.4%,

an estimated USD 10 million in savings. The case study also spurred other similar programs serving different pillars, such as water consumption and waste management.

Thus, the chapter concludes the contextualization of smart cities in the context of the IoT and the connected world, going through all the key points and ending with examples that illustrate the application of contexts in different regions.

References

1. An, J., Le Gall, F., Kim, J., Yun, J., Hwang, J., Bauer, M., ... Song, J. (2019). Toward global IoT-enabled smart cities interworking using adaptive semantic adapter. *IEEE Internet of Things Journal, 6*(3), 5753–5765.
2. Arroub, A., Zahi, B., Sabir, E., and Sadik, M. (2016, October). A literature review on Smart Cities: Paradigms, opportunities and open problems. In *2016 International Conference on Wireless Networks and Mobile Communications (WINCOM)* (pp. 180–186). New York: IEEE.
3. Aujla, G. S., & Jindal, A. (2020). A decoupled blockchain approach for edge-envisioned IoT-based healthcare monitoring. *IEEE Journal on Selected Areas in Communications, 39*(2), 491–499.
4. Aujla, G. S., Jindal, A., & Kumar, N. (2018). EVaaS: Electric vehicle-as-a-service for energy trading in SDN-enabled smart transportation system. *Computer Networks, 143*, 247–262.
5. Aujla, G. S., Kumar, N., Singh, M., & Zomaya, A. Y. (2019). Energy trading with dynamic pricing for electric vehicles in a smart city environment. *Journal of Parallel and Distributed Computing, 127*, 169–183.
6. Aujla, G. S. S., Kumar, N., Garg, S., Kaur, K., & Ranjan, R. (2019). EDCSuS: Sustainable edge data centers as a service in SDN-enabled vehicular environment. IEEE Transactions on Sustainable Computing. https://doi.org/10.1109/TSUSC.2019.2907110
7. Babbar, H., & Rani, S. (2020). Software-defined networking framework securing internet of things. In *Integration of WSN and IoT for Smart Cities* (pp. 1–14). Cham: Springer.
8. Bibri, S. E. (2018). The IoT for smart sustainable cities of the future: An analytical framework for sensor-based big data applications for environmental sustainability. *Sustainable cities and society, 38*, 230–253.
9. Bogatinoska, D. C., Malekian, R., Trengoska, J., & Nyako, W. A. (2016, May). Advanced sensing and internet of things in smart cities. In *2016 39th International Convention on Information and Communication Technology, Electronics and Microelectronics (MIPRO)* (pp. 632–637). New York: IEEE.
10. Campbell, T. (2009). Learning cities: Knowledge, capacity and competitiveness. *Habitat International, 33*(2), 195–201.
11. Chamoso, P., González-Briones, A., Rodríguez, S., & Corchado, J. M. (2018). Tendencies of technologies and platforms in smart cities: a state-of-the-art review. In *Wireless Communications and Mobile Computing* (Vol 2018).
12. Deloitte (2015). *Smart Cities: How rapid advances in technology are reshaping our economy and society*. Deloitte, 1, 1–86. https://www2.deloitte.com/tr/en/pages/public-sector/articles/smart-cities.html
13. Deloitte (2019). *Making Smart Cities Cybersecurity, Deloitte Center for Government Insights*. https://www2.deloitte.com/us/en/insights/focus/smart-city/making-smart-cities-cyber-secure.html
14. Dijcks, J. P. (2012). *Oracle: Big Data for the Enterprise*. Oracle White Paper, 16.
15. Fang, X., Misra, S., Xue, G., & Yang, D. (2011). Smart grid—The new and improved power grid: A survey. *IEEE Communications Surveys and Tutorials, 14*(4), 944–980.

16. Fukuoka City Government (2014). *Fukuoka City Visit and Training*, Fukuoka City Government. http://www.city.fukuoka.lg.jp/data/open/cnt/3/19077/1/00guideenglishall.pdf?20161109140138
17. Giffinger, R., Fertner, C., Kramar, H., & Meijers, E. (2007). *City-ranking of European medium-sized cities*. Centre of Regional Science, Vienna UT, 1–12.
18. Harrison, C., Eckman, B., Hamilton, R., Hartswick, P., Kalagnanam, J., Paraszczak, J., & Williams, P. (2010). Foundations for smarter cities. *IBM Journal of Research and Development, 54*(4), 1–16.
19. Jessen, J. (2015). *How to Create a Smart City? Co-Creation of a Smart City with Citizens* (Doctoral dissertation, Master Thesis, Eindhoven University, Holland. http://www.digitalbydel.dk/wp-content/uploads/2015/01/MA_Guenter_final.pdf)
20. Jindal, A., Aujla, G. S., Kumar, N., Prodan, R., & Obaidat, M. S. (2018, December). DRUMS: Demand response management in a smart city using deep learning and SVR. In *2018 IEEE Global Communications Conference (GLOBECOM)* (pp. 1–6). New York: IEEE.
21. Jindal, A., Aujla, G. S., Kumar, N., & Villari, M. (2019). GUARDIAN: Blockchain-based secure demand response management in smart grid system. *IEEE Transactions on Services Computing, 13*(4), 613–624.
22. Kirimtat, A., Krejcar, O., Kertesz, A., and Tasgetiren, M. F. (2020). Future trends and current state of smart city concepts: A survey. *IEEE Access, 8*, 86448–86467.
23. KPMG, (2019). *Internet of Things in Smart Cities*. KPMG, https://assets.kpmg/content/dam/kpmg/in/pdf/2019/05/urban-transformation-smart-cities-iot.pdf
24. KPMG (2020), *Smart Cities – Adoption of Future Technologies*, KPMG, 1-24, https://worldengineeringday.net/wp-content/uploads/2020/03/Smart-City-IOT-WFEO-Version-1.pdf
25. Kumar, N., Aujla, G. S., Das, A. K., & Conti, M. (2019). ECCAuth: A secure authentication protocol for demand response management in a smart grid system. *IEEE Transactions on Industrial Informatics, 15*(12), 6572–6582.
26. Lynn Jr, L. E., Heinrich, C. J., & Hill, C. J. (2000). Studying governance and public management: Challenges and prospects. *Journal of Public Administration Research and Theory, 10*(2), 233–262.
27. Marsal-Llacuna, M. L., Colomer-Llinàs, J., & Meléndez-Frigola, J. (2015). Lessons in urban monitoring taken from sustainable and livable cities to better address the Smart Cities initiative. *Technological Forecasting and Social Change, 90*, 611–622.
28. Minerva, R., Biru, A., & Rotondi, D. (2015). Towards a definition of the Internet of Things (IoT). *IEEE Internet Initiative, 1*(1), 1–86.
29. Perera, C., Zaslavsky, A., Christen, P., & Georgakopoulos, D. (2014). Sensing as a service model for smart cities supported by internet of things. *Transactions on Emerging Telecommunications Technologies, 25*(1), 81–93.
30. Poon, C. C., Zhang, Y. T., & Bao, S. D. (2006). A novel biometrics method to secure wireless body area sensor networks for telemedicine and m-health. *IEEE Communications Magazine, 44*(4), 73–81.
31. Rana, N. P., Luthra, S., Mangla, S. K., Islam, R., Roderick, S., & Dwivedi, Y. K. (2019). Barriers to the development of smart cities in Indian context. *Information Systems Frontiers, 21*(3), 503–525.
32. Ribeiro, P. F., Polinder, H., & Verkerk, M. J. (2012). Planning and designing smart grids: Philosophical considerations. *IEEE Technology and Society Magazine, 31*(3), 34–43.
33. Sahoo, K. S., Sahoo, B., & Panda, A. (2015, December). A secured SDN framework for IoT. In *2015 International Conference on Man and Machine Interfacing (MAMI)* (pp. 1–4). New York: IEEE.
34. Said, O., & Masud, M. (2013). Towards Internet of Things: Survey and future vision. *International Journal of Computer Networks, 5*(1), 1–17.
35. Salles, R. S., de Souza, A. Z., Ribeiro, P. F. Exploratory research of social aspects for Smart City development in Itajubá. In *2020 IEEE International Smart Cities Conference (ISC2)* (pp. 1–8). New York: IEEE.

36. Sezer, S., Scott-Hayward, S., Chouhan, P. K., Fraser, B., Lake, D., Finnegan, J., ... Rao, N. (2013). Are we ready for SDN? Implementation challenges for software-defined networks. *IEEE Communications Magazine, 51*(7), 36–43.
37. Smart City Council (2021). *Examples and Case Studies.* https://smartcitiescouncil.com/smart-cities-information-center/examples-and-case-studies
38. Sproull, L., & Patterson, J. F. (2004). Making information cities livable. *Communications of the ACM, 47*(2), 33–37.
39. Sun, J., & Poole, M. S. (2010). Beyond connection: Situated wireless communities. *Communications of the ACM, 53*(6), 121–125.
40. Verkerk, M. J., Ribeiro, P. F., Basden, A., & Hoogland, J. (2018). An explorative philosophical study of envisaging the electrical energy infrastructure of the future. *Philosophia Reformata, 83*(1), 90–110.
41. O. Vermesan & P. Friess (Eds.) (2013). *Internet of things: Converging technologies for smart environments and integrated ecosystems.* Denmark: River Publishers.
42. World Economic Forum - Global Future Council on Cities and Urbanization (2020), *Smart at Scale: Cities to Watch - 25 Case Studies*, World Economic Forum.
43. Zeadally, B. H. R. K. S., & Khoukhi, A. F. L. (2017). *Internet of Things (IoT) Technologies for Smart Cities.* https://doi.org/10.1109/ISTAFRICA.2016.7530575
44. Zheng, J., Simplot-Ryl, D., Bisdikian, C., & Mouftah, H. T. (2011). The Internet of Things [Guest Editorial]. *IEEE Communications Magazine, 49*(11), 30–31.
45. Zikopoulos, P., & Eaton, C. (2011). *Understanding big data: Analytics for enterprise class Hadoop and streaming data.* New York: McGraw-Hill Osborne Media.

Part II
Software-Defined Networking

Chapter 3
Challenges of Traditional Networks and Development of Programmable Networks

Fanglin Liu, Godfrey Kibalya, S. V. N. Santhosh Kumar, and Peiying Zhang

3.1 Introduction

The increasing development of science and technology is gradually changing our lives. Among them, the most closely connected with people's lives is the Internet [18, 20]. The emergence and development of the Internet has not only subverted the traditional media industry but also has a revolutionary impact on the basic structure and standards of the entire society and economy [8, 19]. The Internet is the abbreviation of Computer Interactive Network, which is a huge network formed by the series connection between the network and the network [15, 22]. These networks are connected by a set of common protocols to form a logically single and huge global network. Computer network is the foundation of the information society. It connects multiple computer systems that are scattered in different locations and have independent functions with communication equipment and lines, communicates with each other under the support of network protocols and software, and finally realizes the sharing of network resources and real-time interaction of information [13, 14, 24]. However, although the existing design principles of simple and easy access to the Internet have brought convenience to its development, it exposes

F. Liu · P. Zhang (✉)
College of Computer Science and Technology, China University of Petroleum (East China), Qingdao, P.R. China
e-mail: 1556447740@qq.com; zhangpeiying@upc.edu.cn

G. Kibalya
Department of Network Engineering, Universitat, Politecnica de Catalunya, Barcelona, Spain
e-mail: Godfrey.mirondo.kibalya@upc.edu

S. V. N. Santhosh Kumar
School of Information Technology and Engineering, Vellore Institute of Technology, Vellore, India
e-mail: anthoshkumar.svn@vit.ac.in

G. S. Aujla et al. (eds.), *Software Defined Internet of Everything*, Technology, Communications and Computing, https://doi.org/10.1007/978-3-030-89328-6_3

inherent drawbacks in the face of massive data transmission requirements and large-scale business application environments [10, 23, 25]. For example, in a traditional network architecture, in order to meet a specific application requirement, it usually needs to include a large number of hardware devices. However, a noteworthy problem is that network devices produced by different manufacturers usually require different ways to debug and configure, and the command-line debugging interfaces used to manage the devices are also different. Therefore, in a network that mixes equipment from multiple different vendors, managing and deploying the network is a very big challenge [12, 16]. Moreover, in a typical distributed architecture, information is transmitted between devices in the form of a "baton," which will lead to the emergence of redundant data traffic. In addition, with the development of various technologies, the volume of network traffic continues to increase, and the volume of the network gradually increases, which leads to the gradual increase of the time for transmitting information between nodes, which is obviously not reasonable. In addition, the inability to perform intelligent flow control and visualized network status monitoring based on network conditions is also a problem that hinders further development.

In general, the problems that the traditional network architecture may face can be summarized as follows: (1) The traditional network lacks a global concept and cannot control traffic from a macro level; (2) The traditional network structure is difficult to deploy and manage; (3) The way of information exchange between current routing equipment will cause unnecessary bandwidth occupation. In general, the core problem is that there is a contradiction between the diverse and changeable network upper-layer applications and business needs and the current stable and rigid traditional network architecture, and an appropriate solution is urgently needed [27–31, 34, 36, 38]. Based on the above problems, Software-Defined Network (SDN) is a better solution [5, 6]. The predecessor of SDN was a project called Ethane, which allowed network administrators to easily define security control strategies based on network traffic through a centralized controller [3]. In addition, by applying these security policies to network devices, the security control of node communication in the entire network is realized. Based on the inspiration of this project, the concept of OpenFlow was born. In this concept, the switch does not have an independent computing center but only has the forwarding function, and all path calculations, security policies, and other tasks are all done by the controller [9]. Generally speaking, SDN is not a specific technology, but an idea and a framework. As long as the hardware in the network can be managed and controlled through centralized software, and the network has programmability, and the control and forwarding levels in the hardware devices are separated, the network can be considered an SDN network [4, 21]. Therefore, SDN in the narrow sense refers to software-defined networking, while the concept of SDN in the broad sense has extensions in more areas [1]. In general, SDN has the following three advantages: (1) SDN can change the tightly coupled architecture of applications and networks under traditional networks and improve the level of network resource pooling; (2) SDN networks can realize automatic network deployment and configuration and support rapid business launches and flexible expansion; (3) By introducing programmable features, automated network services and protocol scheduling can be realized [11, 17, 26]. At present, the market share of SDN is increasing year by year, and the trend is improving, and it has applications in many fields. However, due to some known challenges, although some feasible solutions are slowly available, SDN is still far from large-scale deployment [2, 7]. These challenges include some technical challenges and non-technical challenges, which will be analyzed in detail later.

Starting from the traditional Internet architecture, this article first reviews the development history of the Internet and hardware devices in the network and gives examples of four common network architectures to analyze and summarize the problems existing in the traditional Internet architecture. Secondly, we introduced the future network architecture-SDN, including determining the definition of SDN, analyzing the necessity of its emergence and the advantages of the architecture, explaining the core technologies it includes, and analyzing the changes and development of the industrial chain under the influence of the architecture. In addition, we further introduced its feasible application scenarios and introduced some existing solutions to reflect the advantages of the SDN architecture compared with the traditional architecture. Finally, we analyzed and summarized the threats and challenges that the SDN architecture may face and looked forward to its future development. This article comprehensively introduces and compares the advantages and disadvantages of the two different architectures and analyzes them with examples, which can lay the foundation for the follow-up research of SDN.

3.2 Traditional Network Architecture

This section discusses the evolution and architecture of network.

3.2.1 Internet Development History

In the 1960s, the US Department of Defense decided to study a distributed command system, the core value of which is that even if several nodes are destroyed, other nodes can still maintain communication. In 1966, Robert Taylor, the third director of IPTO, believed that a compatible protocol should be established to allow all terminals to communicate with each other. In the same year, the new communication network project (named ARPANET) completed the internal project, and since then, the entire project was actually initiated. In the first phase of the project, a network of four nodes was established, and its geographic location is shown in Table 3.1.

The packet switching technology is adopted between the four nodes, and they are connected through a special IMP device and a communication line with a rate of 50kbps. Among them, the role of IMP includes connection, scheduling, and management and is generally regarded as the prototype of a router. Two years later, the original communication protocol, also known as the Network Control Protocol

Table 3.1 The table of the geographic location of the four nodes

Node	Place	Host	OS
Node 1	Network Measurement Center, UCLA	SDS SIGMA 7	SEX
Node 2	The Network Information Center at SRI	SDS 940	Genie
Node 3	The Culler-Fried Interactive Mathematics Center at the UCSB	IBM 360/75	OS/MVT
Node 4	The School of Computing at UTAH	DEC PDP - 10	Tenex

(NCP), was born. However, with the continuous increase of network nodes, it has brought a lot of pressure to the NCP protocol, and further optimization is urgently needed. For this reason, in the end, a stable (fourth generation) TCP/IP protocol was born under continuous development and gradually became the mainstream.

In the next phase, ARPANET was replaced by NSFnet and became the connection between the new supercomputer research centers, and its speed was more than 25 times that of the ARPANET network. In the late 1980s, the number of computers connected to NSFnet far exceeded ARPANET. Therefore, in the early 1990s, ARPANET was formally dismantled. Then, concepts and technologies such as the World Wide Web, Hypertext Transfer Protocol (HTTP), Hypertext Markup Language (HTML), and web browsers were proposed and developed. Since then, the Internet has truly has become the global Internet and has begun to enter people's lives.

The development of the network is inseparable from the upgrade of hardware. With the appearance of personal computers, smart phones, and the development of mobile communication technology, the Internet has gradually entered a new stage of development, namely the era of mobile Internet. Mobile Internet is the product of the integration of mobile and Internet, inheriting the advantages of mobile anytime, anywhere, portable and Internet open, sharing, and interactive. It is a new generation of open telecommunications infrastructure network with high-quality telecommunications services. Compared with the traditional Internet, the mobile Internet emphasizes that it can be used anytime, anywhere and can access the Internet and use application services in a high-speed mobile state. Mobile Internet related technologies are generally divided into three parts, namely mobile Internet terminal technology, mobile Internet communication technology, and mobile Internet application technology. The advantages of mobile Internet, such as interactivity, portability, privacy, positioning, and entertainment, have led to rapid growth in people's demand for the Internet. At this stage, user experience has gradually become the supreme pursuit of the development of terminal operating systems. In addition, with the rapid development of mobile communication technology, unlike the previous technologies that only provide bandwidth mobile communication between people, 5G, as a mobile communication system for the needs of human information society after 2020, will penetrate more Fields, such as the Internet of Things, industrial networks, medical and rescue, transportation, etc., to achieve a comprehensive interconnection of all things. Therefore, in the future, the Internet will penetrate into all areas of social life and will be more closely related to people's lives.

3.2.2 *Equipment Development History*

Network cable, network card, and protocol cable are the three elements that make up the smallest unit network. Among them, the network cable provides the physical medium, carries the bit stream and electrical signals, the network card performs data processing, and the protocol cable can realize data analysis, addressing, flow control, etc. in the communication process. However, once the distance between the terminals is too far, exceeding the upper limit of the physical transmission distance of the network cable, data will begin to be lost. For this reason, a repeater is born. The repeater can relay and amplify information, and its appearance enables long-distance transmission between devices. After that, considering the limitation of the repeater interface, in order to solve the problem that it could not realize the long-distance data communication between multiple hosts, the hub was born. The hub can be described as a multi-interface repeater. Data received from any interface can be transmitted to all other interfaces by flooding.

However, because the hub cannot identify the addressing information and upper-layer content of the data packet, it cannot isolate the end host. This will lead to a reduction in bandwidth utilization if multiple hosts are in the same collision domain. Based on this problem, the bridge provides a solution. The bridge is a link layer product. By recording the MAC address of the terminal host and generating a MAC table, the data flow between the hosts can be forwarded on this basis. The bridge can isolate the conflict domain, and because the data between different interfaces will not conflict with each other, this will improve the bandwidth utilization. However, the limited interface of the bridge made its ability to isolate network conflicts relatively limited, so the switch was born. The switch has been extended and upgraded on the basis of the network bridge. Compared with the network bridge, it has several main advantages, such as: (1) The number of interfaces is more dense, and the bandwidth utilization rate is greatly improved. (2) Adopt dedicated ASIC hardware chip to realize high-speed forwarding. (3) Not only can the conflict domain be isolated, but also the broadcast domain can be isolated through VLAN. On the basis of the switch, in order to solve the long-distance wide area network communication problem, the router was born.

The router has the function of judging the network address and selecting the IP path. It can construct a flexible link system in multiple network environments, and link each subnet through different data packets and media access methods. In addition, in order to solve the shortcomings of limited communication, wireless AC/AP came into being. And, in order to further improve network security and performance, firewalls and flow control devices were born. To sum up, in order to achieve different goals and solve different problems, there are a large number of different types of devices in the network. In the next section, we will introduce

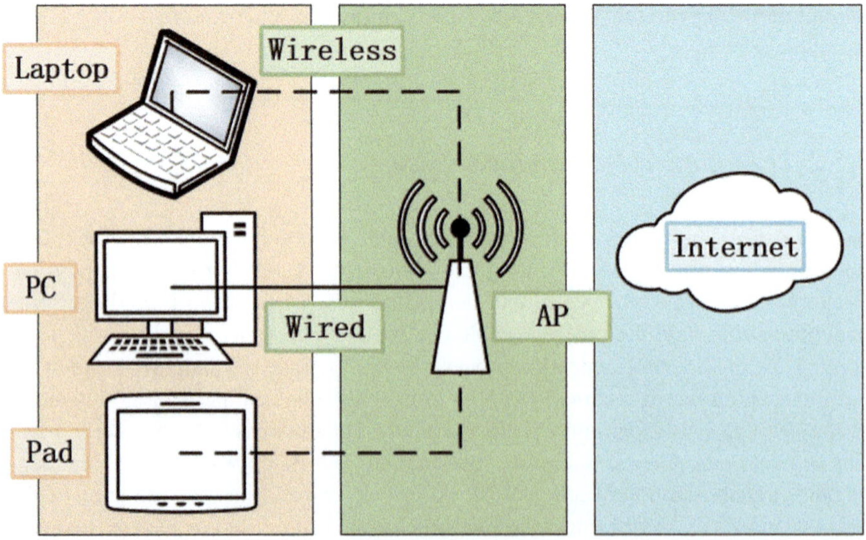

Fig. 3.1 The diagram of a typical home network architecture

several common network architectures and analyze the architectures, technologies used, and the equipment involved.

3.2.3 Typical Architecture

The typical network architecture is discussed in the subsequent sections.

3.2.3.1 Home Network

Figure 3.1 shows a typical home network that provides WiFi hotspot access through a wireless router and connects to the external network through a router, which usually includes a wireless router. The technologies used include WiFi, NAT, PPPOE, DHCP, etc.

3.2.3.2 Campus Network

Figure 3.2 shows the most common campus network architectures such as large and medium-sized enterprise networks or campus networks, which usually use access layer, aggregation, core layer three-layer architecture, and dual-core network. According to different needs, it is usually divided into user area, internal server,

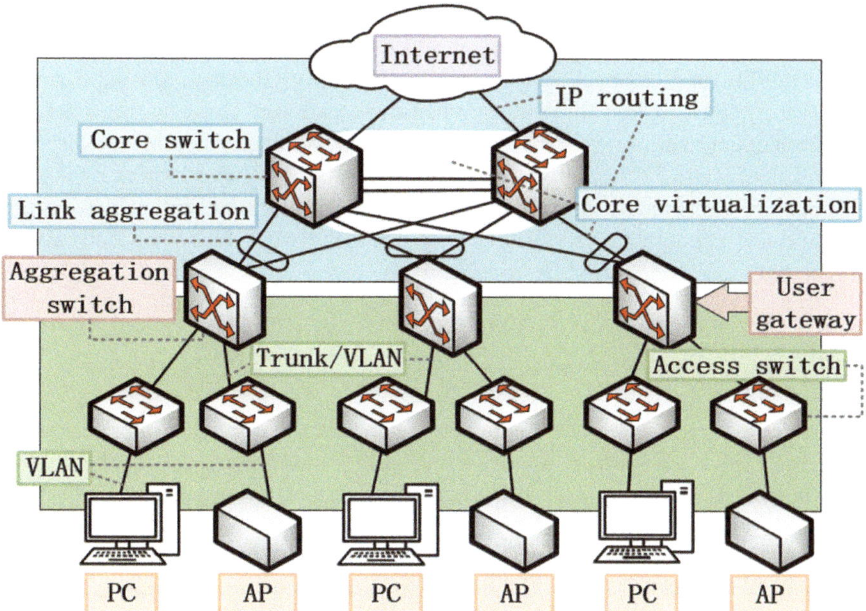

Fig. 3.2 The diagram of a typical campus network architecture

external service area, management area, Internet area, etc., and they are connected and isolated through core switches and firewalls. In addition, the Internet uses multi-outlet connections, dial-up and NAT through routers, and load balancing through flow control products.

In the campus network architecture, the core layer usually does not include any processing of data packets/frames, because this will reduce the speed of packet switching. The main function of the core layer is to provide high-speed connections between the various convergence layer devices in the campus network. The function of the convergence layer is to define the boundaries of the network, and the processing of data packets/frames is also completed at this layer. In addition, in the campus network environment, the access layer usually includes the following functions: shared bandwidth; exchange bandwidth; MAC layer filtering; and micro-segmentation. This type of architecture usually includes routers, switches, wireless AC/AP, firewalls, load balancers, and servers. The technologies involved usually include VLAN, TRUNK, MSTP, HSRP/VRRP, Etherchannel, WLAN, NAI, ACL, SNMP, etc.

3.2.3.3 Government Affairs Network

The government network usually includes the government, electric power, public security, etc., adopts the metropolitan area network architecture, and is designed through MPLS technology. Different regions and cities are connected to the CE through the convergent PE and divided into different VRFs. The core equipment acts as a P/PE for high-speed forwarding. In addition, VRF is used to isolate access between different regions and cities. When it is necessary to access the Internet, government extranet, servers, etc., moderate mutual visits can be achieved through the design of RD/RT. The equipment it contains usually includes routers, switches, load balancers, firewalls, intrusion prevention, DDOS attack prevention equipment, etc. And the technologies involved usually include VLAN, TRUNK, OSPF, BGP, MPLS, QoS, AVL, NAT, SNMP, and security.

3.2.3.4 Data Center Network

In traditional large-scale data centers, the network is usually a three-tier structure, including the access layer, the convergence layer, and the core layer. The access layer is sometimes called Edge Layer. Access switches are usually located at the top of the rack, so they are also called ToR (Top of Rack) switches, and they are physically connected to the server. Aggregation Layer is sometimes called Distribution Layer. The aggregation switch connects to the Access switch and provides other services, such as firewall, SSL offload, intrusion detection, network analysis, etc. The core switch of the Core Layer provides high-speed forwarding for packets entering and leaving the data center and provides connectivity for multiple aggregation layers. The core switch provides a flexible L3 routing network that usually provides the entire network. In addition, in the early data centers, most of the traffic was north-south traffic. However, with the development of technology, the content and form of data have also changed. For example, most of the traffic in a traditional data center is communication between clients and servers. However, with the gradual rise of technologies such as distributed computing and big data, some applications will generate a large amount of traffic between servers in the data center. Therefore, the east-west traffic is increasing significantly. In addition, the software-defined data center requires that the computing storage network of the data center can be software-defined, while the traditional three-tier network architecture did not consider SDN at the beginning of the design. In general, technological development requires new data centers to have smaller over-subscription and needs to provide higher east-west traffic bandwidth and support for SDN. In addition, the largest data center corresponds to the network equipment with the largest volume and the highest performance. Not all network equipment vendors can provide equipment of this scale, and the corresponding capital costs and operation and maintenance costs are also high. Therefore, the use of traditional three-tier network architecture makes enterprises face the dilemma of cost and scalability. This means that we need a new way to resolve this contradiction.

3.2.4 Conclusion of Issues

The life cycle of a network system usually includes four stages: demand investigation, planning and design, deployment and implementation, and operation and maintenance. Based on this cycle, a huge network architecture has now been formed, effectively realizing multiple applications between people and people and data, which has played an important role in promoting economic and social development. However, with the vigorous rise of technologies such as big data, cloud computing, Internet of Things, and mobile Internet, Internet applications are becoming increasingly diversified and business volumes are increasing. Therefore, the current network architecture is gradually unable to meet the demand, and the existing problems are becoming more prominent. For example, (1) The traditional network lacks a global concept and cannot control traffic from a macro perspective. Each router calculates the next hop according to its own dynamic routing protocol, but due to the lack of a global concept, it will cause a lot of waste of resources; (2) Due to many network equipment manufacturers, equipment types, inconsistent control commands, etc., the traditional network structure is difficult to deploy and manage; (3) The current information exchange between routing devices is carried out layer by layer, which will cause unnecessary bandwidth occupation. In general, the core problem is that there is a contradiction between the diverse and changeable upper-layer applications and business needs of the network and the current stable and rigid traditional network architecture, and an appropriate solution is urgently needed, as shown in Fig. 3.3.

3.3 SDN Network Architecture

3.3.1 Development Path

In 2007, Dr. Martin Casado, a member of the Clean Slate project team led by Professor Nick McKeown of Stanford University, proposed a solution and network architecture for decoupling the control plane and the data forwarding plane, which is considered to be the prototype of today's SDN technology. In 2008, Professor Nick McKeown and others published a paper "OpenFlow: Enabling Innovation in Campus Networks" at the SIGCOMM conference, and first proposed the OpenFlow protocol based on the SDN architecture. In the same year, his team released the first open source SDN controller NOX-Classic. In 2011, Internet companies such as Google, Facebook, and Yahoo initiated the establishment of the Open Networking Foundation (ONF) to promote the standardization and development of SDN architecture and technology. In 2013, network vendors such as Cisco, Juniper, Broadcom, and IBM initiated the open source platform project OpenDaylight (ODL), with the goal of launching a universal enterprise-level SDN controller. In 2014, ONOS was born. Facebook openly released the details of the Wedge switch

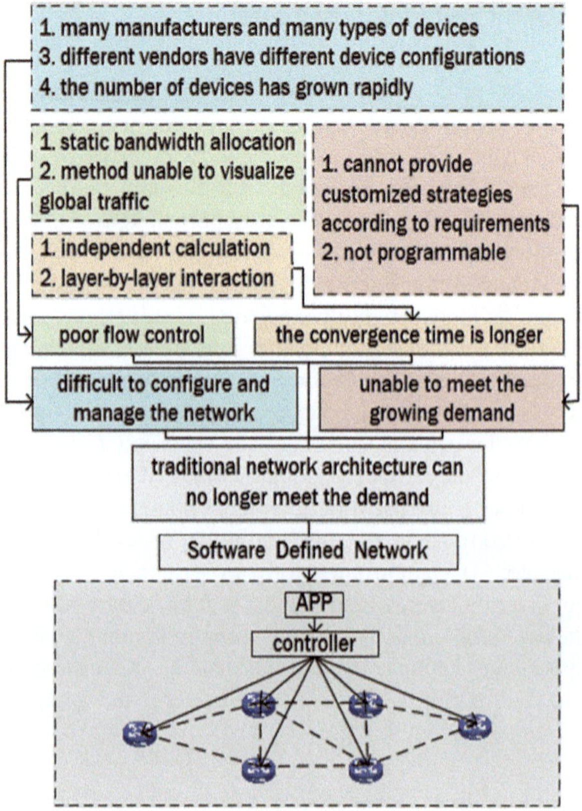

Fig. 3.3 The diagram of problems in the traditional network architecture

design in the OCP project, and the white box switch became the main theme of the year. Cavium acquired SDN startup Xpliant, and Broadcom released the OF-DPA framework compatible with the OpenFlow protocol. In 2015, ONF released an open source SDN project community, and SD-WAN became the second mature SDN application market. The integration of SDN and NFV has become a trend, and this year is a hot year for NFV. In 2016, SDN startups VeloCloud, Plexxi, Cumulus, and BigSwitch received a new round of financing, IEEE held the NFV-SDN conference, and the research on network programming language received the focus of academic circles. The SDN-IoT academic seminar was successfully held.

Since then, with the increase in attention, research is no longer limited to the traditional narrow SDN, and more and more projects tend to move from the original narrow SDN to the broad SDN. That is, it supports rich southbound protocols, which can realize flexible programmability and flexible deployment, as well as intelligent analysis and scheduling. Broad SDN has more powerful vitality, especially with the rapid development of cloud computing and big data. In the future, SDN-based cloud network integration will become one of the main demands of the development of SDN, a new generation of data centers and backbone network infrastructure.

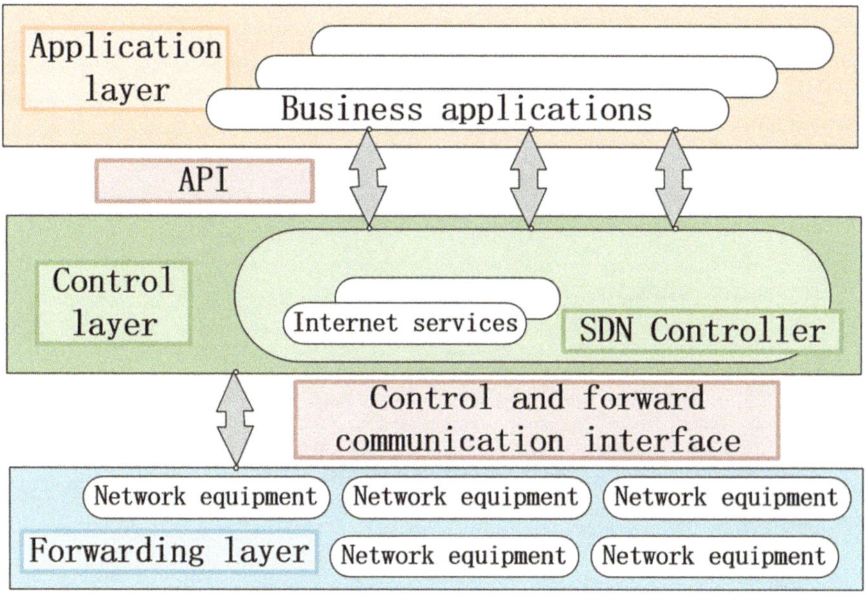

Fig. 3.4 The diagram of problems in the basic architecture of SDN

Refactoring and how to better support applications will be an important area of SDN development.

3.3.2 Definition and Architecture

The core idea of SDN is to separate the control plane and the data plane and use a centralized controller to complete the programmable tasks of the network. The controller interacts with the upper-layer application and the lower layer forwarding device through the northbound interface and the southbound interface protocol, respectively. It is this characteristic of separation (decoupling) of centralized control and data control that SDN has powerful programmability. This powerful programmability enables the network to be truly defined by software, which in turn makes network operation, maintenance, management, and scheduling easier. At the same time, in order to enable SDN to achieve large-scale deployment, it is necessary to support the collaboration between multiple controllers through the east-west interface protocol. Figure 3.4 shows the basic architecture of SDN.

In the SDN architecture, the control plane centrally controls the network equipment through the control-forward communication interface. This part of the traffic occurs between the controller and the network equipment, independent of the data traffic generated by the communication between the terminals. In addition, the network device generates a forwarding table by receiving the control signaling

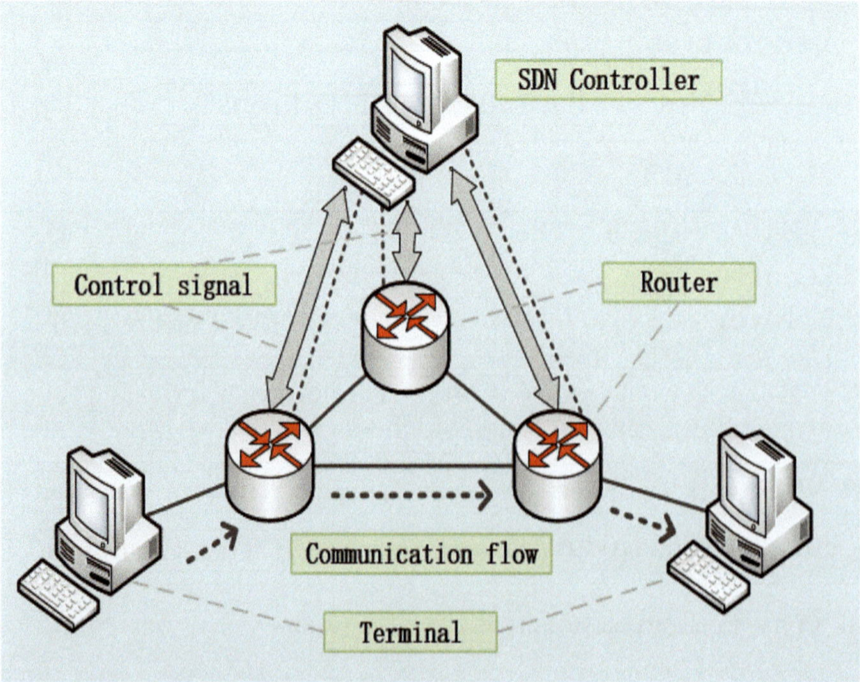

Fig. 3.5 The diagram of data forwarding process between devices

and determines the processing method of the traffic accordingly. Therefore, in this mode, it is no longer necessary to use a complex distributed network protocol for data forwarding, as shown in Fig. 3.5.

3.3.3 Core Technology and Advantages

3.3.3.1 Core Technology

The SDN architecture includes four planes and two interfaces, as shown below:

1. Data plane: It consists of several network elements, and each network element can contain one or more SDN Datapaths. Each SDN Datapath is a logical network device. It has no control capability and is only used to forward and process data. It logically represents all or part of the physical resources. An SDN Datapath includes three parts: control data plane interface agent, forwarding engine table, and processing function.
2. Control plane: The SDN controller is a logically centralized entity. It is mainly responsible for two tasks. One is to convert SDN application layer requests to

SDN Datapath, and the other is to provide SDN applications with an abstract model of the underlying network (which can be states, events). An SDN controller includes three parts: northbound interface agent, SDN control logic, and control data plane interface driver. The SDN controller is only required to be logically complete, so it can be composed of multiple controller instances or a hierarchical controller cluster; geographically speaking, it can be that all controller instances are in the same location, or it can be Multiple instances are scattered in different locations.

3. Application plane: This plane consists of several SDN applications. It can interact with the SDN controller through the northbound interface, that is, these applications can submit the requested network behavior to the controller in a programmable manner. An SDN application can contain multiple northbound interface drivers (using multiple different northbound APIs). In addition, SDN applications can also abstract and encapsulate their own functions to provide a northbound proxy interface to the outside, and the encapsulated interface forms a more advanced northbound interface.
4. Management plane: Responsible for a series of static tasks, which are more suitable for implementation outside the application, control, and data planes, such as configuring network elements, specifying SDN Datapath controllers, and at the same time defining the SDN controller and the control scope of SDN application.
5. SDN control data plane interface (CDPI): SDN CDPI is the interface between the control plane and the data plane. The main functions it provides include: control of all forwarding behaviors, device performance inquiries, statistical reports, and event notifications.
6. SDN Northbound Interface (NBI) SDN NBI is a series of interfaces between the application plane and the control plane. It is mainly responsible for providing an abstract network view and enabling applications to directly control the behavior of the network, which includes abstraction of the network and functions from different layers.

3.3.3.2 The Main Advantages

The emergence of SDN has promoted the transformation of traditional network construction and operation methods, which will effectively improve the service efficiency of cloud service providers, Internet applications, and cloud enterprises, and reduces the cost and complexity of operation and maintenance. The main advantages of SDN can be summarized as the following four:

1. The SDN architecture allows application development, cloud service provision, and network service teams to collaborate on a common platform. In addition, the open interface can provide users with self-service capabilities, which will greatly reduce the time spent on manual processes.

2. With the help of SDN, data centers, cloud providers, and network providers will significantly improve efficiency in many areas. SDN can provide centralized and visualized operation and maintenance of network resource usage, and elastically expand resources when resource usage is high or there are sudden business needs. In addition, when the resource utilization rate is low, idle resources can be released, thereby improving the overall utilization of resources.
3. Combined with the open interface API, the centralized control of SDN and the unified strategy deployment method makes end-to-end application guarantee possible. The original connection-oriented IP network is upgraded to a connection-oriented network service, which can increase new SLA services such as end-to-end application delay and resource availability.
4. Based on the SDN architecture, end users or application providers can customize and adjust network resources according to their needs and can pay according to actual use, and what you see is what you get; data centers or cloud service providers can simplify network deployment The complexity of the network improves the utilization of network resources and reduces the proportion of network costs; network service providers can provide new network connection services, iteratively design new network operation models, and attract and increase potential customers.

In general, the introduction of SDN is not only a promotion of technological development but also an inevitable trend to replace traditional networks. It will bring changes that cannot be underestimated in the communication circle, the Internet circle, and the IT circle.

3.3.4 Industry Chain Analysis

At present, the SDN industry chain can be temporarily divided into six camps, as follows:

1. Traditional equipment vendors: Because the switch function is simple and homogeneous under the SDN architecture, it lacks market value. Therefore, for traditional equipment vendors, the emergence of SDN will make their current dominant position face huge challenges. At the same time, since SDN represents the inevitable trend of network virtualization, traditional equipment vendors cannot refuse or avoid it. Therefore, they often adopt the "walking on two legs" approach: on the one hand, they closely follow the development of SDN by acquiring SDN startups and upgrading their original equipment; on the other hand, they actively launch their own SDN strategy and try to use the existing dominant position to grasp the dominant power of SDN development and integrate it into the existing network architecture.
2. Startups: For startups headed by Nicira (VMware) and Big Switch, the emergence of SDN has created a rare opportunity for them to subvert Cisco's dominance and enter the network equipment industry. In order to save development

costs and enhance versatility at the same time, they are active promoters in the development of a common SDN architecture based on the OpenFlow protocol. At present, these companies are mainly focusing on combining OpenFlow and virtualization technology in a certain field to provide customers with network virtualization solutions.
3. IT service provider: The emergence of SDN also allows IT service providers such as IBM and Hewlett-Packard to see the possibility of entering the network equipment industry and creating new business models. Therefore, they are basically the same as startups in their attitude towards SDN. The difference is that IT service providers also use customized hardware equipment and self-developed SDN operating system to quickly provide a full set of solutions to seize the market space of traditional network equipment manufacturers.
4. Chip manufacturers: Since the SDN architecture has changed the traditional traffic processing method, standardized SDN equipment requires a new generation of SDN-oriented communication switching chips. At present, major international chip manufacturers are actively introducing network processing chip solutions that implement SDN.
5. Internet content providers: For Internet content providers such as Google, Facebook, and Tencent, they pay more attention to the tight coupling between applications and network control brought about by the openness of SDN. If the northbound interface of the SDN controller is fully opened, the Internet company will indirectly gain the dominance of network control, thereby enabling it to integrate the operation and maintenance of its own application network with the operation and maintenance of the underlying transmission network.
6. Operator: Compared with Internet companies, operators are more cautious about SDN. On the one hand, some related technologies are not yet mature enough; on the other hand, although the introduction of SDN may reduce the cost of operators, it may also weaken their profit margins. Therefore, at present, operators are mainly exploring and experimenting with SDN in their data centers.

3.4 Application Scenario Analysis

3.4.1 Application of SDN in Data Center Network

In recent years, with the rapid development of the Internet, more and more applications and data have been concentrated in cloud data centers for processing. Nearly two-thirds of the world's total workload will be processed in the cloud, and data centers will become the source or destination of most Internet traffic. Among all the traffic generated by the data center, internal traffic accounted for 76% of the total traffic in the data center, mainly for data exchange between storage and virtual machines. Therefore, the future Internet will be a network with cloud computing data centers as its core.

Table 3.2 The table of the respective functions and requirements of the five-layer network

Level	Functional	Requirements
Network 1	The interconnection network between VMs of the same server	The access switch can perceive the virtual machine situation in the server
Network 2	The interconnection network between VM and storage	The business network and storage network need to be integrated
Network 3	The interconnection network between different servers within the DC site	Non-blocking network, support virtual machine drift and network configuration migration, and high horizontal expansion capability
Network 4	The interconnection network between servers across DCs	Support virtual machine drift
Network 5	The interconnection network between DC and Client	According to business needs, build a proprietary virtual transport network

Under this trend, the data center has entered the peak of development. However, the traditional data center network architecture has become increasingly difficult to meet the needs of market development. As mentioned in the previous section, in a traditional data center, each business occupies resources independently of each other and has no resource mobility requirements. The internal traffic of the data center is basically north-south. Therefore, each business network is physically isolated, self-contained, and cannot be reused. The network built in this way has limited scalability and can only be replaced vertically but cannot be extended horizontally. In order to solve the problem of low resource utilization, virtualization technology was created. The data center is mainly composed of three types of resources: computing, storage, and network. At present, only network resources have not yet been virtualized, which has become a bottleneck restricting the efficiency improvement of data centers. According to different communication subjects, the data center network can be divided into five layers. Communication between different subjects has different requirements for the network, as shown in Table 3.2.

According to Table 3.2, the current data center network usage requirements can be summarized as follows: (1) Non-blocking network, and possesses approximately unlimited high scalability; (2) Able to perceive virtual machines, and support the drift of virtual machines within a single data center and between multiple data centers, and ensure that related network policies are migrated accordingly; (3) Support multi-service and multi-tenant. On the same physical network, freely construct a business network according to business requirements and ensure network security; (4) Unified network operation and maintenance, highly automated, and intelligent management.

In order to meet the above requirements, a variety of technical solutions have emerged, such as Trill and SPB technologies, which are mainly aimed at multi-path forwarding and flexible deployment requirements of data centers. However, this technology has problems such as low link utilization, poor network stability, and complicated deployment of Layer 3 paths. In addition, the introduction of EVB technology meets the needs of virtual machine deployment and migration but

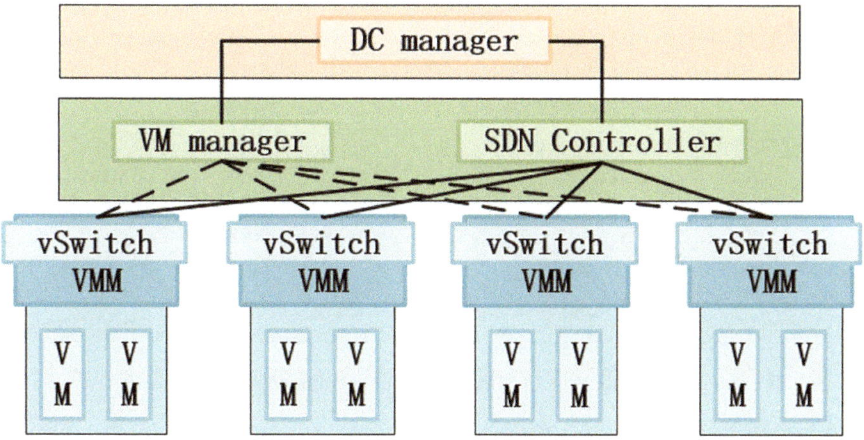

Fig. 3.6 The diagram of the future data center network deployment plan based on SDN

cannot meet the needs of centralized management, multi-path forwarding, virtual multi-tenancy, and IaaS in the data center. On top of this, NVGRE technology uses GRE encapsulation to achieve virtual multi-tenancy, but the considerations are still not comprehensive enough. It can be seen that the current mainstream data center network technology is mainly designed for a specific requirement of the data center, and there is still room for improvement.

In contrast, the SDN architecture has the characteristics of separation of forwarding and control, centralized control logic, network virtualization, and open network capabilities. Therefore, SDN technology can better meet the requirements of centralized management of data center networks, flexible multi-path forwarding, virtual machine deployment, and intelligent migration. The future data center network deployment plan based on SDN is shown in Fig. 3.6.

Among them, the architecture in this solution is mainly composed of three parts, namely the SDN controller, the VM manager, and the DC manager. Among them, the SDN controller is mainly used to implement centralized management and control of network devices, and the VM manager is mainly used to implement VM management, including creation, deployment, and migration. The DC manager is used to achieve overall coordination and control. In general, separation of control and forwarding, centralized logic control, and open network programming API are regarded as the three main characteristics of SDN that distinguish it from traditional network technologies. It is these characteristics that enable SDN to well meet the needs of data center networks.

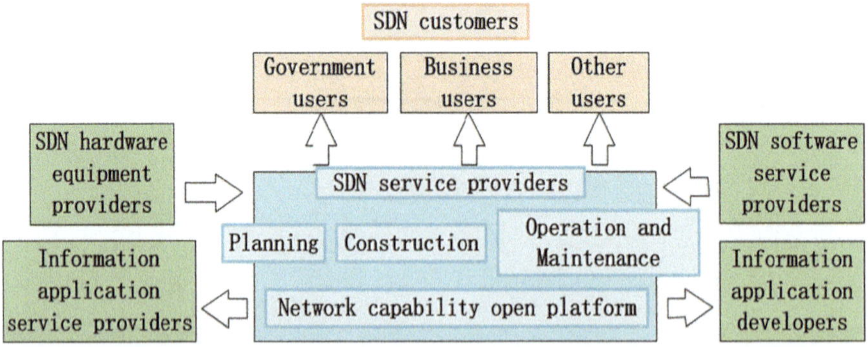

Fig. 3.7 The diagram of application of SDN in telecom operator network

3.4.2 Application of SDN in Government and Enterprise Networks

There are many types of services in government and enterprise networks, and the functions and types of network equipment are complex, which require high network security. Moreover, in this environment, the need for centralized management and control is even more urgent. In addition, there is also a certain demand for network flexibility and customization. However, after the network in the traditional architecture is deployed and launched, if the business requirements change, it is very cumbersome to re-modify the network device configuration. With the gradual development of the Internet, it can no longer meet the needs. However, SDN makes up for this very well. As shown in Fig. 3.7, on the basis of resource support provided by the SDN software and hardware providers, the SDN controller performs unified management, planning, and operation of resources. This method realizes not only centralized control but also the needs of network service providers in an open platform. It can be seen that the software-defined model significantly reduces the difficulty of network maintenance, shortens the network deployment cycle, and saves operation and maintenance costs. Moreover, with the help of existing network resources, by transforming traditional application development service providers into SDN service providers, it is also effective, feasible, and beneficial to provide large government enterprises with exclusive SDN network construction and maintenance services.

SDN strips out complex business functions, which not only reduces equipment hardware costs but also makes enterprise networks more simplified and clearer. At the same time, the logical concentration of SDN control can realize the centralized management and control of the enterprise network, the centralized deployment and management of enterprise security policies, and the flexible customization of network functions in the controller or upper-layer application to better meet the needs of the enterprise network.

3.4.3 Application of SDN in Telecom Operator Network

The separation of forwarding and control of SDN can effectively realize the gradual integration of equipment and reduce the cost of equipment hardware. In addition, the centralized feature page of SDN's control logic can gradually realize centralized management and global optimization of the network and effectively improve operational efficiency. In addition, providing end-to-end network services is also conducive to the intelligent and open development of telecom operators' networks, and the development of richer network services can increase revenue.

Nowadays, both NTT and Deutsche Telekom have begun to test deployment of SDN. NTT has set up test environments in Japan and the United States, while Deutsche Telekom has tried to use SDN in cloud data center, wireless, and fixed access environments. However, because SDN technology is not yet mature enough, the degree of standardization is not high enough. In addition, the management issues of a wide range of network equipment, the security and stability issues of ultra-large-scale SDN controllers, the coordination and interoperability issues of multiple vendors, and the coordination and docking issues of different network levels or standards all need to be resolved as soon as possible. Therefore, the current large-scale application of SDN technology in telecom operator networks is still difficult to achieve, but it can be gradually used in local networks or specific application scenarios, such as mobile backhaul network scenarios, packet and optical network collaboration scenarios, etc.

3.4.4 Application of SDN in Network Virtualization Technology

Network virtualization is an important network technology that can virtualize multiple isolated virtual networks on a physical network, so that different users can use independent network resource slices, thereby improving network resource utilization and realizing network flexibility [32, 33, 35, 37, 39–41]. The emergence of SDN makes the realization of network virtualization more flexible and efficient. At the same time, network virtualization has also become a heavyweight application in the SDN architecture.

It should be noted that SDN is not equivalent to network virtualization. SDN is a centralized control architecture, while network virtualization is a network technology. Traditional network virtualization requires manual configuration (such as VLAN) one by one, with high efficiency and low cost. In today's DC scenarios, in order to achieve rapid and flexible deployment and dynamic adjustment, automated deployment solutions must be used. The emergence of SDN has brought new solutions to network virtualization. Through the separation of forwarding and control, the deployment of automated services has been realized, and the deployment time of services has been significantly shortened.

Generally speaking, network virtualization through SDN includes the following four parts:

1. Network virtualization platform: an agent between the control plane and the data plane to realize the creation and management of virtual networks.
2. Network resource virtualization: including topology virtualization, node virtualization, and link resource virtualization.
3. Network isolation: including control plane isolation (so that tenant controllers do not affect each other), and data plane isolation (isolation of nodes' CPU, flow table, link bandwidth and other resources).
4. Address isolation: ensure that tenants can use any address space to achieve address isolation, mainly through address mapping. In a physical network, by using different physical addresses, address isolation can be achieved.

SDN changes the control mode of the traditional network architecture and divides the network into a control plane and a data plane. The network management authority is handed over to the controller software of the control layer, and commands are uniformly issued to the data layer devices through the OpenFlow transmission channel. Data layer equipment only relies on control layer commands to forward data packets. Due to the openness of SDN, third parties can also develop corresponding applications and place them in the control layer, which can make the deployment of network resources more flexible. In contrast, network administrators only need to issue commands to data layer devices through the controller, without logging in to the devices one by one, which saves labor costs and improves efficiency. It can be said that SDN technology has greatly promoted the development of network virtualization.

3.5 Future and Challenges

3.5.1 Existing Challenges

3.5.1.1 Security Issues

Today, when network security is receiving increasing attention, SDN technology cannot actually meet this demand. For traditional router switch firewalls, the operating system is a highly embedded Unix system, which has passed various tests and tests by manufacturers to ensure its security. Despite this, equipment is still not secure enough. For example, many manufacturers regularly release security patches or risk warnings for network equipment. However, when SDN really arrives, security issues will rely more on application-level prevention. Therefore, if the SDN controller as the core of the network greatly enriches the open interfaces for network customization, it also opens more doors for illegal access and malicious attacks. Therefore, the existence of the SDN controller may expose the network to more security risks.

3.5.1.2 Standardization Issues

At present, ONF only defines the southbound interface between the controller and the switch, but has not yet defined the interface between the controllers and the northbound interface that the controller opens to applications. The reason is that the organization believes that it is too early to standardize these interfaces and may stifle innovation in key components of the network infrastructure. But this has undoubtedly increased the difficulty of intercommunication among devices of various manufacturers and has delayed the commercialization of SDN to a certain extent.

3.5.1.3 Performance Issues

Under the SDN architecture, the controller needs to formulate an optimized routing strategy for each flow. The computational pressure can be imagined, and this pressure will increase geometrically as the number of control network elements increases. In addition, because different applications will establish different logical networks in the SDN system, each application will hinder each other's functions, and resource competition will be very fierce. From the perspective of the development history of computer programs, in order to coordinate the operation of various programs and improve resource utilization efficiency, the complexity and computational complexity of resource allocation algorithms often increase exponentially, which may become a system bottleneck. At the same time, in order to achieve the programmability of the network, applications will be given a lot of control over the environment, which can easily lead to system crashes. Therefore, how to strike a balance between software complexity and computing efficiency is a major challenge facing SDN.

3.5.2 Future Development

With the continuous development and transformation of science and technology, the current SDN technology has become a hot technology leading the network transformation. Many companies in the world have also made in-depth research and predictions on the commercial process of SDN. In the future, narrow SDN will move towards broad SDN, and broad SDN has more powerful vitality, especially with the rapid development of cloud computing and big data, cloud-network integration based on SDN will become one of the main demands of the development of SDN, a new generation of data restructuring of the center and backbone network infrastructure and how to better support applications will be important areas for the development of SDN. In addition, SDN may simplify the exchange of traffic in the data center, thereby enabling traffic to be routed and forwarded more efficiently. SDN allows traffic processing policies to follow virtual machines and containers, so that this information can be moved within the data center to minimize traffic and deal with bandwidth bottlenecks.

In addition, the main experience in the development of SDN in the past two years has been continuous attempts to integrate commercial deployment of its applications, mostly in the field of data centers. Most large-scale and ultra-large-scale data centers have adopted flat architecture, SDN and storage management, and adopted SDN / NFV (Network Function Virtualization), and they are developing very rapidly. By 2021, more than two-thirds of data centers will adopt SDN in whole or in part. As part of the traffic in the data center, SDN/NFV is already transmitting 23% of the data, which will increase to 44% by 2021. SDN technology will gradually develop towards network infrastructure and will experience a long process of change and development. However, due to the traditional SDN technology deployment problems such as complex management, long configuration cycles, difficult business migration, low quality, reliability, etc., in the future development, the problems existing in the SDN architecture should also be solved one by one.

3.6 Conclusion

With the joint development of multiple technologies, the Internet has become inseparable from people's lives. However, although the existing distributed network architecture has the advantages of anti-attack, simple, and easy access, it still exposes some problems in the face of massive data transmission needs. For example, network devices produced by different manufacturers usually require different ways to debug and configure. Therefore, it is a very big challenge to manage and deploy a network with multiple devices. In addition, with the increase in QoS and security requirements, the inability to customize network services according to customer needs, and the inability to perform intelligent flow control and status supervision based on network conditions are also problems that hinder further development. In addition, the current way of information exchange between routing devices will also cause unnecessary bandwidth occupation. Based on the above problems, the introduction of SDN architecture is a better solution. In general, SDN architecture has the following three advantages: (1) SDN can change the tightly coupled architecture of applications and networks under traditional networks and improve the level of network resource pooling; (2) SDN networks can realize automatic network deployment and configuration, and support fast business Go online and expand flexibly; (3) By introducing programmable features, automated network services and protocol orchestration can be realized. At present, the market share of SDN is increasing year by year, and the trend is improving.

This article analyzes and summarizes the problems existing in the traditional Internet architecture by giving examples of four common network architectures. Secondly, we introduced the SDN architecture and analyzed its necessity and advantages of the architecture. And, we further introduced its feasible application scenarios and existing solutions. Finally, we analyzed and summarized the threats and challenges that the SDN architecture may face, laying the foundation for the follow-up research of SDN.

Acknowledgments This work is partially supported by the Major Scientific and Technological Projects of CNPC under Grant ZD2019-183-006, partially supported by Shandong Provincial Natural Science Foundation under Grant ZR2020MF006, and partially supported by "the Fundamental Research Funds for the Central Universities" of China University of Petroleum (East China) under Grant 20CX05017A.

References

1. Aujla, G. S., & Kumar, N. (2018). SDN-based energy management scheme for sustainability of data centers: An analysis on renewable energy sources and electric vehicles participation. *Journal of Parallel and Distributed Computing, 117*, 228–245.
2. Aujla, G. S., Jindal, A., Kumar, N., & Singh, M. (2016). SDN-based data center energy management system using res and electric vehicles. In *2016 IEEE Global Communications Conference (GLOBECOM)* (pp. 1–6).
3. Aujla, G. S., Chaudhary, R., Kumar, N., Kumar, R., & Rodrigues, J. J. P. C. (2018). An ensembled scheme for QoS-aware traffic flow management in software defined networks. In *2018 IEEE International Conference on Communications (ICC)* (pp. 1–7). IEEE.
4. Aujla, G. S., Jindal, A., & Kumar, N. (2018). EVaaS: Electric vehicle-as-a-service for energy trading in SDN-enabled smart transportation system. *Computer Networks, 143*, 247–262.
5. Aujla, G. S., Chaudhary, R., Kaur, K., Garg, S., Kumar, N., & Ranjan, R. (2018). SAFE: SDN-assisted framework for edge–cloud interplay in secure healthcare ecosystem. *IEEE Transactions on Industrial Informatics, 15*(1), 469-480.
6. Aujla, G. S. S., Kumar, N., Garg, S., Kaur, K., & Ranjan, R. (2019). EDCSuS: Sustainable edge data centers as a service in SDN-enabled vehicular environment. *IEEE Transactions on Sustainable Computing*, 1–1.
7. Aujla, G. S., Singh, A., & Kumar, N. (2020). Adaptflow: Adaptive flow forwarding scheme for software-defined industrial networks. *IEEE Internet of Things Journal, 7*(7), 5843–5851.
8. Aujla, G. S., Singh, A., Singh, M., Sharma, S., Kumar, N., & Choo, K. R. (2020). Blocked: Blockchain-based secure data processing framework in edge envisioned v2x environment. *IEEE Transactions on Vehicular Technology, 69*(6), 5850–5863.
9. Aujla, G. S., Singh, M., Bose, A., Kumar, N., Han, G., & Buyya, R. (2020). Blocksdn: Blockchain-as-a-service for software defined networking in smart city applications. *IEEE Network, 34*(2), 83-91.
10. Budhiraja, I., Kumar, N., Tyagi, S., Tanwar, S., & Obaidat, M. S. (2020). URJA: Usage jammer as a resource allocation for secure transmission in a CR-NOMA-based 5g Femtocell system. *IEEE Systems Journal*, 1–10.
11. Garg, S., Singh, A., Aujla, G. S., Kaur, S., Batra, S., & Kumar, N. (2020). A probabilistic data structures-based anomaly detection scheme for software-defined internet of vehicles. *IEEE Transactions on Intelligent Transportation Systems*, 1–10.
12. Kumar, N., & Kumar, M. (2015). Closely spacified wide dual-band microstrip band pass filter using coupled stepped-impedance resonators. In *2015 2nd International Conference on Electronics and Communication Systems (ICECS)* (pp. 865–867).
13. Kumar, N., & Tripathi, M. M. (2017). Evaluation of effectiveness of ANN for feature selection based electricity price forecasting. In *2017 International Conference on Emerging Trends in Computing and Communication Technologies (ICETCCT)* (pp. 1–5).
14. Kumar, N., Chilamkurti, N., Zeadally, S., & Jeong, Y. (2014). Achieving quality of service (QoS) using resource allocation and adaptive scheduling in cloud computing with grid support. *The Computer Journal, 57*(2), 281–290.

15. Kumar, N., Vinoy, K. J., & Gopalakrishnan, S. (2018). Improved well-conditioned model order reduction method based on multilevel Krylov subspaces. *IEEE Microwave and Wireless Components Letters, 28*(12), 1065–1067.
16. Neeraj, N., Naresh, M., Yadav, A. K., & Mathew, L. (2019). Effect of statcom on integration of renewable energy generation with the main grid. In *2019 Innovations in Power and Advanced Computing Technologies (i-PACT)* (vol. 1, pp. 1–5).
17. Singh, A., Aujla, G. S., & Bali, R. S. (2020). Intent-based network for data dissemination in software-defined vehicular edge computing. *IEEE Transactions on Intelligent Transportation Systems*, 1–9.
18. Singh, A., Aujla, G. S., Singh Bali, R., Chahal, P. K., & Singh, M. (2020). A self organised workload classification and scheduling approach in IoT-edge-cloud ecosystem. In *2020 IEEE 92nd Vehicular Technology Conference (VTC2020-Fall)* (pp. 1–5).
19. Singh, M., Aujla, G. S., & Bali, R. S. (2020). A deep learning-based blockchain mechanism for secure internet of drones environment. *IEEE Transactions on Intelligent Transportation Systems*, 1–10.
20. Singh, P., Kaur, A., Aujla, G. S., Batth, R. S., & Kanhere, S. (2020). Daas: Dew computing as a service for intelligent intrusion detection in edge-of-things ecosystem. *IEEE Internet of Things Journal*, 1–1.
21. Singh, A., Batra, S., Aujla, G. S., Kumar, N., & Yang, L. T. (2020). BloomStore: dynamic bloom-filter-based secure rule-space management scheme in SDN. *IEEE Transactions on Industrial Informatics, 16*(10), 6252–6262. https://doi.org/10.1109/TII.2020.2966708.
22. Sood, K., Karmakar, K. K., Varadharajan, V., Kumar, N., Xiang, Y., & Yu, S. (2021). Plug-in over plug-in (pop) evaluation in heterogeneous 5g enabled networks and beyond. *IEEE Network*, 1–7.
23. Vangala, A., Bera, B., Saha, S., Das, A. K., Kumar, N., & Park, Y. H. (2020). Blockchain-enabled certificate-based authentication for vehicle accident detection and notification in intelligent transportation systems. *IEEE Sensors Journal*, 1–1.
24. Vangala, A., Das, A. K., Kumar, N., & Alazab, M. (2020). Smart secure sensing for IoT-based agriculture: Blockchain perspective. *IEEE Sensors Journal*, 1–1.
25. Verma, G. K., Kumar, N., Gope, P., Singh, B. B., & Singh, H. (2021). Scbs: A short certificate-based signature scheme with efficient aggregation for industrial internet of things environment. *IEEE Internet of Things Journal*, 1–1.
26. Wen, Z., Garg, S., Aujla, G. S. S., Alwasel, K., Puthal, D., Dustdar, S., Zomaya, A. Y., & Rajan, R. (2020). Running industrial workflow applications in a software-defined multi-cloud environment using green energy aware scheduling algorithm. *IEEE Transactions on Industrial Informatics*, 1–1.
27. Zhang, P., Zhang, Z., & Zhang, W. (2013). An approach of semantic similarity by combining HowNet and Cilin. In *2013 IEEE International Conference on Green Computing and Communications and IEEE Internet of Things and IEEE Cyber, Physical and Social Computing* (pp. 1638–1643).
28. Zhang, P., Yao, H., & Liu, Y. (2016). Virtual network embedding based on the degree and clustering coefficient information. *IEEE Access, 4*, 8572–8580.
29. Zhang, P., Wu, S., Wang, M., Yao, H., & Liu, Y. (2018). Topology based reliable virtual network embedding from a QoE perspective. *China Communications, 15*(10), 38–50.
30. Zhang, P., Yao, H., & Liu, Y. (2018). Virtual network embedding based on computing, network, and storage resource constraints. *IEEE Internet of Things Journal, 5*(5), 3298–3304.
31. Zhang, P., Huang, X., & Li, M. (2019). Disease prediction and early intervention system based on symptom similarity analysis. *IEEE Access, 7*, 176484–176494.
32. Zhang, P., Hong, Y., Pang, X., & Jiang, C. (2020). VNE-HPSO: Virtual network embedding algorithm based on hybrid particle swarm optimization. *IEEE Access, 8*, 213389–213400.
33. Zhang, P., Li, C., & Wang, C. (2020). Smarttext: Learning to generate harmonious textual layout over natural image. In *2020 IEEE International Conference on Multimedia and Expo (ICME)* (pp. 1–6).

34. Zhang, P., Pang, X., Bi, Y., Yao, H., Pan, H., & Kumar, N. (2020). Dscd: Delay sensitive cross-domain virtual network embedding algorithm. *IEEE Transactions on Network Science and Engineering, 7*(4), 2913–2925.
35. Zhang, P., Pang, X., Kumar, N., Aujla, G. S., & Cao, H. (2020). A reliable data-transmission mechanism using blockchain in edge computing scenarios. *IEEE Internet of Things Journal*, 1–1.
36. Zhang, P., Wang, C., Aujla, G. S., Kumar, N., & Guizani, M. (2020). IoV scenario: Implementation of a bandwidth aware algorithm in wireless network communication mode. *IEEE Transactions on Vehicular Technology, 69*(12), 15774–15785.
37. Zhang, P., Wang, C., Aujla, G. S., & Pang, X. (2020). A node probability-based reinforcement learning framework for virtual network embedding. In *2020 IEEE 21st International Symposium on "A World of Wireless, Mobile and Multimedia Networks" (WoWMoM)* (pp. 421–426).
38. Zhang, P., Wang, C., Jiang, C., & Benslimane, A. (2020). Security-aware virtual network embedding algorithm based on reinforcement learning. *IEEE Transactions on Network Science and Engineering*, 1–1.
39. Zhang, P., Huang, X., Wang, Y., Jiang, C., He, S., & Wang, H. (2021). Semantic similarity computing model based on multi model fine-grained nonlinear fusion. *IEEE Access, 9*, 8433–8443.
40. Zhang, P., Jiang, C., Pang, X., & Qian, Y. (2021). Stec-IoT: A security tactic by virtualizing edge computing on IoT. *IEEE Internet of Things Journal, 8*(4), 2459–2467.
41. Zhang, P., Li, C., & Wang, C. (2021). Viscode: Embedding information in visualization images using encoder-decoder network. *IEEE Transactions on Visualization and Computer Graphics, 27*(2), 326–336.

Chapter 4
Architecture and Deployment Models-SDN Protocols, APIs, and Layers, Applications and Implementations

Bhawana Rudra and Thanmayee S.

4.1 Introduction

SDN has emerged recently as it addresses the lack of programmability issues in the existing network and promotes network management [6, 35]. The programmable term is used to make the network management and the reconfiguration concept as simple. This allows the encapsulation of the wider ideas by focusing on different planes and achieving the goal for various means. The concept of Programmability emerged in the mid-90s, when the Internet was successful for its use by users all over the world. As the spread was large, experts were interested to experiment with new ideas and protocols that can provide various services [8, 23, 36]. For the support of wide networks, the support for the specific protocols was important for the better outputs without vendor interoperability. The modification of the control logic in the network devices is not possible, restricting the evolution of the network [9, 17, 37, 45]. Many researchers have focused to find open, flexible, extensible, and programmable network devices. OpenSignaling (OpenSig) and Active Networking were the two initiatives that were developed for handling the underlying hardware issues and provide an open interface for the control and management of the network [12, 41, 42]. Open Signaling was emerged in 1995 by focusing on the concept of programmability in the networks. The main idea is to separate the control plane and data plane in the network allowing the open interfaces to interact between them. It is easy to control and program the switches remotely, making the entire network into a distributed platform by simplifying the deployment of new services. Towards the direction of research, Tempest framework allowed multiple switches to control and manage the multiple partitions of the switch, allowing multiple architectures to run

B. Rudra (✉) · Thanmayee S.
Department of Information Technology, National Institute of Technology, Mangalore, Karnataka, India

G. S. Aujla et al. (eds.), *Software Defined Internet of Everything*, Technology, Communications and Computing, https://doi.org/10.1007/978-3-030-89328-6_4

over the physical ATM network [1, 10, 14]. This gave much freedom to the network operators to define the Unified control architecture by controlling the requirements of future services provided by the network.

Another architecture that came into existence was DCAN (Devolved Control of ATM networks) [14]. The main focus was to control and manage the network switches and assign the work to the external workstations. The networks are distributed inherently, allocate resources across the network to provide QoS. Minimalistic protocols like OpenFlow were designed to manage the communication between the network and the management entity. This allows to add the functionality of synchronization streams in the management domain. The main goal was to develop programmable networks and promote innovations. The network nodes are introduced through network APIs and allow network providers to actively participate in controlling the nodes by executing some arbitrary code. This allowed the development of customized services along with dynamic configuration at run time. This architecture consists of a three layer stack on the active nodes. The bottom layer is on an operating system (NodeOS) which multiplexes the communication between the nodes. The next layer is the execution environment and allows the writing environment for the active network applications like ANTS [43], PLAN [21]. At the top layer, the applications that are actively executed where the code was developed by the network providers.

The demand for more flexible and dynamic services allows us to add more new features into the network. Two models fall into this category and they are Capsule model and Programmable switch model. In the capsule model, the code is included in the data packets itself. In the switch model, the code is executed at the network nodes through out-of-band mechanism. Capsule model allows for more innovation and is associated with active networking as it offers a different approach for network management and provides a simple installation for the new data plane across the network paths [16, 29, 31]. The concept of various approaches was envisioned for the programmable networks that can allow innovation and open networking experimental environments. None of them was successful and not widespread due to the lack of compelling problems. Out of all these, OpenSignaling and Active networking were not successful as they focused on the wrong user group [31, 40]. One more reason behind the programmable network failure is due to their focus on innovative architectures, models and not concentrating on the issues like security and performance of the network. Although there are many theoretical advantages, the reasons like network performance and security did not allow for adoption into the network. The attempts were clearly defined that the networks can be perceived towards the research ideas for the development of flexible programmable architectures. These shortcomings have high significance and addressed the deficiencies and paved a path to the development of the accepted path of SDN [5, 24, 39, 44].

4.2 SDN Architecture

In the traditional network architecture shown in Fig. 4.1, SDN Architecture comprises two major components: Control Plane and Data Plane. The control plane is where the decisions about traffic flow take place. The data plane is where the forwarding of traffic takes place and is bound together in the underlying network devices. The data plane is also called the forwarding plane. The decision about forwarding happens at the network switches based on their configured routing tables. Any update to the control plane requires the network engineer/administrator to program the network switches individually with required network policies. This results in a monotonous job that can be impractical in some applications, for

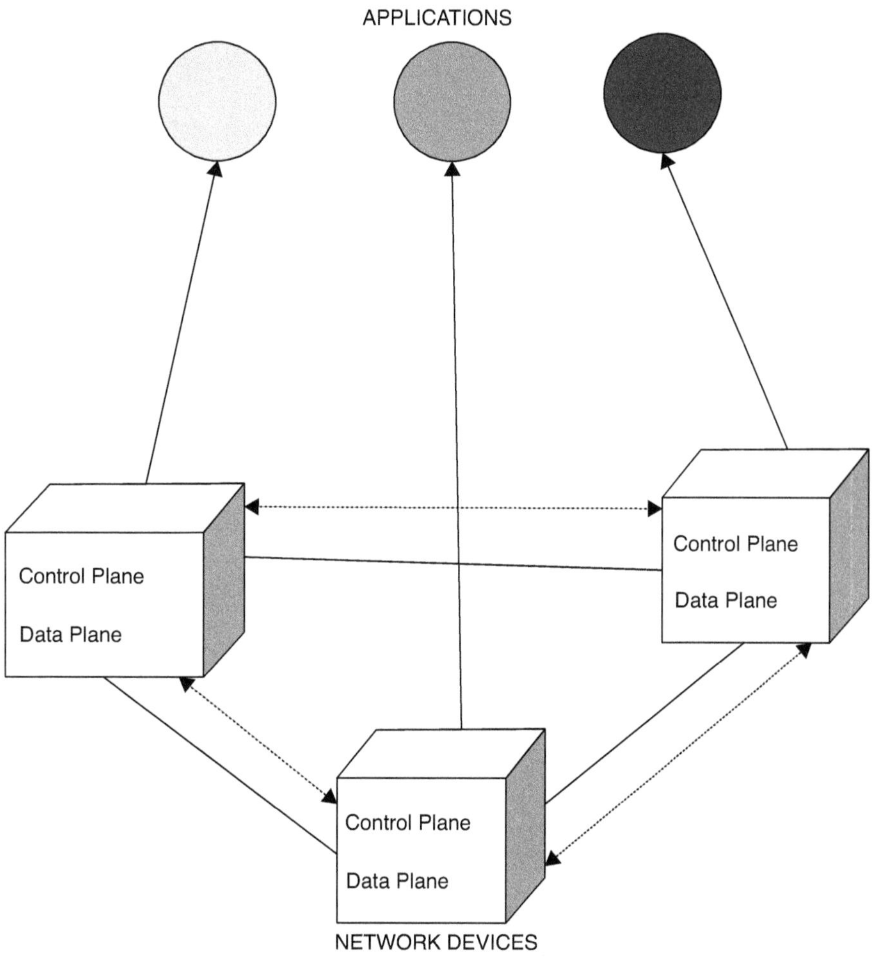

Fig. 4.1 Traditional network architecture

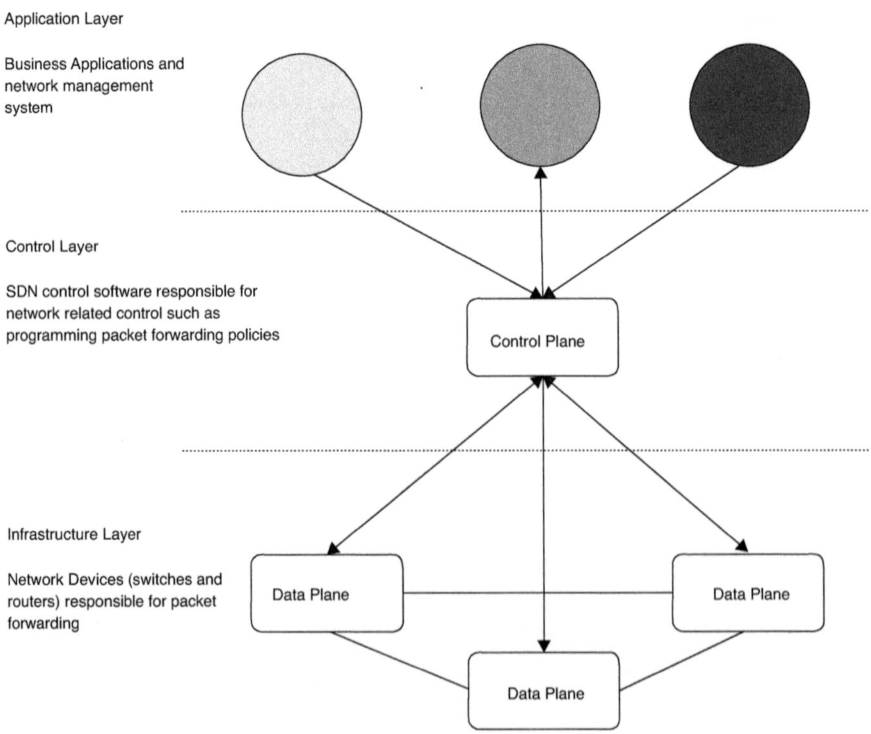

Fig. 4.2 SDN architecture

example, IoT environments [15, 20, 27, 34]. To waive away the challenges involved in traditional network architecture, SDN has plotted down a captivating technique; that is, SDN separates the control plane and data plane illustrated in Fig. 4.2. Thus, the system that makes decisions about where the traffic has to flow and is separated from the system that actually forwards traffic to the desired destination. This introduces flexibility, manageability, adaptability, scalability, agility features as an added advantage for IoT network management [2, 11, 19, 22].

Infrastructure Layer

It is the lowest layer in the SDN architecture. It comprises switching devices. These devices are interconnected to form a network. They could be connected through a wired or wireless transmission media. It is important to make switching devices operate efficiently and also make good utilization of transmission media at the Infrastructure layer. This will result in improved performance of the Infrastructure layer. Switching devices have a control plane and data plane. In the data plane, the switching devices perform data forwarding with the help of its processors. Some of the examples of network processors include XLP processor family (MIPS64 architecture) from Broadcom, XScale processor (ARM architecture) from Intel, NP-x NPUs from EZChip, PowerQUICC Communications Processors

(Power architecture) from freescale, NFP series processors (ARM architecture) from Netronome, Xelerated HX family from Marvell, OCTEON series processors (MIPS64 architecture) form Cavium and general purpose CPUs from Intel and AMD. The responsibility of the control plane is to receive rules such as packet forwarding rules from the control layer and link it with the rules in the data link layer. It then stores this rule in the local memory. This new design for switching devices shows how the operations of these devices have been simplified to gain advantages in various application areas [20, 21, 38].

There are wide varieties of transmission media that are used to connect the switching devices. There can be wired connection or wireless connections, resulting in heterogeneous connectivity between the devices. Each transmission media will have their own characteristics and thus there must be specific configuration and management technologies. These technologies must be integrated with SDN controllers in order to have good control over the network. Software Defined Radio is one of the advanced wireless transmission technologies. It allows software controlled wireless transmission. Thus it is easy to integrate SDR and SDN. Another well known transmission media is Optical Fiber which is known for its high capacity and low power consumption. The technology, called Reconfigurable Optical Add/Drop Multiplexers (ROADMs), enables the integration of optical fiber connectivity technology into SDN control planes [20, 21, 38].

Control Layer

Control Layer connects the application layer and the infrastructure layer. Its working principle or strategy can directly affect the overall SDN network performance. It provides an abstract view of the network infrastructure. It simplifies the task of applying custom policies/protocols on the network hardware. The network operating system (NOX) controller is the most common controller that is used widely. SDN controllers have to do network controlling and network monitoring [28, 30]. It is responsible to translate the application policy into the infrastructure layer's packet forwarding rules. As we can see in Fig. 4.2 that SDN controller is having interfaces with the application layer and the infrastructure layer. The interface between the controller and the interface is called south-bound interface. This interface collects network status, updates rules for packet forwarding at the switching devices in the infrastructure layer. The interface between controller and application layer is called north-bound interface. It helps in providing an overall view of the network status. It takes policies defined in high-level languages from the application layer and uses it to define rules at the infrastructure layer. This basically translates the requirements of any SDN applications into packet forwarding rules [3, 4, 20, 21, 38].

4.3 SDN Protocols

The SDN concept was emerged in 2005, when the experts came up with a 4D approach for network control. Later, Ethane architecture was developed to control the network using centralized policies for the control flow of routing. Ethane switches were used to forward the packets from controller to the destination based on the instructions. DATALOG language was used to design the policies based on the security. An experiment was conducted by installing the Ethane in the Stanford computer science lab to serve 300 systems and for a small business of 30 systems This was to derive the working of the network management and proved that a single controller can handle 10,000 flow requests per second for small business and set of distributed controllers can be deployed for larger topologies [8, 18, 32, 33, 39]. Ethane is not suitable for the present traditional networks techniques as it requires to know about the users and the nodes along with the control over the routing. These limitations were addressed by NOX by allowing access to the source and destination for each and every event that occurs. This architecture will allow to build a scalable network with flexible control as it uses the intermediate granularity in the flow [19, 31, 39].

4.4 Principles of SDN Architecture

SDN applications can be network aware or we can say as the network application aware. Traditional applications describe the network requirements indirectly for the implementation by involving the several processing steps to negotiate and support the execution of the applications based on policy controls. They will not support the dynamic user requirements like throughput, delay, or availability [7, 13, 31]. In this, the network service providers will not trust the users for traffic markings for the priority based packet headers. So in order to overcome this, some networks will try to support the user requirements by incurring some additional cost which may lead to misclassification. These networks do not allow the user to know about the information and the state of the network SDN will allow the user to specify the need in a trusted environment which can be monitored. SDN applications can monitor and adapt accordingly [25, 26, 40]. The use of controllers will allow the summarization of the network state and translates the requirements towards the lower level rules. Logically centralized SDN is distributed for the corporation between the physical controllers to achieve better performance, scalability, and reliability. The control decisions are up-to-date on a global view instead of on each distributed network so that the behavior should not change in network hops. Control plane acts as a single centralized network operating system for scheduling and solving the resource conflicts [2, 11, 26, 31].

The controller will control the Data paths with limited capabilities by not competing with other control elements that simplify the scheduling of the jobs and resource allocation. The SDN networks will run with network resource utilization based on the complex and follow the specified policies based on the information model that is defined by OpenFlow. In the traditional network architecture, the control plane, where the decisions about traffic flow take place, and the data plane also called a forwarding plane, where the forwarding of traffic takes place, are bound together in the underlying network devices. The decision about forwarding happens at the network switches based on their configured routing tables. Any update to the control plane requires the network engineer/administrator to program the network switches individually with required network policies. This results in a monotonous job that can be impractical in an IoT environment. To waive away the challenges involved in traditional network architecture, SDN has plotted down a captivating technique; that is, SDN separates the control plane and data plane. Thus, the system that makes decisions about where the traffic has to flow and is separated from the system that actually forwards traffic to the desired destination. This introduces flexibility, manageability, adaptability, scalability, agility features as an added advantage for IoT network management. In the near future SDN is expected to become a crucial part of IoT the agile and flexible architecture that it provides [18, 20, 21, 44].

4.5 SDN Tools and Languages

Various tools and languages were being used to implement and monitor the architecture of SDN. Many focused on the platform like Onix for the implementation of the controllers in distributed networks and its management. Veriflow is capable of finding the errors in the application rules by avoiding the disruption of the network performance. Routeflow is another routing architecture that was designed based on the SDN concepts, used to provide the interaction between hardware and the open source routing stacks. This paved a path for the migration towards SDN from Traditional IP. Later, physical SDN prototypes were introduced which paved a path for the SDN innovation like Mininet. It is a virtual emulator that allows any SDN prototype evaluation. If the evaluation is positive, the SDN services are deployed for the general and research purpose else again the prototypes are developed and tested. Mininets performance is poor at high loads and its lightweight virtualization is also not suitable. Another design is Frenetic, high-level programming for OpenFlow architecture. It uses the SQL syntax for the queries, stream processing language, and a specific language for packet forwarding. These three languages will make the programmers task as simple by allowing them to produce the high-level forwarding policies. It addresses some issues related to consistency and synchronization between the arrival of the packet time and the installation of the rule time. It consists of two abstraction levels, i.e., one for traffic control and the other for installation of the rules in the switches. Other programming languages like ProceraNettle came into existence for the reactive programming and facilitate the management of the network along the event-driven networks. A list of simulators and emulators like Mininet, NS-3, and so on that are supported by for the real time experiments are as follows [18, 33, 39].

Mininet

It is based on the OF protocol which runs the end hosts, routers, and related links on the Linux kernel using a lightweight virtual network. The components present in Mininet act as real network components and allow us to check possible bandwidth, node connectivity, and deepest nodes along with speed. It supports various tools and real view of the network traffic. This is being used by researchers and developers due to its easy interaction with the network using API and CLI features and also allows for the development of the various real hardware. This is mainly used because it is fast, supporting, allowing packet forwarding, running real programs available on laptops, servers, open source and active. It does not allow huge amounts of data in a single system, support OF controllers, support only Linux platform, NAT is out of box, shares host file system and virtual time notion absence.

NS-3

It is a simulator suited for research and educators. The library is split into modules. The OF is switchNet which acts as a switch. The objects present in NS-3 implement the flow table for the packets received, give a connection, and behave like a controller as in SDN. DropController and Learning Controller are the two controllers which are available in the package. This allows the addition of new protocols, distance between real and simulated networks reduced with the integration and customizable simulator. The disadvantages of this simulator are loss of models, creation of interfacing topology, and loss of visible capability.

EstiNet

This is a simulator and an emulator that support OF and switches. In the simulation of EstiNet, POX, NOX, and so on controllers act as SDN controllers. Controllers will run on external machines in emulation mode. It allows the controller with a dedicated hardware using an Ethernet cable resulting in remote controlling of the device. The other advantages include accuracy, repetitions, fast, and scalability. This allows "Kernel-reentering simulation Methodology" for testing whether the novelty of the controller programs by the researcher is effective and simple to adopt or not.

4.6 SDN Benefits and Application Domains

The inheritance of the decoupling of the control plane and data plane will offer a great control over the network with the help of programming. This combined feature can benefit the system by improving the configuration, performance and encourage the innovation for the network architecture and its operations. SDN allows a real time centralized control due to its ability to fetch the status of the network instantaneously as well the user defined policies. This helps in network optimization thus allows improving the network performance. SDN offers the platforms for the innovations and experimentation of the new techniques for the network design and has the ability to define the isolated network which is virtual through a control plane [40, 44].

Enhanced Configuration

Configuration plays a major role when the new equipment is added to the existing network which is required to achieve the coherent network operations. Manual processing is required at a certain level due to the heterogeneity of the device manufacturers and also the interface configuration. The manual processing is prone to errors and tedious as well it requires some troubleshoot for the configuration errors. The unification of the control plane will allow all kinds of network devices, which include switches, routers, firewalls, and load balancers, to configure network devices from a single point of failure automatically. The entire program can be configured and optimized dynamically based on network status. Performance Improvement: The key objective is to maximize the utilization of the network infrastructure. The existence of various technologies and the stakeholders in the network optimization of the network is a difficult task. The approaches which are available focus on performance optimizations of subnets for the quality experience of the user towards the network services. This optimization will be performed based on the local information of the network. SDN provides the optimization globally as it allows the centralized control with the global network along with a feedback control that has the information which is being exchanged within the layers of the architecture. With the centralized algorithms, the optimizations issues are manageable. It can deal with traffic scheduling, end-to-end congestion, energy efficient operations, and Quality of service which can be easily deployed and tested for effectiveness in the network.

Encourage Innovation

The architecture must be able to continue the evolution of the network applications in terms of innovation rather than only to predict the requirements of the future. The main problem will arise due to the proprietary hardware used in the conventional networks widely. When new services are developed, tested on separate testbed rather than on the network, it will not provide confidence for the adoption of the technology by the industry. The community like PlanetLab and GENI was enabled for experiments, those efforts did not solve the issue completely. SDN provides a programmable network by enabling the platform for innovation in terms of new revenue generation, deployment of new ideas, flexibility, and so on. This architecture provides a clear separation between the virtual and real environment.

SDN Applications

SDN is applied in a wide range of networks which includes Data centers, WLANs and heterogeneous networks, Optical Networks, Cellular and Internet of Things [18, 19, 31, 33, 40].

Data Centers

Data centers need to scaleup for the support of servers and the virtual machines that exist in the entire large network. From the network point of view, scaling up of these devices is an issue. The forwarding table size will increase as the number of servers increases, raising the requirement of sophisticated forwarding devices. With this increment of devices, traffic management and policy enforcement will become critical as data centers need to achieve high levels of performance. With careful design and configuration, the aforementioned requirements can be met in the traditional networks. In most of the cases this will be achieved using the preferred routes and by placing the middleboxes at choke points on the physical network. This concept will be contradictory for the requirement of scalability as the manual configuration will lead to error prone as the network size increases and it will not be able to adapt to the application requirements. The aforementioned gaps were filled with SDN with the help of decoupling concept which made the forwarding of the services from device to device much simpler. The control logic was delegated to one control entity which is centralized by allowing the dynamic management of the flow of packets, balance of the traffic, and allocation of the resources by adjusting the data center operations. This concept will increase the performance of the network and eliminate the concept of middleboxes in the network.

Cellular Networks
There is an increase in the rise of cellular devices from the decade and pushed cellular networks to their limits. The interest of integrating the SDN into the cellular networks raised the development of 3G and 4G communications which we are using today. The main issue with the cellular architecture was the centralized data flow, allowing the traffic through specialized equipment with multiple network functions from the routing and billing increasing the architectural cost because of the complexity of the devices. The cell size tends to become small in order to cover the demands of the ever increasing traffic over the network. This leads to the interference in the neighboring base stations, rendering the allocation of the resources statistically no longer adequate. SDN in cellular networks will solve some of the aforementioned deficiencies. The decoupling of the control plane and data plane introduces the centralized controller which has a complete view of the network thus allowing equipment to be simpler by recusing the architectural cost. The operations like routing, mobility management, policy enforcement are assigned to various cooperating controllers by making them more flexible and manageable. The centralized controller simplifies the various operations of the load as well as the noise management and does not require any direct interaction and coordination among the base stations. Controller makes the decision and instructs the dataplane to operate for various services. Introducing the virtual operators using SDN is simple and easy into the telecommunication market. Providers will be responsible for the management of the flow of the subscribers and the controllers without paying extra amounts.

4.7 Research Challenges

SDN is a promising technology for the communication between the IT and cloud providers and its enterprises; some challenges are still unsolved like the performance of the cloud in the wireless networks, security challenges raised due to the introduction of the programmable concept. The dynamics of the network will change with the attacks like DDoS, spam, Malware, phishing, and so on which need to be addressed for the development of secure SDN. Mobile networks are more vulnerable as the broadcast channels allow the eavesdropping and injection attack. The common challenges that arise with the SDN are listed below [7, 13].

1. Reliability: Network topologies need to be configured intelligently for the prevention of the errors by increasing the network availability that may occur manually. There is a possibility of single point failure because of brain-split problem. Networks need to be routed to alternative nodes when the network devices fail for the flow of continuation. Controller will be responsible for the failure of the entire network in the absence of a stand-by-controller. To overcome this issue, many organizations have come forward with many solutions like the implementation of the multipaths for the reroute of the traffic towards the active links, support of various technologies like Virtual Router Redundancy Protocol (VRRP), Multi-Chassis Link Aggregation Group (MC-LAG) for the development of network availability. The concept of clustering has emerged in order to overcome the network failure with the help of stand by controllers. Memory synchronization needs to be maintained between the active and stand by controllers. Many have proved that the centralized controller concept will interrupt the traffic flow in the network and may lead to network failure. Many have suggested various solutions but still the problem exists as a research challenge.
2. Scalability: SDN is distinguished with traditional networks with the decoupling of the data plane and control plane from the architecture. The planes can evolve independently until unless APIs are connecting them. The changes are accelerated in the control plan with the help of the centralized view of the network. Although SDN supports decoupling, it has its own disadvantages like standard APIS need to be defined which may arise the scalability issue. It has been noted that the number of switches in the network increases along with the end hosts then the controller will become a key bottleneck to be considered. When the bandwidth is increased, the end users will increase and more requests will be queued towards the controller which will fail to handle. The flow-setup process contains some limitations that lead to scalability issues. SDN network causes limited visibility of the traffic which does not allow troubleshooting on this platform. When the network team will find the network slowdown, they will immediately reschedule the backup.
3. Performance under Latency Constraints: SDN performance is measured on flow-setup time and the number of flows that occurs per second where the controller can handle. The flow setup will be proactive and reactive whereas the proactive will occur before the packet arrival, so the switch will understand how to deal with it. This will help in removing the limit of the number of flows per second which will be handled by the controller. Reactive mode will come into existence when the packet arriving the switch does not match with the existing entries of the flow table. Here, the controller will decide how to process the packet and those

related instructions will be cached onto that switch. This will consume more time compared to the proactive mode as it is the sum of the time taken to process at the controller and the updation of the switch about the change of flow. This reactive mode will introduce an overhead that can be used to limit network scalability and introduce the flow-setup delay. Many experts have suggested solutions but still remain an issue. Some solutions based on DevoFlow and McNettle architectures are used to overcome this issue.

4. Use of Low-level Interfaces between the controller and Network Device: Although Network management is made easy with the help of simple interfaces and the control applications which determines the high-level network policies. The SDN has to translate these policies to low-level configurations present on the switch. The current available controllers support the event-driven model, imperative. When the network packet flows, the interfaces will react to the arrival of the packet and for the link status updates with the installation or the uninstallation of the individual packet processing rules specified in the low-level interfaces along with rule-by-rule and switch by switch. The programmers continuously need to uninstall or install the policies which will affect the future events which will be monitored by the controller. The interfaces need to coordinate continuously between the multiple asynchronous events to perform a simple task. This will increase the burden on the controller and may generate time-absorption issues and may slow down the entire network. Solutions were suggested based on a high-level programming language that contains operators which can allow or deny the flows maintaining QoS.
5. Controller Placement Problem: This problem arises from the decoupling of the control and data plane to the flow-setup latency towards the reliability, fault tolerance to the performance of the system. We can consider an example of delay due to the use of wide area networks (WANs), availability limit of the network. The practical implications with this are from the software design, which affects the controller's response towards the events in the real time. This even includes the network topology and the number of controllers required for the smooth communication.
6. Security: The studies performed by the IT professional on the security challenges are the lack of integration with the available technologies that are unable to poke around each and every packet. The controller vulnerability will increase with the increase of intelligence of the controllers. Once the hacker or the attacker gains the controller access, they can damage part or full network. The SDN has to incorporate the authentication and authorization services for the various classes of the network administrators. The system should be able to alert the network providers in case of sudden attack and limit the communication of the controller.

SDN is a promising technology but lacks the standard policies. The current architecture still lacks the standard topology, delay, and loss of packets in the network. It does not even support horizontal communications among the nodes and collaborating devices. It still has experience with the absence of OpenFlow drivers and a standard high-level programming language. The other concerns include interoperability, performance and privacy concerns, and lack of technical experts for the support. It gained popularity due to the proposed prototypes, development of the tools and a particular language for the OpenFlow and for Controllers along with the Cloud computing networks [25, 26, 31, 40].

References

1. Alsmadi, I., Alazzam, I., & Akour, M. (2017). A systematic literature review on software-defined networking. In *2021 International Conference on Information Technology (ICIT)*. https://doi.org/10.1007/978-3-319-44257-0_14
2. Aujla, G. S., & Kumar, N. (2018). SDN-based energy management scheme for sustainability of data centers: An analysis on renewable energy sources and electric vehicles participation. *Journal of Parallel and Distributed Computing, 117*, 228–245.
3. Aujla, G. S., Jindal, A., Kumar, N., & Singh, M. (2016). SDN-based data center energy management system using RES and electric vehicles. In *2016 IEEE Global Communications Conference (GLOBECOM)* (pp. 1–6). New York: IEEE.
4. Aujla, G. S., Jindal, A., & Kumar, N. (2018). EVaaS: Electric vehicle-as-a-service for energy trading in SDN-enabled smart transportation system. *Computer Networks, 143*, 247–262.
5. Aujla, G. S., Chaudhary, R., Kumar, N., Kumar, R., & Rodrigues, J. J. (2018). An ensembled scheme for QoS-aware traffic flow management in software defined networks. In *2018 IEEE International Conference on Communications (ICC)* (pp. 1–7). New York: IEEE.
6. Aujla, G. S, Singh, A., & Kumar, N. (2019). Adaptflow: Adaptive flow forwarding scheme for software-defined industrial networks. *IEEE Internet of Things Journal, 7*(7), 5843–5851.
7. Braun, W., & Menth, M. (2014). Software-defined networking using OpenFlow: Protocols, applications and architectural design choices. *Future Internet 2014, 6*, 302–336.
8. Cabaj, K., Wytrębowicz, J., Kuklinski, S., Radziszewski, P., & Dinh, K. (2014). *SDN Architecture Impact on Network Security*. https://doi.org/10.15439/2014F473
9. Cai, Z., Cox, A. L., Ng, T. S. E. (2010). *Maestro: A System for Scalable OpenFlow Control*, Rice University Technical Report TR10-08, December 2010.
10. Cai, Z., Cox, A. L., & Ng, T. S. E. (2011). *Maestro: Balancing Fairness, Latency, and Throughput in the OpenFlow Control Plane*. Rice University Technical Report TR11-07, December 2011.
11. Cao, H., Wu, S., Aujla, G. S., Wang, Q., Yang, L., & Zhu, H. (2019). Dynamic embedding and quality of service-driven adjustment for cloud networks. *IEEE Transactions on Industrial Informatics, 16*(2), 1406–1416.
12. Campbell, A. T., et al. (1999). Open signaling for ATM, internet and mobile networks (OPENSIG'98). *ACM SIGCOMM Computer Communication Review, 29*(1), 97–108.
13. Conti, M., Chong, S., Fdida, S., Jia, W., Karl, H., Lin, Y., Mähönen, P., Maier, M., Molva, R., Uhlig, S., & Zukerman, M. (2011). Research challenges towards the Future Internet. *Computer Communications 2011, 34*(18), 2115–2134.
14. Devolved Control of ATM Networks (2013). Available from http://www.cl.cam.ac.uk/research/srg/netos/old-projects/dcan/
15. Feghali, A., Kilany, R., & Chamoun, M. (2015). SDN security problems and solutions analysis. In *2015 International Conference on Protocol Engineering (ICPE) and International Conference on New Technologies of Distributed Systems (NTDS)*, Paris, 2015 (pp. 1–5). https://doi.org/10.1109/NOTERE.2015.7293514.
16. Fei, H., Hao, Q., & Bao, K. (2013). A Survey on software-defined network (SDN) and OpenFlow: From concept to implementation. *IEEE Communications Surveys & Tutorials, 16*(4), 2181–2206 (2013)

17. Ferro, G. (2012). *OpenFlow and software-defined networking*. http://etherealmind.com/software-defined-networking-openflow-so-farand-so-future/.
18. Foster, N., Freedman, M. J., Harrison, R., Rexford, J., Meola, M. L., & Walker, D. (2010). Frenetic: A highlevel language for OpenFlow networks. In *Proceedings of the Workshop on Programmable Routers for Extensible Services of Tomorrow (PRESTO '10)*, Philadelphia, PA (2010) (article no. 6)
19. Foster, N., Harrison, R., Freedman, M. J., Monsanto, C., Rexford, J., Story, A., & Walker, D. (2011). Frenetic: A network programming language. In *ACM SIGPLAN Notices—ICFP '11* (Vol. 46, pp. 279–291).
20. Hakiri, A., Gokhale, A., Berthou, P., Schmidt, D. C., & Gayraud, T. (2014). Software-defined networking: Challenges and research opportunities for Future Internet. *Computer Networks, 75*(Part A), 453–471. ISSN:1389-1286.
21. Hicks, M., et al. (1998). PLAN: A packet language for active networks. *ACM SIGPLAN Notices, 34*(1), 86–93 (1998).
22. Hu, F. (2014). *Network innovation through OpenFlow and SDN: Principles and design*. Boca Raton: CRC Press (2014). http://dx.doi.org/10.1201/b16521
23. Jammal, M., Singh, T., Shami, A., Asal, R., & Li, Y. (2014). Software-defined networking: State of the Art and research challenges. *Computer Networks, 72*. https://doi.org/10.1016/j.comnet.2014.07.004
24. Kim, E.-D., Lee, S.-I., Choi, Y., Shin, M.-K., & Kim, H.-J. (2014). A flow entry management scheme for reducing controller overhead. In *2014 16th International Conference on Advanced Communication Technology (ICACT)* (pp. 754–757).
25. King, D., Rotsos, C., Aguado, A., & Georgalas, N. (2016). *The Software Defined Transport Network: Fundamentals, Findings and Futures*. https://doi.org/10.1109/ICTON.2016.7550669
26. Khan, S., Shah, M., Khan, O., & Wahab Ahmed, A. (2017). *Software Defined Network (SDN) Based Internet of Things (IoT): A Road Ahead* (pp. 1–8). https://doi.org/10.1145/3102304.3102319
27. Kreutz, D., Ramos, F. M. V., Verissimo, P. E., Rothenberg, C. E., Azodolmolky, S., & Uhlig, S. (2015). Software-defined networking: A comprehensive survey. *Proceedings of the IEEE, 103*(1), 14–76.
28. L. Foundation, *Opendaylight: An Open Source Community and Meritocracy for Software-Defined Networking*. A Linux Foundation Collaborative Project (April 2013).
29. Lara, A., Kolasani, A., & Ramamurthy, B. (2014). Network innovation using OpenFlow: A survey. *IEEE Communications Surveys & Tutorials, 16*(1), 493–512 (2014). First Quarter.
30. Liu, D., & Deng, H. (2013). *Mobility Support in Software Defined Networking*, Tech. Rep.
31. Marina, M. K., & Kontovasilis, K. (2015). *Software Defined Networking Concepts*. 19 June 2015 https://doi.org/10.1002/9781118900253.ch3
32. Monsanto, C., Reich, J., Foster, N., Rexford, J., & Walker, D. (2013). Composing software-defined networks. In *Proceedings of the 10th USENIX Symposium on Networked Systems Design and Implementation*.
33. NSDI'13 (2013). *Proceedings of the 10th USENIX conference on Networked Systems Design and Implementation (NSDI '13)*, Lombard, IL (pp. 1–14).
34. Nunes, B. A. A., Mendonca, M., Nguyen, X.-N., Obraczka, K., & Turletti, T. (2014). A survey of software-defined networking: Past, present, and future of programmable networks. *IEEE Communications Surveys and Tutorials, 16*(3), 1617–1634.
35. Open Network Foundation (2013). *SDN Architecture Overview*, version 1.0.
36. *OpenFlow Components*, http://archive.openflow.org/wp/openflowcomponents/,2011
37. Rexford, J. (2012). *Software-defined networking*. COS 461: Computer networks lecture. http://www.cs.princeton.edu/courses/archive/spring12/cos461/docs/lec24-sdn.pdf
38. Rowshanrad, S., Namvarasl, S., Abdi, V., Hajizadeh, M., & Keshtgary, M. (2014). A survey on SDN, the future of networking. *Journal of Advanced Computer Science and Technology, 3*, 232–248. https://doi.org/10.14419/jacst.v3i2.3754

39. Sezer, S., Scott-Hayward, S., Chouhan, P. K., Fraser, B., Lake, D., Finnegan, J., Viljoen, N., Miller, M., & Rao, N. (2013). Are we ready for SDN? Implementation challenges for software-defined networks. *IEEE Communications Magazine, 2013*, 36–43.
40. Singh, S., & Jha, R. K. (2017). A survey on software defined networking: Architecture for next generation network. *Journal of Network and Systems Management, 25*, 321–374 (2017). https://doi.org/10.1007/s10922-016-9393-9
41. Tennenhouse, D. L., et al. (1997). A survey of active network research. *IEEE Communications Magazine, 35*(1), 80–86.
42. Van der Merwe, J. E., et al. (1998). The tempest—A practical framework for network programmability. *IEEE Network, 12*(3), 20–28.
43. Wetherall, D. J., Guttag, J. V., & Tennenhouse, D. L. (1998). ANTS: A toolkit for building and dynamically deploying network protocols. In *IEEE Open Architectures and Network Programming* (pp. 117–129).
44. Xia, W., Wen, Y., Foh, C. H., Niyato, D., & Xie, H. (2015). A survey on software-defined networking. In *IEEE Communications Surveys and Tutorials* (Vol. 17, no. 1) (pp. 27–51). Firstquarter 2015. https://doi.org/10.1109/COMST.2014.2330903
45. Yeganeh, S.H., Tootoonchian, A., & Ganjali, Y. (2013). On scalability of software-defined networking. *IEEE Communications Magazine, 51*(2), 136–141.

Chapter 5
Network Policies in Software Defined Internet of Everything

Rashid Amin, Mudassar Hussain, and Muhammad Bilal

5.1 Introduction

Software Defined Network [5, 8, 9] is an emerging concept that has been deployed in various domains. There are several mechanisms for network policy management that exist in the literature [6].

5.1.1 What are the Network Policies?

Network policies are the set of rules which instruct network devices to function as per requirements of users, applications, and/or organizations. The network managers implement these policies to restrict/allow specific communication to a certain resource or group of users. The correct implementation of network policies helps to protect the integrity, confidentiality, and availability of precious data while providing efficient and effective access to information systems. The scope the network policies depend upon the category of networks, i.e., Local Area Network (LAN), Metropolitan Area Network (MAN), and Wide area Network (WAN). Network policies in LAN consist of a collection of goals for an organization regarding rules for users/administrators and system/management requirements to achieve a

R. Amin (✉)
University of Engineering and Technology, Taxila, Pakistan

M. Hussain
University of Wah, Wah Cantt, Pakistan

M. Bilal
Department of Computer Engineering, Hankuk University of Foreign Studies, Yongin-si, Gyeonggi-do, South Korea
e-mail: mbilal@hufs.ac.kr

G. S. Aujla et al. (eds.), *Software Defined Internet of Everything*, Technology, Communications and Computing, https://doi.org/10.1007/978-3-030-89328-6_5

certain access level. Network policy is a "living document" that is continuously updated as per the requirements of users or evolution in technology [38]. In this way, the network policy can be configured that assigns users to a Virtual Local Area Network (VLAN). When VLANs are configured by using internetworking devices, e.g., routers, switches, bridges, etc., then the network policy restricts communication between different VLAN users or severs. Moreover, VLANs help to cluster multiple network resources logically which is quite beneficial to design , implement, and manage networks. Similarly, all remote access to the MAN and WAN shall be authenticated, logged, and restricted to minimize the risk to the valuable assets. The access control in this case normally implemented by using layer 3 or layer 4 devices, for example, routers and firewalls.

Internet of Everything (IoE) is the superset of Internet of Things (IoT) introduced by Cisco and is appeared as an innovation in IoT. In IoE, people, processes, data, and things are being connected to the Internet that comprises the wider concepts in the field of connectivity. So, this connection would be producing a massive data volume. There are wide application areas of IoE and such kind of networks can be implemented in healthcare, construction, factories, agriculture, logistics, etc. It enables network connectivity between diverse kind of devices for automatic processing without human intervention. Recently, the efforts are being carried out to interconnect IoE infrastructure and cloud computing platforms. It is increasing the complexity of networks which results in various security problems. In such environment, where there are involved numerous actors for data communication, the implementation of policies for an effective data flow is quite difficult to implement [22].

In traditional computer networks, the routers/firewalls are configured based on network policies to filter network traffic. Although, these networks are broadly adopted, however, these are complex and hard to manage due to the distributed nature where network control and management are implemented in every device [13]. In addition, network policy implementation in these networks is also quite tedious and often takes weeks to months to translate and configure the policies at the networking devices. The network policies are sequential collection of permit and deny conditions that state a router or firewall to forward/permit packets or drop/deny. The data communication is controlled by the routing protocols or by installing network policies on the interfaces of internetworking devices. In addition, the network policies are configured manually using device specific commands. Moreover, the network policies are normally implemented based on destination IP addresses or in tuple form that comprises source IP, destination IP addresses, ports, and protocol. The below Access List 112 shows that the network traffic from network 192.168.12.0 is allowed for communication to network 10.10.1.0 and otherwise denied. Similarly, Access List 122 shows that TCP traffic for telnet application is allowed from host 192.168.13.1 to host 10.10.1.2.

```
Access List 112 Permit IP 192.168.12.0 0.0.0.255 10.10.1.0 0.0.0.255
Access List 112 Deny IP Any
Access List 122 Permit TCP Host 192.168.13.1 Host 10.10.1.2 Eq Telnet
```

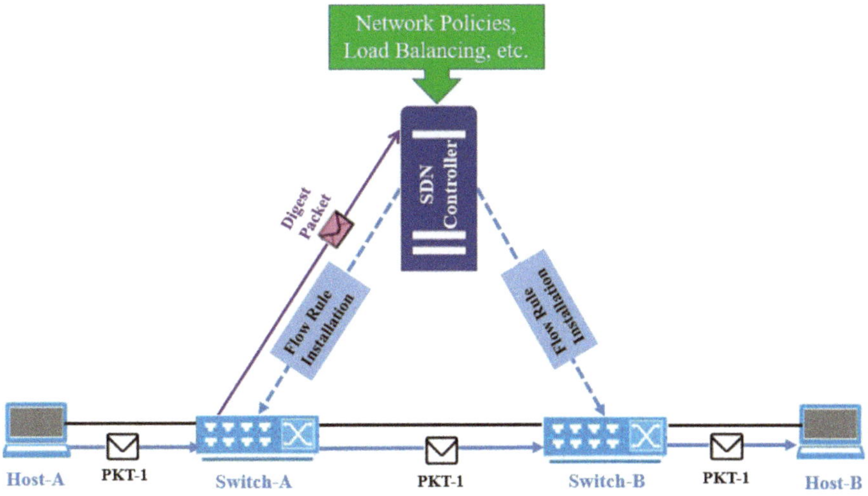

Fig. 5.1 Network policy implementation in SDN

Software Defined Networking (SDN) [3, 18, 33, 37, 45] is an emerging network architecture which helps to solve limitations of traditional networking by separating data plane from network control and management planes of forwarding devices. In SDN, the network policies are configured at the control plane which computes best path between source and destination for data flow as shown in Fig. 5.1. Based on specified network policies, the controller computes flow rules and installs at the switches along the path between source and destination.

For example, a network policy based on destination IP address is represented via pyretic language [35] as (Match (Switch = SW1) and Match (Dest_IP = '192.168.1.3') >> FWD(2)). This policy states that when a packet is received at Switch-1 (SW1) whose destination IP address is 192.168.1.3 will be forwarded to Port 2 of switch-1. Similarly, a complex network policy that is expressed as Employees (Faculty and Staff) has access to the Servers via Transmission Control Protocol (TCP) and Destination Port Numbers 20, 25, 80. This policy is more complex than the destination IP address based policy, as it includes multiple parameters, like Source Employees, Destination Servers, Protocol TCP, Port Numbers 20, 25, 80. Such policies are implemented at the central controller which computes best path between source and destination in addition to installing flow rules at the along the best path as per policy. This centralized policy implementation mechanism helps to reduce complexity and network management in SDN.

5.1.2 Role and Importance of Network Policies

The network policies in communication networks play a vital role with respect to the effective flow of data between a valid source and destination. A communication network in which network policies are implemented can be automated more easily and can react quicker in case of any change in the policies of an organization. In such networks the internetworking devices, users, and applications need to follow the instruction from defined network policies. In IoEs, the access privileges are provided to all stakeholders, i.e., users, devices, and applications and updated as networks expand. The access rights enable users to perform certain actions and can get more access to the information resources. In this way, more devices, users, applications can be added in the network for information exchange. In addition, network policies help to implement access control to the resources in order to protect the sensitive data. It helps to provide consistent services and make performance dependable and verifiable. The following are very important benefits of implementing network policies in communication networks, such as, IoTs and IoEs.

5.1.2.1 Business Intent and Agility

Network policies are building block of any organization and implemented in a way to reflect the business intent to achieve the desired business outcomes. The policies are implemented at the internetworking equipment for accessing a certain data and without implementation of policies, the organizations fail to deliver the output in an optimal way.

5.1.2.2 Consistent Services

The well implemented network policies offer consistent services throughout the network and provide seamless mobility, fault tolerance, and predictable data delivery. So, the users and things can access the network resources remotely without affecting the access privileges while maintaining quality of services [7].

5.1.2.3 Network Automation

Network automation is the process in which internetworking devices are configured by implementing security and management policies. It helps to increase network performance and efficiency. Network virtualization is often helpful to automate communication networks.

5.1.2.4 Performance Monitoring

Performance monitoring is an important aspect which reveals that the implementation of policies, goals, and metrics are meeting the organization requirements or not [15]. It helps to measure the network performance and ensures that the policies are meeting the goals and objectives of an organization.

5.1.2.5 Network Security

With policies in place, any violations can be easier to detect. Security is more easily enforced, threats more quickly contained, and risk rapidly reduced with security-related policies [10, 11, 16].

5.2 Types of Network Policies for IoE

There are several types of network policies, i.e., access control, load balancing, archiving, failover, quality of service, traffic engineering, etc. [47]. Some common network policies are depicted in Fig. 5.2. Since network policies specify how the network must function in different circumstances, there is no set list of policies [17, 54]. A network's policies depend on what is necessary to achieve business objectives. These network policies manage different types of services, applications, and tasks in the IoE network.

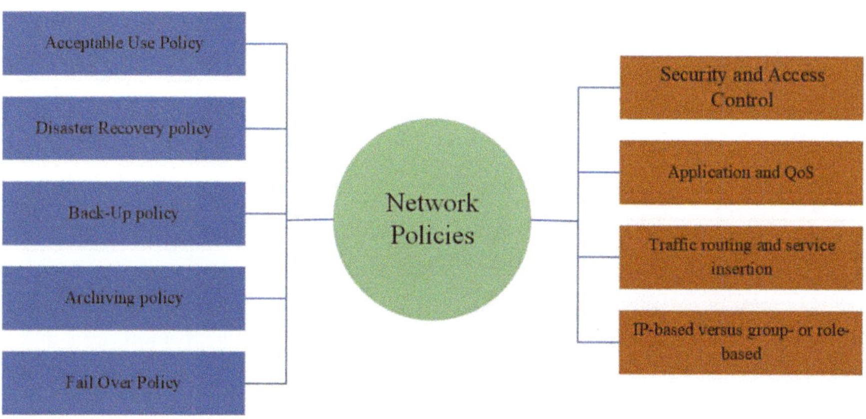

Fig. 5.2 Type of network policies in IoE

5.2.1 Security and Access Control

This policy determines when a user or object can join the network and what services they can access. Since data and application protection is based on access and security policies, they may be the most critical types of policies.

5.2.2 Application and QoS

These policies determine the relative priority of different applications and how traffic can be prioritized for each. Hackers can gain access to your networking environment from any computer or software product you use. As a result, it is important to keep all systems up to date and fixed in order to avoid cyberattackers from manipulating bugs to gain access to confidential data [28]. The mixture of hardware, applications, and best practices you use to track problems and close holes in your protection coverage are referred to as application security and it enhances the QOS.

5.2.3 Traffic Routing and Service Insertion

These policies specify how traffic from specific groups of users or hosts, such as guest traffic, can be routed via a firewall and other security modules. Routing protection guarantees that the routing protocol is working properly . It involves putting in place mechanisms to ensure that state modifications on devices and network elements are monitored, whether they are dependent on external or internal inputs (physical security of the device itself and parameters maintained by the device, including, e.g., clock) [51].

5.2.4 IP-Based Versus Group- or Role-Based

Policies may be specified at the IP address or function stage. Role-based policies are more complex, flexible, and easy to automate, and they promote consumer and system mobility. IP-based policies are inflexible, do not scale, and are ideally tailored to a stable climate.

A network that adheres to the well-defined policy will quickly fulfill the market requirements for which it was built. The network cannot be set up to deliver optimally without specific targets, and its efficiency cannot be calculated without priorities. Network policies can also be classified as follows:

5.2.5 Standard Usage Policy

Employees who request access to a network must sign a nondisclosure document prior to being given access. They have to promise not to use it for non-work-related purposes, such as sharing copyrighted content, watching pornography, or social networking. The arrangement is referred to as an Appropriate Usage Policy [46]. It relieves the company from worry over responsibility for the network use. The result is that an organization may legitimately terminate anyone who uses the network for work-related reasons but only with a prior warning about what will occur.

5.2.6 Disaster Recovery Policy

There is a possibility that both of the networks would go down in flames. This may be anything from a building containing a network on fire or flooding to malicious destruction caused by an intruder or a dissatisfied employee. When a tragedy strikes, an organization's network must be restored as soon as possible or risk going out of business [14]. A disaster management strategy is a written plan that seeks to get administrators in a company to think ahead of what they intend to do in the case of a disaster. Companies would realize what to do easily and calmly in the case of a major situation if they plan accordingly to prepare for the worst.

5.2.7 Backup Policy

In all moments, all information is at risk because of a number of reasons; an intruder or a dissatisfied employee can erase data, a fire may occur, data may be compromised by software or an update, a piece of machinery such as a hard disc may malfunction, an earthquake or other natural catastrophe may occur, or a device containing data may be stolen, and so on. Regular backups are conducted to ensure the data can be retrieved [30].

5.2.8 Archiving Policy

One does not like to keep all the files on the system, however, may not be able to erase them. For example, after a year group graduate, you no longer need to have them on the school's admin system. It is doubtful that the tax office would use these reports again because the corporation has done with them (companies have to keep tax records for seven years by law). After a plane accident, they would have to consult the archives for parts [55]. They do not want to know the latest machine specifications but might want to include them in the future. So, in all these cases, archiving policies help to manage the entire system smoothly.

5.2.9 Failover Policy

In the event that the system fails, businesses often have replacement systems for essential network infrastructure. If the first system fails, these replacement systems are programmed to start automatically [31]. It shows that the IoE network is not having some "down period." "Down time" happens anytime a corporation is reluctant to use the network, considering the reality that it is vital to its activities. When an IoE network was offline for even a few hours, it might cost them thousands, if not millions, of pounds. Servers and routers are common examples of infrastructure that is duplicated in this way [39]. A written protocol is known as a Failover policy, which specifies which items of equipment can be duplicated, how they will be set up, and how long customers can anticipate downtime to be, among other things.

5.3 Automation of Network Policies

The automation of network policies is extremely important to increase the network performance. The high-level programming languages, such as, Pyretic [35], Frenetic [19], and Maple [53] help to specify these policies as per the application environment. These languages provide parallel and sequential composition operators for effective implementation of policies. These help to implement multiple policies via composition operators to process network packets in series or parallel. These languages provide network programmers a platform where they can build network applications based of network policies to test the behavior of whole network.

In [24, 25], the problem of changing or modifying the network policies at controller is discussed in SDN. This changing or modification of network policies leads to packet violations due to already installed flow rules at switches. They solve this problem by detecting network policy change with the help of matrices and multi-attributed graphs. After detection of policy change the proposed approaches detect policy conflicts/overlapping and the conflicting flow rules are deleted from the switches. In addition, the controller computes best path and new flow rules which are cached at controller and installed along the path. In this way the proposed approaches automate network policy change which improves network performance and efficiency. In research work [52], the authors detect irregularities in the network policies before the implementation in the network. It means that the anomalies are fixed before the installation of flow rules at switches. To achieve the desired goal, the forwarding policies are formally represented and set of anomalies are detected against set of flow rules for the respective policies. In addition, it also provides provision for network administrators to specify their own anomalies.

In [43], SDN based proactive flow rules installation approach is proposed for efficient communication in Internet of Things (IoT). It resolves the problem of flow installation delay as well as congestion due to packet-in messages. This saves energy and other potential resources of network nodes. The results reveal that the proposed mechanism reduces congestion between controller and network nodes and reduces average flow rule installation delay by 90%. In SDN, flow rules are installed in switches on the basis of exact matching [43] or wildcard-based matching [34]. The wildcard-based matching improves reusability of flow rules and reduces packet-in messages. This improves scalability both at data and control planes. However, in case of exact matching almost every flow passing the switch will generate a packet-in message to the controller which exhausts precious resources. To resolve this problem, some researchers suggest using load balancing mechanism by installing proactive flow rules on multiple switches [36] or reactive caching flow rules in each switch. In [56], SDN based wildcard rule caching mechanism namely CAching in Buckets (CAB) is proposed for efficient flow rules placement during network policy implementation process. It suggests partitioning the field space into logical structures called "buckets" and cache buckets along with all the associated flow rules. The CAB helps to solve the flow rule dependency problem with quite less overhead. In addition, it significantly reduces flow setup time, saves bandwidth and flow setup requests. There are different other research works which are helpful to debug networks [1, 32] and ensure network consistency [42]. These are quite effective to automate network policies as during automation; network consistency is really important, and testing/debugging ensures efficient implementation process.

5.4 Network Policies in SDN

SDN is a networking model in which a single software program called a controller dictates the actual network activity. SDN transforms network equipment into basic packet forwarding devices (data plane), with the controller operating as the "brain" or control logic (control plane) [57] . As opposed to traditional network approaches, this paradigm change has many advantages. First, utilizing a software program to bring innovative concepts into the network management is far simpler than using a preset series of commands in proprietary network devices because the software is far easier to modify and control. Second, SDN introduces the advantages of a centralized approach to network design over distributed management: instead of needing to customize all network equipment separately to alter network behavior, operators may make networkwide traffic forwarding policies in a theoretically single location, the dispatcher, who has global awareness of the network state.

Network policies are defined and implemented in an enterprise network for achieving better network management, ensuring network security and access to resources [44]. In a traditional network, network policies are implemented on devices interfaces using low-level commands as follows, (i) all switches are identified where network policies are to be implemented. (ii) Then the respective interfaces of switches are selected manually and policies are configured using specific commands. Moreover, these network device control and management commands vary from one forwarding device manufacturing company to another. For example, to check the current device configurations, HP uses command as display current-configuration, Cisco uses command show running-config, and Juniper uses command as show configuration/display set. Variation of these commands creates many problems for the network administrator and requires skills according to multiple operating systems to smoothly manage the entire network.

In an SDN network, network policies are implemented at the controller, and according to these policies, the controller installs rules on the switches and IoE devices. However, there is a need to identify the switches where the policies are to be implemented [4]. For example, creating a network policy that does not allow traffic from a source node to the destination on router "R3", it can be done in pyretic language as follows: $match(switch = R3, inport = 2, srcip =' 10.0.1.1')[drop]$ In addition to this, high-level names can be added to the corresponding ranges of end-user IP addresses, a communication application, protocols, and/or port number, as shown below. $match(Faculty, LMS, TCP, 20)[Permit]$ NOX SDN controller provides the following format for policy implementation: $Install < switch, pattern, priority, timeout, actions >$. The above command deploys a network policy by specifying pattern, priority, timeout, and actions. This information is passed to the flow table of an OpenFlow switch. There are three attributes in the OpenDayLight controller: i.e., Type, ID, Permissions (actions) for a network policy. Type refers to the user or group for which NETWORK is being created, ID is the unique identity of the user and Permissions are the operations/actions to be performed.

5.5 Conflict and Overlapping Among the Network Policies

The network policies need to be composed in an effective manner to avoid conflicts and overlapping. The conflicts and overlapping in network policy implementation process normally occur due to the misconfiguration within a single policy or between policies in different devices. To avoid such conflicts, the in-depth understanding of the causes of these conflicts, automated inspection of policy rules and minimum manual implementation is desired.

Consider an example of three network policies (P1, P2, P3) and discuss conflicts and overlapping in these policies. It is noted that in IoEs and IOTs, where there are different kinds of devices, tenants, and connectivity parameters, such conflicts and overlapping may increase. In SDN based networking environment where these policies are configured at control plane are explained below:

Policy 1 (P1) states that:
Faculty can communicate with Learning Management System (LMS) Server through TCP Port
Policy 2 (P2) states that:
Employees (Faculty and Management) can communicate with Servers through TCP Ports 20, 25, 80
Policy 3 (P3) states that:
Faculty and Management can communicate with each other through TCP Ports 20, 25, 80, 587, 993
The network policies (P1, P2, P3) in tuple form are represented below:
P1 = "Faculty, LMS, TCP, 20, Permit"
P2 = "Employees, Servers, TCP, (20, 25, 80), Permit"
P3 = "Faculty, Management, TCP, (20, 25, 80, 587, 993), Permit"

The policies P1 and P2 have conflict with each other because P1 says that only Faculty is permitted to communicate with LMS Server and others access is denied. However, policy P2 states that Employees can communicate with all Servers. So, there is a conflict in network policies (P1 and P2), because P1 is denying access of Management to LMS and policy P2 is permitting access of all Employees including Management to all servers which also includes LMS as well. In addition there is an overlapping in network policies (P1 and P2), as P1 is permitting access to LMS through TCP Port 20 and P2 states that Employees can access LMS including all Servers through TCP Ports 20, 25, 80 in which access of Faculty to LMS is again permitted, because Employees also include Faculty. The third policy P3, however, has no conflict and no overlapping. These policies can be correctly implemented with the help of human operator who can manually compose these policies into a composite policy. This composition can be done through any programming language in IF-THEN-ELSE program as follows:

```
If_Match (SOURCE = "Faculty", DESTINATION = "LMS", PROTOCOL =
"TCP", PORT = 20)
    PERMIT
Else If_match (DESTINATION = "LMS")
    DENY
Else If_Match (SOURCE = "Employees", DESTINATION = "Servers", PROTO-
COL = "TCP", PORT = "20,25,80")
    PERMIT
Else If_Match (SOURCE = "Faculty", DESTINATION ="Management", PROTO-
COL = "TCP", PORT = "20,25,80,587,993")
    PERMIT
```

These network policies can also be sketched with the help of Policy Whiteboarding and Set Operator [29] which is shown in Fig. 5.3. It shows that Faculty is part of Employees and LMS is part of Servers. So, Faculty is subset of Employees and LMS is subset of Servers.

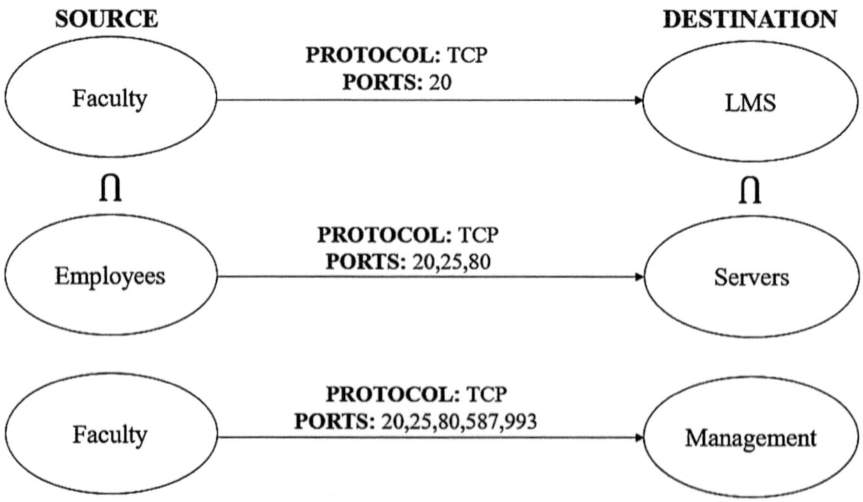

Fig. 5.3 Network policy whiteboarding

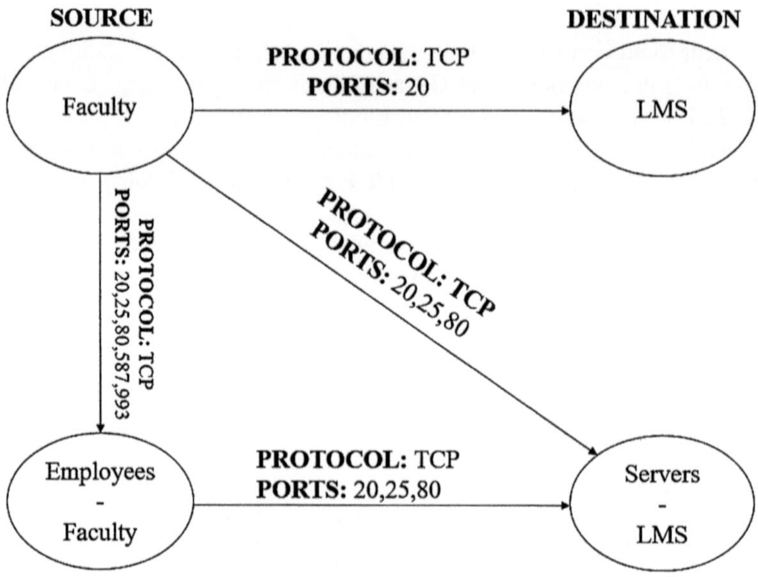

Fig. 5.4 Graph composition of network policies

The network policies (P1, P2, P3) can be composed correctly with the help of graph composition as shown in Fig. 5.4. It presents that the faculty can communicate with LMS server via TCP port 20. We can observe that because there is no relationship between Management and LMS, so graph composition reflects exclusive access of faculty to LMS which is the intent of network policy P1. In addition, the remaining Servers excluding LMS are represented by Difference ('-') Set Operator that allows desired communication of Servers with the Employees as well as the Faculty.

It means that graph composition is quite helpful to present conflict free network policies and play a vital role in automatic policy composition process. Policy Graph Abstraction (PGA) [41] presents a very interesting research work that helps users to easily specify and implement conflict free network policies automatically. With the help of PGA, users, tenants, admins, and applications can produce network policies as graphs which are forwarded to the graph composer through a PGA User Interface (UI) that can collect further information from external sources for the effective policy implementation. Afterwards, the graph composer offers conflict free graph by fixing all Errors.

5.6 Network Policies Optimization

A large number of network policies, i.e., monitoring, load balancing, routing, security, are deployed in the IoE to manage and control the devices. Network policies optimization is a set of best practices used to improve network performance. A variety of algorithms and techniques can monitor and improve network performance, such as: global load balancing, minimize latency, packet loss monitoring, bandwidth management, etc. For the efficient functioning of IoE networks, a variety of network management strategies, i.e., data management, routing, network mobility, heterogeneous node interoperability, and data protection, become crucial concerns.

An IoE network has strategic design objectives; policy execution is typically broken down into several parts in the form of policy artifacts. Each policy object handles a particular form of setup, such as IPsec for all machines, or fine-tunes a policy for a specific category of individuals, such as setting the network proxy server for all salespeople. As the decision management system progresses, the policy to manage gets more complex [21]. Administrators must configure with a range of configurations in current networks, including access control, program installation, device choice, and so on. As a result, creating a comprehensive strategy that covers anything is difficult, if not impossible. Instead, constructing decomposed and specialized regulation artifacts is far more achievable, enabling the management burden to be spread among several managers.

While today's IoE policy management systems still depend on the definition and incorporation of policy artifacts, this methodology has a range of drawbacks: (1) When several options are open, conflict mediation is a process that decides the final policy. While this is a general policy management problem, it is especially important to use policy artifacts since policy setups in separate objects that conflict, necessitating a resolution method. (2) When many objects work together to achieve a policy design objective, a transactional update is required. Such objects must be changed atomically without disclosing intermediate states to reflect a design shift.

Some analysis [49, 50] has centered on business networks in particular. They use the oversimplified "divide-and-conquer" approach, and network architecture is performed step by stage. Although these studies have an advanced state of the art in systematic network architecture, their models may result in overly complex configurations. Zhang et al. [58] investigate how to build effective, shared data structures for multiple packet filters, such as the HyperCuts decision tree. They discovered various important factors that can impact the efficiency of the mutual HyperCuts trees they created. Delignated is an innovative solution to clustering packet filters into joint HyperCuts decision trees. Memory use can be significantly decreased by allowing several packet filters to share HyperCuts decision trees, according to the assessment using both actual and synthetic packet filters.

5.6.1 Detection and Settlement of Policy Variations

Policy disagreement is a common phenomenon that involves all policy languages. According to [48], there are two kinds of policy disputes: intra-policy conflicts and inter-policy conflicts, with the former being exacerbated by an ill-defined policy and the latter by several rules being implemented, resulting in contradictory behavior. When the requirements of two or more laws are fulfilled simultaneously, but policy compliance cannot conduct their acts in these policies at the same time [40], an intra-policy dispute arises. For example, in an access management situation, one application policy accepts the access request and the other rejects it. These two policies cause intra-policy tension. Furthermore, an inter-policy dispute occurs when two or more relevant regulations produce opposing configuration commands and mechanisms for networked devices. The opposing policies in the inter-policy dispute situation do not conflict objectively, but they do conflict when allocated to framework elements at run time [23].

As a consequence, there are two approaches to resolving this problem: the first is to prevent setting competing rules for a network or device in advance, which is a static strategy of avoiding intra-policy conflicts; the second is to select a result or merge different outcomes when the policy deployment encounters disagreement results at run time. This is a complex approach for dealing with intra-policy and inter-policy disputes. The first proposal, above all, could immediately detect policy clashes. To this end, proposals should take the shape of a formal phrase or be able to be converted into one. ASL, for example, is a first-order logic language [20], while PDL can be converted into logic programs indirectly . When in logical representation, techniques for dispute checking that have already been established in this field can be conveniently extended. Policymakers may prevent setting competing rules by changing policy requirements and/or behavior before adding policies to the mechanism using this dispute detection [27].

5.6.2 Fine-Tuning of Goal Policies

High-level policies, which are mainly goal policies from the usage case perspective, and low-level policies, which are mostly action policies, are the two types of network policies. To finally realize management in action policy-driven processes, a mapping is needed to delegate behavior to specific artifacts and machines [2]. However, in goal policy-driven schemes, an additional refining phase is required before mapping into the actual framework components. A Target policy does not have detailed instructions as low-level system behavior, so the goal policy must be simplified to low-level ones about how to achieve the aim. Goal policy refinement's key challenges include determining how to dynamically extract low-level policies from a high-level policy and ensuring that they are compatible with the initial high-level policy [26]. Two foundations are needed to address these issues: a formal foundation for both the structure and the policy model and several refinement techniques to align the target priorities and concrete system behaviors.

Bandara et al. [12] propose a policy refinement strategy focused on the EC and goal-based criteria elaboration (Event Calculus). This approach aims to minimize high-level Goal policies to low-level Goal policies, which can then be mapped to concrete framework components. First and foremost, the UML-modeled framework should be converted into EC, a standardized language for representing the system. Abductive logic is then used to refine high-level objectives to low-level organizational policies, picking suitable methods to elaborate operations goals. Finally, the device artifacts are delegated to these operations for enforcement. It is worth noting that the term "strategy" applies to the process through which a device will accomplish a certain objective. Abductive logic may be used to infer methods because both the scheme and the target have a structured definition.

5.7 Conclusion

Network policies play a vital role in communication networks to implement access control. In traditional networking as the network expands in size then administration and management challenges also increase with the increase in complexity. However, SDN offers a promising approach to resolve these challenges by implementing network policies at the central controller. In this chapter, we have discussed network policies in detail along with the types, roles, and importance in IoTs and IoEs. In addition, we discussed the network policies with respect to the automation and implementation of conflict free polices. Finally, we discuss optimization of network policies for effective communication in IoTs and IoEs.

References

1. Al-Shaer, E., & Al-Haj, S. (2010). Flowchecker: Configuration analysis and verification of federated openflow infrastructures. In *Proceedings of the 3rd ACM Workshop on Assurable and Usable Security Configuration* (pp. 37–44), 2010.
2. Alvarez-Campana, M., López, G., Vázquez, E., Villagrá, V. A., & Berrocal, J. (2017). Smart CEI moncloa: An IoT-based platform for people flow and environmental monitoring on a smart university campus. *Sensors, 17*(12), 2856.
3. Alvizu, R., Maier, G., Kukreja, N., Pattavina, A., Morro, R., Capello, A., & Cavazzoni, C. (2017). Comprehensive survey on T-SDN: Software-defined networking for transport networks. *IEEE Communications Surveys & Tutorials, 19*(4), 2232–2283.
4. Amin, R., Shah, N., Shah, B., & Alfandi, O. (2016). Auto-configuration of ACL policy in case of topology change in hybrid SDN. *IEEE Access, 4*, 9437–9450.
5. Amin, R., Reisslein, M., & Shah, N. (2018). Hybrid SDN networks: A survey of existing approaches. *IEEE Communications Surveys & Tutorials, 20*(4), 3259–3306.
6. Amin, R., Shah, N., & Mehmood, W. (2019). Enforcing optimal ACL policies using k-partite graph in hybrid SDN. *Electronics, 8*(6), 604.
7. Aujla, G. S., Chaudhary, R., Kumar, N., Kumar, R., & Rodrigues, J. J. P. C. (2018). An ensembled scheme for QoS-aware traffic flow management in software defined networks. In *2018 IEEE International Conference on Communications (ICC)* (pp. 1–7). New York: IEEE.
8. Aujla, G. S., Singh, A., & Kumar, N. (2019). Adaptflow: Adaptive flow forwarding scheme for software-defined industrial networks. *IEEE Internet of Things Journal, 7*(7), 5843–5851.
9. Aujla, G. S., Kumar, N., Garg, S., Kaur, K., & Ranjan, R. (2019). EDCSuS: Sustainable edge data centers as a service in SDN-enabled vehicular environment. *IEEE Transactions on Sustainable Computing*. https://doi.org/10.1109/TSUSC.2019.2907110
10. Aujla, G. S., Singh, A., Singh, M., Sharma, S., Kumar, N., & Choo, K.-K. R. (2020). Blocked: Blockchain-based secure data processing framework in edge envisioned v2x environment. *IEEE Transactions on Vehicular Technology, 69*(6), 5850–5863.
11. Aujla, G. S., Singh, M., Bose, A., Kumar, N., Han, G., & Buyya, R. (2020). Blocksdn: Blockchain-as-a-service for software defined networking in smart city applications. *IEEE Network, 34*(2), 83–91.
12. Bandara, A. K., Lupu, E. C., Moffett, J., & Russo, A. (2004). A goal-based approach to policy refinement. In *Proceedings of the Fifth IEEE International Workshop on Policies for Distributed Systems and Networks, 2004. POLICY 2004* (pp. 229–239). New York: IEEE.
13. Benson, T., Akella, A., & Maltz, D. A. (2009). Unraveling the complexity of network management. In *NSDI* (pp. 335–348).
14. Berke, P., Cooper, J., Aminto, M., Grabich, S., & Horney, J. (2014). Adaptive planning for disaster recovery and resiliency: An evaluation of 87 local recovery plans in eight states. *Journal of the American Planning Association, 80*(4), 310–323.
15. Cao, H., Wu, S., Aujla, G. S., Wang, Q., Yang, L., & Zhu, H. (2019). Dynamic embedding and quality of service-driven adjustment for cloud networks. *IEEE Transactions on Industrial Informatics, 16*(2), 1406–1416 (2019)
16. Cisco (2010). *What is network policy?* Available at https://www.cisco.com/c/en/us/solutions/enterprise-networks/what-is-network-policy.html (2021/03/10)
17. Damianou, N., Bandara, A., Sloman, M., & Lupu, E. (2002). *A survey of policy specification approaches*. Department of Computing, Imperial College of Science Technology and Medicine, London (Vol. 3, pp. 142–156).
18. Feamster, N., Rexford, J., & Zegura, E. (2014). The road to SDN: An intellectual history of programmable networks. *ACM SIGCOMM Computer Communication Review, 44*(2), 87–98.
19. Foster, N., Harrison, R., Freedman, M. J., Monsanto, C., Rexford, J., Story, A., & Walker, D. (2011). Frenetic: A network programming language. *ACM Sigplan Notices, 46*(9), 279–291.
20. Gabillon, A., Gallier, R., & Bruno, E. (2020). Access controls for IoT networks. *SN Computer Science, 1*(1), 1–13.

21. Gusmeroli, S., Piccione, S., & Rotondi, D. (2013). A capability-based security approach to manage access control in the internet of things. *Mathematical and Computer Modelling, 58*(5–6), 1189–1205.
22. Hameed, S., Khan, F. I., & Hameed, B. (2019). Understanding security requirements and challenges in internet of things (IoT): A review. *Journal of Computer Networks and Communications, 2019*, 2019. https://doi.org/10.1155/2019/9629381
23. Huang, D., Chowdhary, A., & Pisharody, S. (2018). *Software-Defined networking and security: From theory to practice*. Boca Raton: CRC Press.
24. Hussain, M., & Shah, N. (2018). Automatic rule installation in case of policy change in software defined networks. *Telecommunication Systems, 68*(3), 461–477 (2018)
25. Hussain, M., Shah, N., & Tahir, A. (2019). Graph-based policy change detection and implementation in SDN. *Electronics, 8*(10), 1136.
26. Keoh, S. L., Kumar, S. S., & Tschofenig, H. (2014). Securing the internet of things: A standardization perspective. *IEEE Internet of things Journal, 1*(3), 265–275.
27. Kolar, M., Fernandez-Gago, C., & Lopez, J. (2018). Policy languages and their suitability for trust negotiation. In *IFIP Annual Conference on Data and Applications Security and Privacy* (pp. 69–84). New York: Springer.
28. Kreibich, C., Handley, M., & Paxson, V. (2001). Network intrusion detection: Evasion, traffic normalization, and end-to-end protocol semantics. In *Proceedings of the USENIX Security Symposium*, Vol. 2001.
29. Lee, J., Kang, J.-M., Prakash, C., Banerjee, S., Turner, Y., Akella, A., Clark, C., Ma, Y., Sharma, P., & Zhang, Y. (2015). Network policy whiteboarding and composition. In *Proceedings of the 2015 ACM Conference on Special Interest Group on Data Communication* (pp. 373–374).
30. Levitin, G., Xing, L., Zhai, Q., & Dai, Y. (2015). Optimization of full versus incremental periodic backup policy. *IEEE Transactions on Dependable and Secure Computing, 13*(6), 644–656.
31. Li, D., Wang, S., Zhu, K., & Xia, S. (2017). A survey of network update in SDN. *Frontiers of Computer Science, 11*(1), 4–12.
32. Mai, H., Khurshid, A., Agarwal, R., Caesar, M., Godfrey, P. B., & King, S. T. (2011). Debugging the data plane with anteater. *ACM SIGCOMM Computer Communication Review, 41*(4), 290–301.
33. McKeown, N. (2011). How SDN will shape networking. *Open Networking Summit.*
34. McKeown, N., Anderson, T., Balakrishnan, H., Parulkar, G., Peterson, L., Rexford, J., Shenker, S., & Turner, J. (2008). Openflow: Enabling innovation in campus networks. *ACM SIGCOMM Computer Communication Review, 38*(2), 69–74 (2008)
35. Monsanto, C., Reich, J., Foster, N., Rexford, J., & Walker, D. (2013). Composing software defined networks. In *10th {USENIX} Symposium on Networked Systems Design and Implementation ({NSDI} 13)* (pp. 1–13).
36. Moshref, M., Yu, M., Sharma, A., & Govindan, R. (2013). Scalable rule management for data centers. In *10th {USENIX} Symposium on Networked Systems Design and Implementation ({NSDI} 13)* (pp. 157–170).
37. Mousa, M., Bahaa-Eldin, A. M., & Sobh, M. (2016). Software defined networking concepts and challenges. In *2016 11th International Conference on Computer Engineering & Systems (ICCES)* (pp. 79–90). New York: IEEE.
38. Paquet, C. (2012). *Implementing Cisco IOS Network Security (IINS 640-554) Foundation Learning Guide: Imp Cisco IOS Netw Sec F _c2*. Indianapolis: Cisco Press.
39. Pashkov, V., Shalimov, A., & Smeliansky, R. (2014). Controller failover for SDN enterprise networks. In *2014 International Science and Technology Conference (Modern Networking Technologies)(MoNeTeC)* (pp. 1–6). New York: IEEE.
40. Pisharody, S. (2017). *Policy conflict management in distributed SDN environments*. PhD thesis, Arizona State University, 2017.

41. Prakash, C., Lee, J., Turner, Y., Kang, J.-M., Akella, A., Banerjee, S., Clark, C., Ma, Y., Sharma, P., & Zhang, Y. (2015). PGA: Using graphs to express and automatically reconcile network policies. *ACM SIGCOMM Computer Communication Review, 45*(4), 29–42 (2015).
42. Reitblatt, M., Foster, N., Rexford, J., Schlesinger, C., & Walker, D. (2012). Abstractions for network update. *ACM SIGCOMM Computer Communication Review, 42*(4), 323–334.
43. Sanabria-Russo, L., Alonso-Zarate, J., & Verikoukis, C. (2018). SDN-based pro-active flow installation mechanism for delay reduction in IoT. In *2018 IEEE Global Communications Conference (GLOBECOM)* (pp. 1–6). New York: IEEE.
44. Sezer, S., Scott-Hayward, S., Chouhan, P. K., Fraser, B., Lake, D., Finnegan, J., Viljoen, N., Miller, M., & Rao, N. (2013). Are we ready for SDN? Implementation challenges for software-defined networks. *IEEE Communications Magazine, 51*(7), 36–43.
45. Shenker, S., Casado, M., Koponen, T., McKeown, N., et al. (2011). The future of networking, and the past of protocols. *Open Networking Summit, 20*, 1–30.
46. Stephen, B., & Petropoulakis, L. (2007). The design and implementation of an agent-based framework for acceptable usage policy monitoring and enforcement. *Journal of Network and Computer Applications, 30*(2), 445–465.
47. Stone, G. N., Lundy, B., & Xie, G. G. (2001). Network policy languages: A survey and a new approach. *IEEE Network, 15*(1), 10–21.
48. Strassner, J., & Schleimer, S. (1998). Policy framework definition language. *draft-ietf-policy-framework-pfdl-00. txt.*
49. Sun, X., Rao, S. G., & Xie, G. G. (2012). Modeling complexity of enterprise routing design. In *Proceedings of the 8th International Conference on Emerging Networking Experiments and Technologies* (pp. 85–96).
50. Sun, X., Sung, Y.-W., Krothapalli, S. D., & Rao, S. G. (2010). A systematic approach for evolving vlan designs. In *2010 Proceedings IEEE INFOCOM* (pp. 1–9). New York: IEEE.
51. Tsao, T., Alexander, R., Dohler, M., Daza, V., Lozano, A., & Richardson, M. (2015). A security threat analysis for the routing protocol for low-power and lossy networks (RPLS). *RFC 7416 (Informational), Internet Engineering Task Force.*
52. Valenza, F., Spinoso, S., & Sisto, R. (2019). Formally specifying and checking policies and anomalies in service function chaining. *Journal of Network and Computer Applications, 146*, 102419.
53. Voellmy, A., Wang, J., Yang, Y. R., Ford, B., & Hudak, P. (2013). Maple: Simplifying SDN programming using algorithmic policies. *ACM SIGCOMM Computer Communication Review, 43*(4), 87–98.
54. Wang, B., & Liu, K. J. R. (2010). Advances in cognitive radio networks: A survey. *IEEE Journal of Selected Topics in Signal Processing, 5*(1), 5–23.
55. Whitlock, M. C., Bronstein, J. L., Bruna, E. M., Ellison, A. M., Fox, C. W., McPeek, M. A., Moore, A. J., Noor, M. A. F., Rausher, M. D., Rieseberg, L. H., et al. (2016). A balanced data archiving policy for long-term studies. *Trends in Ecology & Evolution, 31*(2), 84–85.
56. Yan, B., Xu, Y., Xing, H., Xi, K., & Chao, H. J. (2014). Cab: A reactive wildcard rule caching system for software-defined networks. In *Proceedings of the Third Workshop on Hot Topics in Software Defined Networking* (pp. 163–168).
57. Yan, Q., Yu, F. R., Gong, Q., & Li, J. (2015). Software-defined networking (SDN) and distributed denial of service (DDOS) attacks in cloud computing environments: A survey, some research issues, and challenges. *IEEE Communications Surveys & Tutorials, 18*(1), 602–622.
58. Zhang, B., & Ng, T. E. (2010) On constructing efficient shared decision trees for multiple packet filters. In *2010 Proceedings IEEE INFOCOM* (pp. 1–9). New York: IEEE.

Chapter 6
Analysis of Load Balancing Techniques in Software-Defined Networking

Gurpinder Singh, Amritpal Singh, and Rohit Bajaj

6.1 Introduction

To introduce the needs of the current and future architecture for the computer network-like high speed routing, centralized control of the network, managing QoS, end-to-end user connectivity, load balancing, centralized security control, cloud computing, a concept has been introduced as by name of SDN [1–3]. SDN is a simple concept that is used to separate the control group from the forwarding group of networking devices. This feature allows separation of control plane (e.g., SDN controller) on computer platforms from network equipment software (e.g., switches/routers) [4, 5]. Another feather in the cap is that it is providing security to the main SDN controller of the network because forwarding a packet is only controlled by the network equipment and all controlling commands of the network are written on SDN controller plane. This chapter is written to discuss the various load balancing and routing techniques of SDN to optimize the available resources in an efficient manner [6].

6.2 Software-Defined Network

The configured network consist of various routers, switches, mainframe computers, servers, and other components to establish communication. Basically, the routers and switches are backbone of the computer network. These components are used for end-to-end user communication, routing the data with maximum quality of service.

G. Singh · A. Singh (✉) · R. Bajaj
Department of Computer Science and Engineering, Chandigarh University, Mohali, India
e-mail: amritpal.cse@cumail.in; rohitbajaj.cse@cumail.in

G. S. Aujla et al. (eds.), *Software Defined Internet of Everything*, Technology, Communications and Computing, https://doi.org/10.1007/978-3-030-89328-6_6

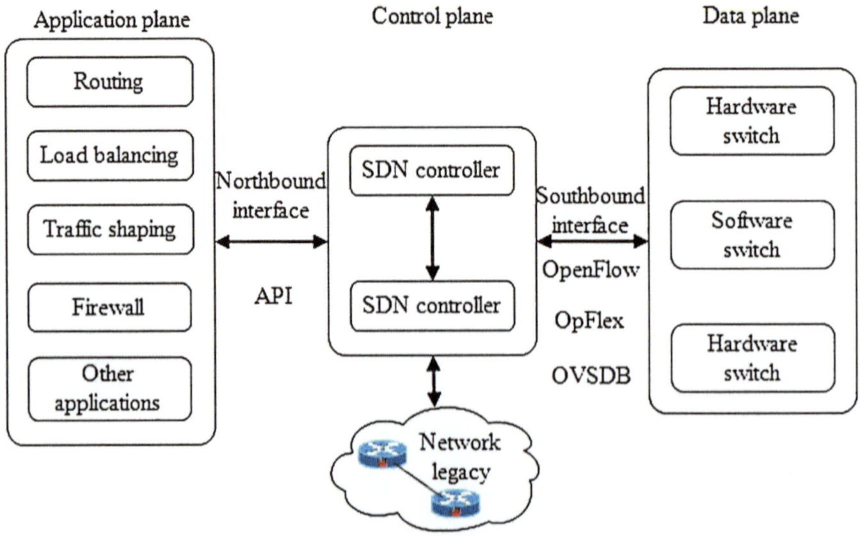

Fig. 6.1 Architecture of Software-Defined Networking (SDN)

However, all these devices are placed over the different geographical locations and act independently to establish the communication. Therefore, the maintenance of such kind of the network is quite difficult without centralized control because each device is making its decision according to their environment without considering the whole network. Therefore, a new concept is introduced, i.e., SDN to control the network traffic. A centralized software-defined control unit is connected to all the available network devices as shown in Fig. 6.1.

As the block diagram depicts, the whole network is divided into two planes. The first part is the control plane, which is a centralized unit to change, configure, and manage the network via software. The second plane is the data plane, which acts as per the directions directed by the control plane and forwarding the data accordingly. Most of the routers act as a forwarding device, and however, the routing decisions are made by SDN controller [7]. The various interfaces of the network are discussed below:

- **Northbound Interface:** It provides an interface between the SDN controller and the various executing applications. Applications can be used by the end users to control the network; however, the issue with the approach is lack of standardization. Therefore, it is called as API-based network.
- **Southbound Interface:** This interface helps to communicate with the lower defined layers, i.e., forwarding layer via different protocols, e.g., OpenFlow and Network Configuration Protocol. The defined protocols are used to instruct the forwarding plane with the help of various commands of the SDN controller.

- **Eastbound Interface:** It is used to connect conventional IP networks with the SDN-enabled networks. By using this interface, the SDN domain can communicate to the routing protocol through the messages for different activities.
- **Westbound Interface:** This interface provisioned the SDN controller to communicate with different configured domain controllers in the network for better performance and data exchange in the geo-distributed environment [8–10].

6.2.1 Types of Software-Defined Network

Typically, three types of Software-Defined Network architecture are used. All the different categorized SDN architectures are defined as per their functionality as compared to the existing network devices. The existing devices cannot replace at once with the SDN controllers due to the compatibility issues of the network. In the below section, the various architectures of the SDN controllers are highlighted with proper details.

6.2.1.1 Centralized SDN

The working of the centralized Software-Defined Network (SDN) is clear by its name only as one Software-Defined Network controller is used to manage all networking devices in the network. It is the model of Open Networking Foundation (ONF) that uses OpenFlow protocol for instructing the forwarding or data plane devices. It is very easy to manage as all the instructions are flowing from one point/controller as shown in Fig. 6.2.

The centralized SDN collects all the networking information from all the configured network devices for efficient load balancing, routing, fault tolerance, and security. However, there are few disadvantages of using this category of architecture as discussed below:

- **Frequent Updation:** All the networking devices are connected to the centralized network controller, and therefore, for smooth functionality of the network, an abrupt updation at each connected devices is required.
- **Simplicity Comes at a Greater Cost:** A single controller manages the entire network, and therefore, a larger storage space, computation power, and extra energy are required. An extra cost is required to scale up the network; otherwise, the entire network can be crashed.
- **Introducing New Flow Entry:** During updation of the new entry in the flow table, the following steps are followed:

    1. *PACKET_IN* message forwarded to the controller.
    2. New flow rule generated by the controller.
    3. *PACKET_OUT* message updation on each connected switch in the network.

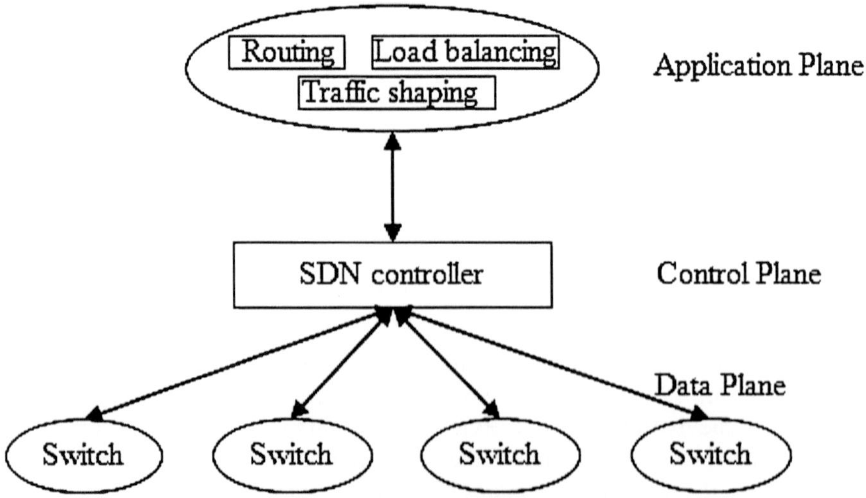

Fig. 6.2 Architecture of Centralized Software-Defined Networking (CSDN)

The abovementioned steps will be followed for each entry, and it will be time consuming for the bigger networks [8, 11].

6.2.1.2 Distributed SDN

In case of Distributed Software-Defined Network (SDN), more than one Software-Defined Network controllers is connected to manage all the configured networking devices in the network. This kind of architecture is more reactive than centralized controller because for each domain there is a dedicated Software-Defined Network controller, and in the similar manner, there are numerous of domain controllers connected to each other as shown in Fig. 6.3.

There can be more than two Software-Defined Network controllers for two domains, and it can be more than two as per the requirements. The Distributed Software-Defined Network quickly updates the changes and makes the network more robust in nature. There are three scenarios of distributed SDN: the first one is parallelism between different switches to send the packets to the controller. The second one is used when synchronization is required between the controllers of different domains. The third scenario is used where at-least two layers are present, the controllers are at the second layer containing different switches/devices as one domain, and other domains also exist at the other layer. At the first layer, controllers of different domains are connected to the dedicated centralized controller [8, 12].

The main disadvantages of Distributed Software-Defined Networking (DSDN) are discussed below:

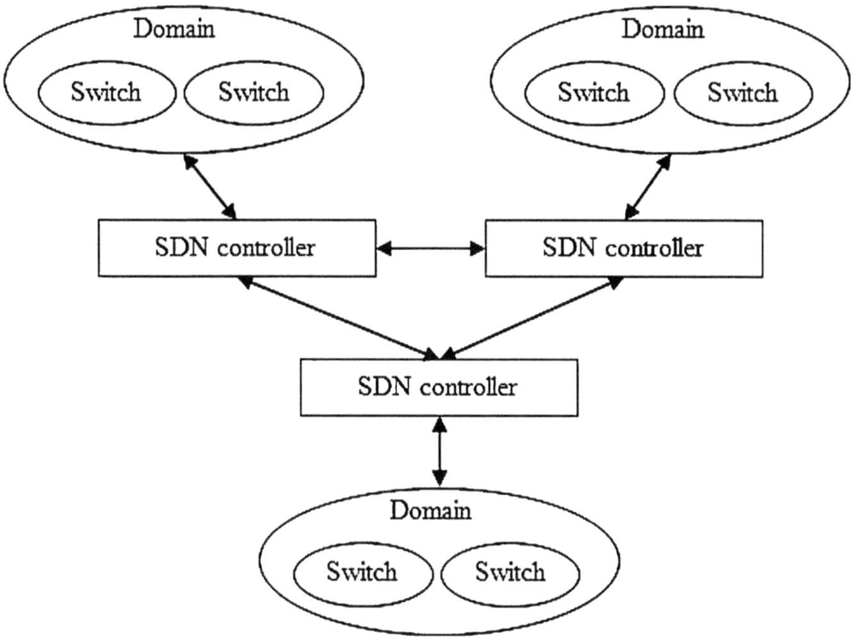

Fig. 6.3 Architecture of Distributed Software-Defined Networking (DSDN)

- **Consistency for Global View:** The network is not controlled by a single controller, and therefore, the knowledge of the entire network is required to make the network consistent.
- **Synchronization at Regular Intervals:** All the configured controllers required synchronization with each other to update the information of each and every connected device to make the network more robust.

6.2.1.3 Hybrid SDN

Hybrid SDN architecture is the combination of traditional infrastructure and SDN-enabled network and communicates together to manage the entire network. As shown in Fig. 6.4, devices of the traditional and SDN-enabled network of data and control planes are visible. The devices of both of the networks conditionally communicate with each other. The SDN-enabled and traditional network switches are present at the data plane, and they can communicate with each other without any bar with the help of LLDP (link layer discovery protocol), etc. of a traditional switch. Communication at the control plane is feasible by considering the advertisements of OSPF (open shortest path first) in order to discover the network topology or to improve the routing performance by SDN controller [13–16].

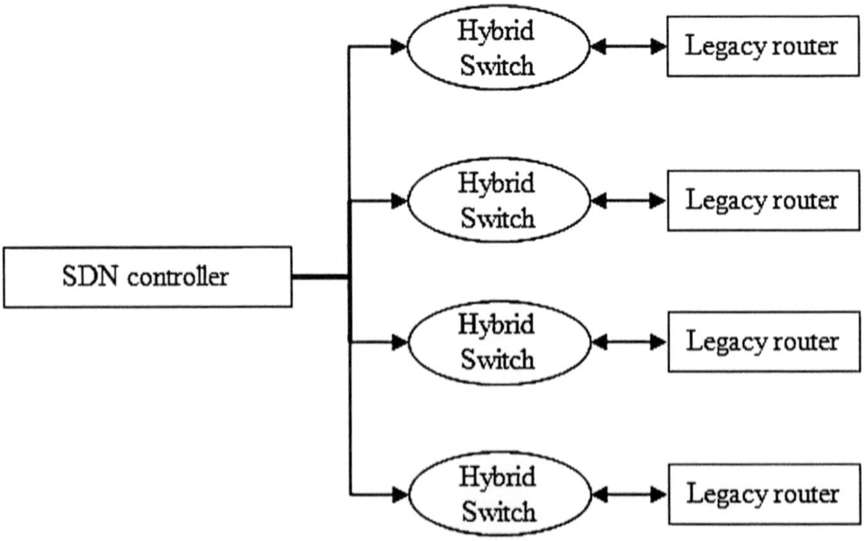

Fig. 6.4 Architecture of Hybrid Software-Defined Networking (DSDN)

6.3 Techniques for Load Balancing in Software-Defined Network

Software-Defined Networking framework is used to manage the network traffic in an efficient manner by using the programmable interface. However, a heavy workload can be noticed on some of the resources available in the network, resulting into usage of extra energy and delay. To overcome this issue, a framework is required to balance the workload on the available nodes in the network. In the following section, the techniques used to manage the workload on the nodes are discussed.

6.3.1 Balance the Load by Filtering the Load Based on TCP and UDP [17]

This approach filters the incoming traffic by identifying the header of the used protocol like, TCP or UDP. The proposed algorithm captures the incoming packets and then analyzes the type of protocol, i.e., TCP or UDP. If the packet is using TCP protocol, it is forwarded to the CT (TCP load balancer controller) and discards the flow of UDP protocol. Furthermore, if the packet belongs to the UDP, then it is forwarded to the CU (UDP load balancer controller) and discards the flow of TCP. The controller follows the standard policies, and they can use the reverse path to avoid the congestion situation. The TCP and UDP controllers also have backup or secondary controllers in case of failure. The ARP_REQ packets are forwarded from secondary to primary controllers and receive ARP_REP for aliveness after a specified interval of time. The secondary controller can act as primary controller when the attempt time fail (Maxf) is greater than the specified timeout of fail (N). Despite the failure of primary, the active controller uses ARP_SYN to send ARP_REQ to check the availability and health of non-available controller. At the end, the collected flow is forwarded to the back-end servers for reliability purpose.

6.3.2 Stable the Network by Shifting the Workload from Overloaded Controller to the Underloaded Controllers [18]

This approach is used whenever a controller is at idle state and it receives the data flow events from the overloaded controller without migration of the switches. However, in a situation where the processing time is more than the threshold time to relieve the overloaded controller, the switches of the overloaded controller are migrated to the underloaded controller. In the case of heavy traffic, the cooperation between controllers is the best option rather to migrate the switches for handling the temporary traffic. The idle controller usually considers the nearby controller as a target; however, if the nearby controller is not accepting the steal request, then it is forwarded to the second nearby controller and this iteration continues for three times. After the third attempt, the controller recalculates the status of its resources and then decides whether to go for stealing or not. The flowchart of flow-stealing method is shown in Fig. 6.5.

The switch migration is the case when cooperation between controllers are not able to handle the load, each controller has its network region containing a number of switches, and regions are further divided into zones, a subset of switches of region without intersection. The subsets of the switches (zones) are configured for alternative controllers near to the zone. The region that belongs to the overloaded controller is named as overloaded region (OR). If a switch in the zone is overloaded, it is migrated to alternate controller by considering the minimum distance condition. The unique thread is created to serve each switch in the controller. In case of overloading scenario in a zone, more than one thread in a controller can be overloaded. Each overloaded switch is considered as per the maximum hit time condition. The list of zones is collected by considering the numbers of hits in each zone and arranged in the ascending order. A zone with maximum hit time is prioritized and selected for migrating the load of the switch to the alternative controller, and the same process continues as per the stored list. Following are the three conditions in case of switch migration.

$$\sum_{i \in P_k} a_i <= c_T, \quad \forall k \in Z, \tag{6.1}$$

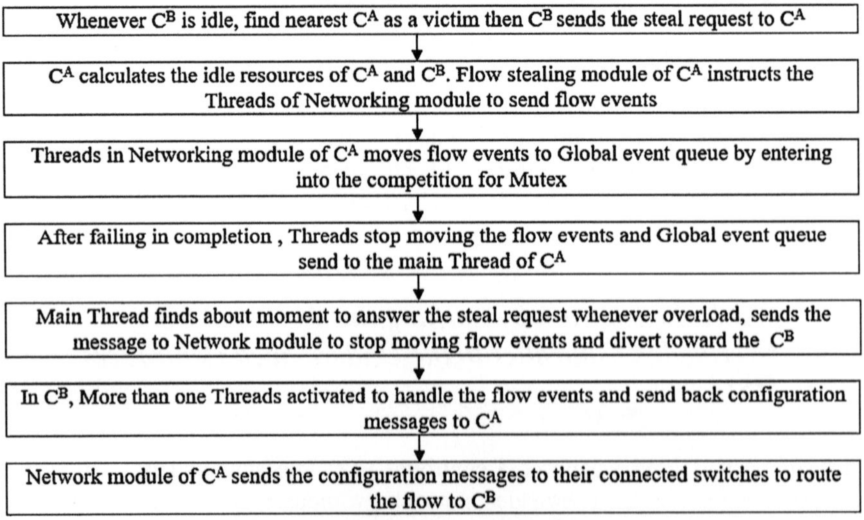

C^A	CONTROLLER A [TRAGET]
C^B	CONTROLLER B [IDLE]
Steal Request	IP address, port and idle computing resources of C^B

Fig. 6.5 Flowchart of flow-stealing method

$$\sum_{i \in P_k} b_i <= m_T, \quad \forall k \in Z, \tag{6.2}$$

$$d_i T <= distance_{Max}, \quad \forall_i \in P_k, \forall k \in Z, \tag{6.3}$$

where a_i is the computing requirement of ith switch, b_i is the memory requirement of ith switch, c_T is the computing resources for Tth controller, m_T is the memory resources for Tth controller, $d_i T$ is the distance between ith switch and Tth controller, P_k is the migrated switches in kth zone, and $distance_{Max}$ is the maximum allowed distance between a switch and a controller.

6.3.3 Balancing of Load by Authentication, Monitoring Load, and Switch Migration [19]

This approach consist of three layers, cooperative, controllers, network region consist of different switches as shown in Fig. 6.6.

Authentication module of the cooperative platform is used to secure the controllers from hacking with the help of hash function. The hash function key is shared

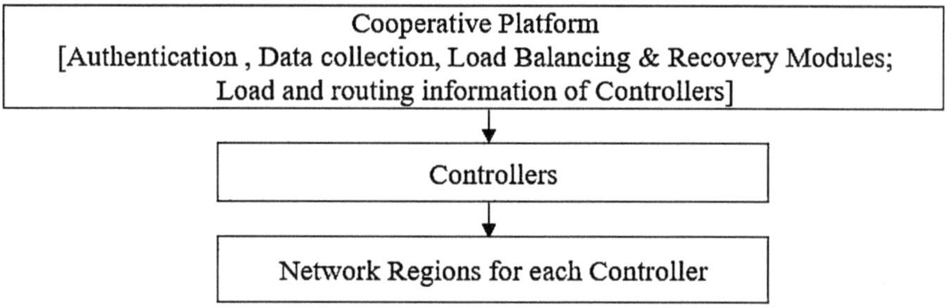

Fig. 6.6 Architecture of load by authentication, monitoring load, and switch migration

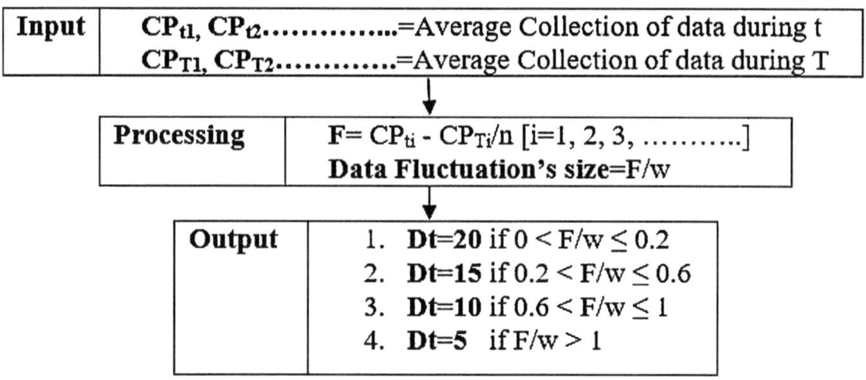

Fig. 6.7 Flowchart of data collection approach

with all the controllers and cooperative platforms. To authenticate, each sender controller is verified by the cooperative platform. The sender controller generates a unique hash message and forwards the message to the cooperative platform. The cooperative platform generates hash from the same message and compared it with the received hash message for authentication. The data collection module stores the load information of each controller from packet-in message received by the controller. The time interval for collection of load information depends upon the traffic's frequency. If the traffic is normal, then the time interval may be longer. However, in case of heavy traffic, lesser time interval can be observed as shown in Fig. 6.7.

Load balancing and recovery is done by migration of switch of overloaded controller to alternate controller. The pseudo-code of load balancing algorithm is shown in Fig. 6.8.

Input	Read load of every Controller(**CP$_i$**) LB=0.8 & FL=0.2[Adjust according to frequency of Load] Select Controller with Maximum Load from all(C **MAX**) Select Controller with Minimum Load from all(C **MIN**)
Processing	m= Average of Load of each Controller/ MAX CP IF m is in range of [LB-FL, LB+FL] then 0; else 1
Output	0= No need of migration Switch's Load 1=Module select switch with maximum Packet-in produced and connected to "C MAX" and migrate load of that switch to the "C MIN"

Fig. 6.8 Pseudo-code of load balancing algorithm

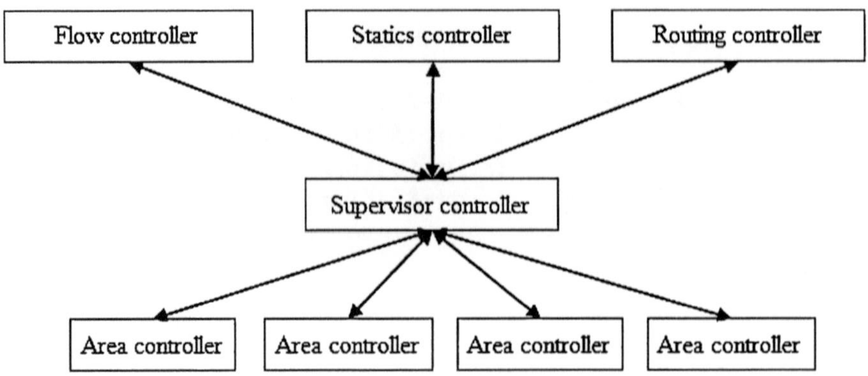

Fig. 6.9 Architecture of scalable network

6.3.4 Different Services Provided by Supervisor Controller to Local Controller for Load Sharing in Scalable Network [20]

This approach discussed about scalable network with uneven load. The supervisor controller which acts as an SDN controller for data collection of the routes leads to other controller's network and some other services. Afterward, there are different area controllers for different switches used to handle the packets from the switches. The controllers attached to the supervisor controller except area controllers are contributory controllers to handle the specific services and connected with the eastbound API and westbound API to the supervisor controller as shown in Fig. 6.9.

Initially, the supervisor controller starts by pre-deciding the number of specific services of the contributory controllers. This collected information is installed on each area controller with the id of contributory controllers. To initiate the communication, a new packet is created and forwarded to the area controller. The packet is matched with the flow table entries for further action, and in case no entry is matched, $PACKET\_IN$ message is forwarded to the supervisor controller and then to a particular contributory controller to create new flow entry. The supervisor controller does not open the packet as unique id of the contributory controller, which is mentioned on the header of the packet. In another scenario, same new entry packet is created and forwarded to the area controller and to the supervisor. It is handled by supervisor, and the entry is stored in the flow table for further reference. In case, the threshold value crosses the handling request capacity of supervisor, and then it activates the contributory controllers for load balancing.

6.3.5 Load Balancing by Making Cluster of Controller by Super Controller [21]

It contains three layers, an extended version of two-tier architecture. The super controller is at the top layer and other controllers are placed into a different cluster. To collect and manage the data/load of all the clusters, super controller uses cluster vector for each cluster to get address of every controller in the cluster. The cluster vector provides the facility to the controllers in the cluster to share their load with respective controller. To migrate the load among various controllers, all the controllers are synchronized to know the status of the underloaded and overloaded controllers, and accordingly the load is migrated from overloaded to underloaded controllers. A super controller (SC) handles the load among all the controllers in the cluster. The master controllers share the cluster load of each controller to the super controller, and by using the details, the SC decides the master controller. Parallelly, the master controller handles the process of load balancing in the cluster. The super controller handles the load balancing in clusters by collecting the data: (i) minimum differences in the load of clusters are considered as a target and (ii) minimum distance is considered between the controllers in the cluster to reduce the traffic transfer time in case of overloaded nodes. The process of master controller is known as reassignment where load balancing is done in the cluster by sharing load of different switches under the controllers in the cluster. Y is the matrix used by SC to gather information about placement of each controller in which cluster C_j represents the controller where j =1,2,3... and G_i represents the cluster where i=1,2,3...

$$Y(t)_{ij} = \begin{cases} 0: & C_j \in G_i \\ 1: & \text{else} \end{cases}$$

The above equation represents that if C_j belongs to the G_i, then 0, else 1.

Calculation of Cluster's Load Difference:

Load of cluster $[\theta(t)i]$ = Sum of controllers' average load

$$\sum_{j=1}^{M} l(t)jY(t)ij \tag{6.4}$$

Difference between a cluster load and other clusters' load [Avg]

$$\sum_{j=1}^{M} l(t)j/k \tag{6.5}$$

Distance:

$$\vartheta(t)_j = |\theta(t)l - Avg| \tag{6.6}$$

Total load difference between clusters' load:

$$\delta(t) = \sum_i = 1^k \vartheta(t)_i / K \tag{6.7}$$

Calculation of Distance Between Controllers in a Cluster The distance between the controllers needs to be minimum to reduce the data travel time in case of overloading, and the group of switches needs to be connected with the nearby controller. The maximal distance between the controllers in the same cluster (compare to all clusters) is given by $\eta(t)$.

$$\text{maxDistance}[\eta(t) = \max_{l<=c<=k} \max_{l<=i,j<=M} d_{ij} Y(t)_{ic}], \tag{6.8}$$

where c is the cluster number and i and j are the controllers in the c cluster.
η(t) needs to be minimal but not too narrow, i.e., minMaxDistance (Cnt). If the distance is too narrow, then the selection of alternate controller is a difficult scenario. To avoid such situation, an offset value is added to the calculated distance.

$$\text{Cnt} = \text{maxDistance} + \text{offset}. \tag{6.9}$$

6.3.6 *Lighten Up the Overloaded Controller by Migration of the Switches to Lightly Load by Broadcasting Load Between Controllers [22]*

This approach is used to migrate the load of the switches of the overloaded controller to the underloaded controllers. The controllers communicate with each other after a regular interval of time to collect the load information. The main components of the approach are discussed in the following sections:

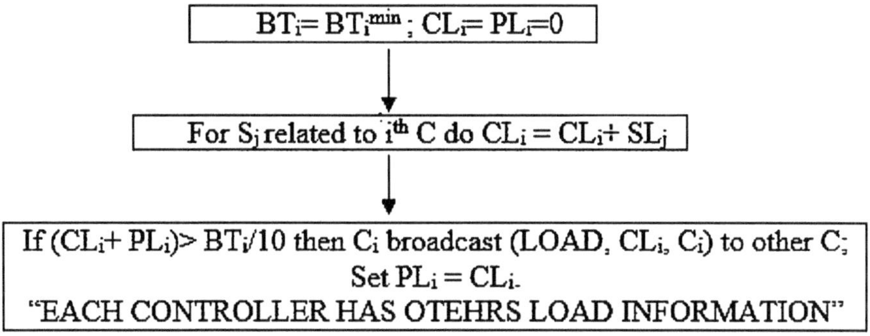

Fig. 6.10 Flowchart of load broadcasting

1. **Load Measurement Component:** Measurement of each controller's Load in network (CL_i).
2. **Load Broadcast Component:** Broadcasting of load to the different controllers in the network as shown in Fig. 6.10.
3. **Load Balancing Component:** Used to evaluate the different load balancing conditions under various load balancing scenarios in the network as shown in Fig. 6.11.
4. **Load Migration Component:** Used to migrate the load of the switch from heavy loaded controller to lightly loaded controller as shown in Fig. 6.12. The migration of S_k from C_i to C_j, where $S_k \epsilon SC_i$, $C_i \epsilon HLC$ and $C_j \epsilon LLC$

   The flowchart of migration of S_k from C_i to C_j is shown in Fig. 6.13. The migration approach reduce the bandwidth utilization and traffic frequency as the migration of the load of various switches saturating the network by migrating the load of the controllers.
5. **Link Reset Component:** It is used to reset the link from (S_k, C_j) to (S_k, C_i) after completing the migration of the overloaded data. This process usually starts at a particular time in a day, and if the network is still overloaded, then the process of reset is postponed till the network is not stable.

6.3.7 Tenant Controllers Are Finding Best Paths by Calculating the Maximum Throughput [23]

This approach uses network hypervisor (NH) for dividing the network into multiple virtual isolated networks that plays an important role between the tenant controllers and physical switches where NH acts as a controller to physical switches and switches to the virtual networks. Whenever a physical switch receives any request and the flow entry is not available in the flow table, then it is forwarded to the NH. The NH translates the address to virtual address and sends it to the tenant

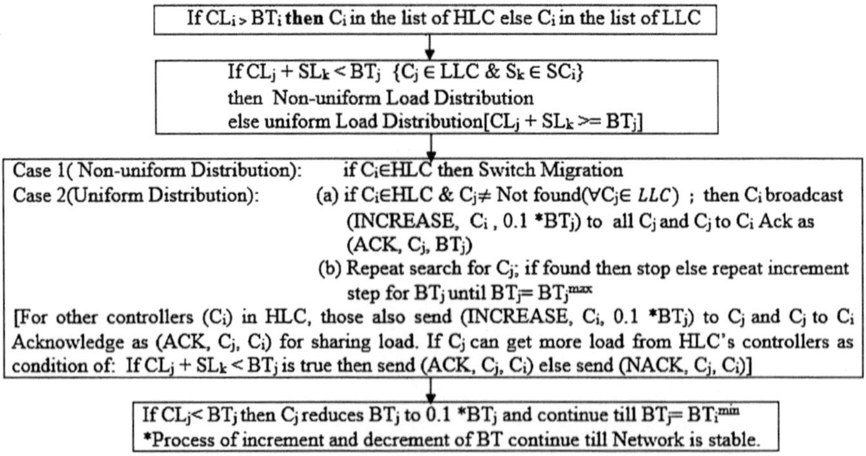

Fig. 6.11 Flowchart of load balancing

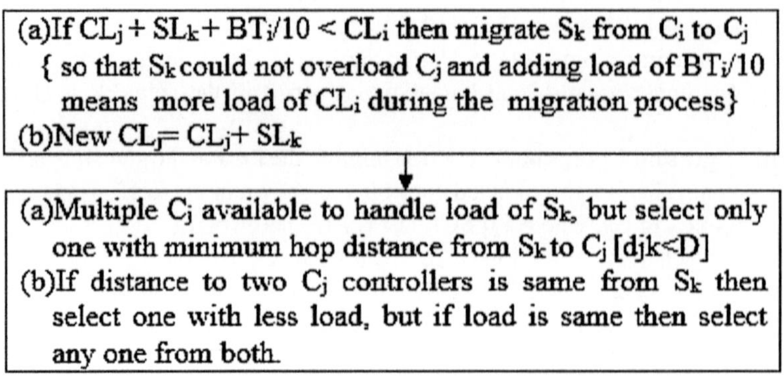

Fig. 6.12 Flowchart of load migration

controller of SDN. The tenant controller finds a suitable path by using the virtual switches and reverts back to the NH. The NH translates the path to physical address and forwards it to the physical switch. This process helps the tenant controller to achieve the required bandwidth to connect the required virtual switch. It is followed by calculating the path/paths of required throughput and finalizing the path/paths accordingly. The above discussed steps are explained as follows:

1. Selection of the path according to the requirement of throughput by calculating the bandwidth:

   a. **Link Manager and Throughput Requirement Manager:** The main purpose of the link manager is to collect the information of maximum allocated bandwidth to the physical links, the tenant controllers, and the switches. SDN usually collects the information when the switches are activated in the network

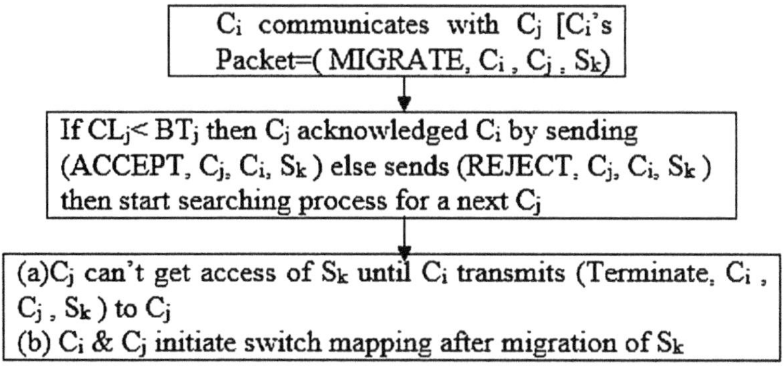

Fig. 6.13 Flowchart of migration of S_k from C_i to C_j

through their ports. The purpose of the throughput requirement manager is to note down the throughput requirement of the tenant and store the same in the forms of tables using hashing method.

b. **Path Extractor:** After getting route from the tenant controller, it is used to create new path. If the flow entry is already exist in the flow table rules otherwise send the message to the controller. After finalizing the path, it is shared with the throughput requirement manager.
c. **Throughput Allocator:** After getting the main path from the tenant using path extractor, it calculates the remaining capacity of the path (CR). If the calculated path satisfy the requirements, finalized as a main path otherwise sub-paths created by other tenants from the main path is deallocated and the complete main path is available for the tenants. Still, the main path is not able to satisfy the new requests of the tenant, and the path splitting task is activated to divide the traffic of tenant through different created paths to fulfill the requirement.

2. **Path Splitting:** This method is used to handle the incoming traffic from various end devices. In this approach, the rule of flow for traffic is divided into three categories.

   a. **Ingress Rule:** The incoming traffic is controlled by directing the flow from one in-port to multiple out-ports.
   b. **Egress Rule:** The traffic is controlled by forwarding the flow from multiple in-ports to a single out-port.
   c. **Core Rule:** This rule is used not to change or split the flow of traffic, and it acts as intermediate rules of flow of Ingress and Egress. It worked with the flow from single in-port to single out-port.

The flowchart to define the flow and group rules in case of splitting the traffic to main path (MP) and sub-path (SP) is shown in Fig. 6.14:

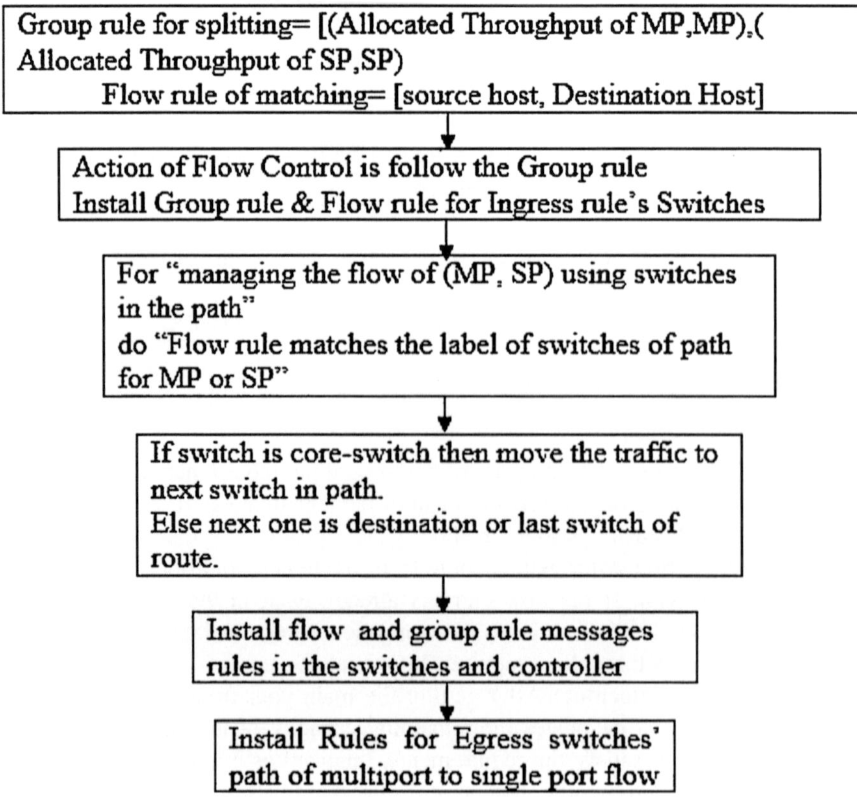

Fig. 6.14 Flowchart of splitting main path (MP) and sub-path (SP)

6.3.8 Balance the Flow of User's Load by a Different Method of Sharing Resources [24]

In SDN-enabled network, the main purpose is to handle the user requests in an efficient manner to improve the processing of the network. An efficient load balancing and high traffic in the network can be managed by transmitting the user requests to the pool of servers according to their requirements. In this approach, a load balancer (OpenFlow Switch) is used with the controller of the SDN. The load balancer has virtual IP address that helps to reroute the traffic of a different user to various connected servers. The user forwards the request to the load balancer, further, it check the flow table to direct the request as per the flow rule with the help of the request header. If the flow entry is not matched as per the load balancer's entry, then, the request is forwarded to the controller to create a path between the user and the server. The load balancer mostly uses two methods to send the request: random and round robin. The flowchart of both the methods is shown in Figs. 6.15 and 6.16.

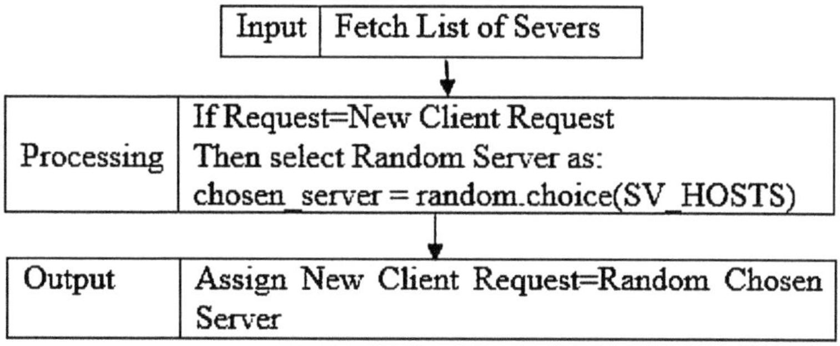

Fig. 6.15 Flowchart of random method

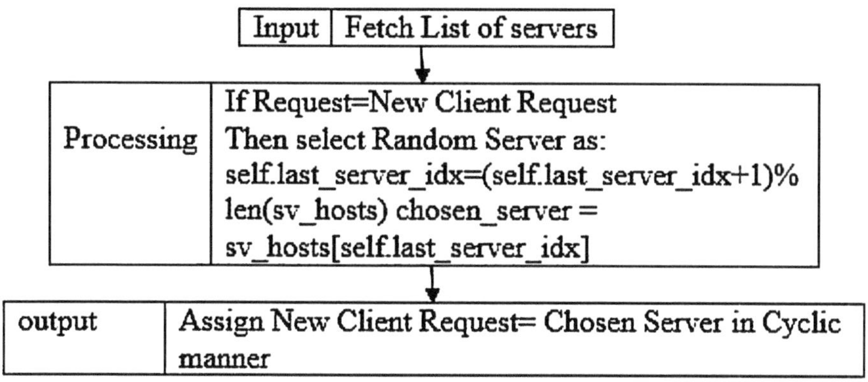

Fig. 6.16 Flowchart of round-robin method

The controller selects one of the available servers from the cluster of servers. The round-robin approach uses the circular method, i.e., the request containing the load of each client is forwarded to the server one by one, e.g., if there are four clients and two servers, then, the first client request is sent to server_1, the second client's request to the server_2, the third client's request to the server_1, and the forth client's request to the server_2 (servers 1 and 2 are of the same specification). In this manner, the load of the different devices is balanced among the different connected devices.

6.3.9 Finding Best Path by Considering Bandwidth and Delay for Cloud Data Centers [18]

The load balancing mechanism is used to manage the traffic of the source and destination data centers. The flowchart of the modules is shown in Fig. 6.17.

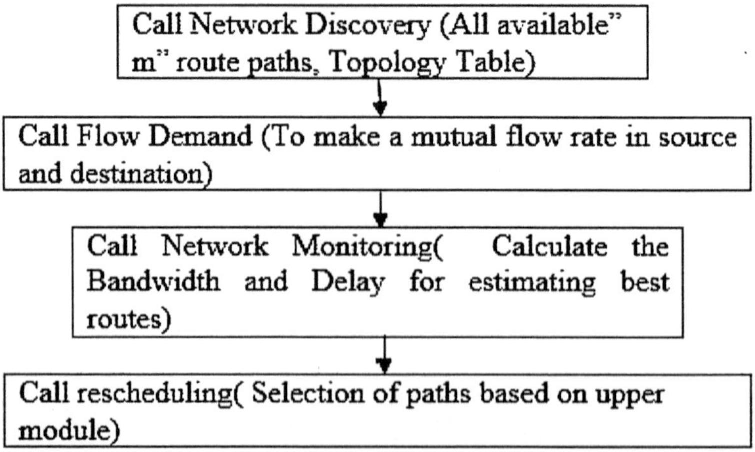

Fig. 6.17 Flowchart of modules of approach

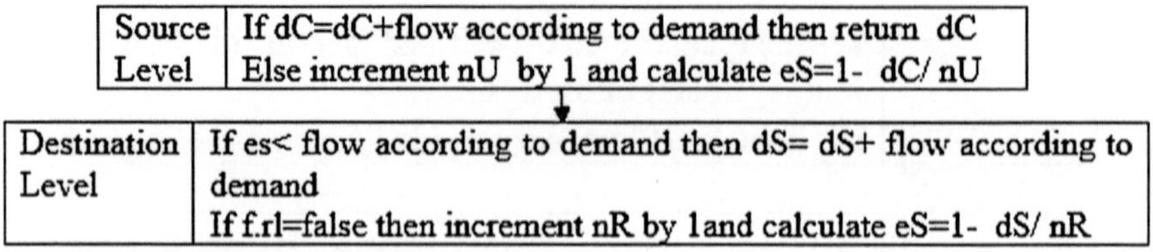

Fig. 6.18 Flowchart of flow demand

1. **Network Discovery:** This module is managed by the SDN controller to calculate the shortest path (m) without any loop and fat-tree topology. The SDN creates two tables; the first table contains m shortest path, and the second contains path except m. Both tables help to provide the links at the time of failure of network links.
2. **Flow Demand:** This module creates fat-free topologies. The source (Sr) and destinations (Dest) are used to configure links the by following intermediate devices and lead to the point of convergence, as explained in the form of flowchart in Fig. 6.18.
3. **Network Monitoring:** This module is used to calculate the bandwidth and link delay between the source and the destination. The bandwidth and delay are calculated to evaluate the load to avoid all the congestion.

$$\text{Latency} = \text{Source's Latency} - \text{Destination's Latency} \tag{6.10}$$

$$\text{Netlatency} = \text{ReplayDelay} - \text{Latency} \tag{6.11}$$

$$\text{Maximum Link Delay} = (\text{RequestDelay} + \text{Netlatency})/2. \tag{6.12}$$

Table 6.1 Notations table

Notation	Description
n	Number of controllers
LB	Load balancing threshold value
FL	Floating parameters
CPi	Load of controllers ($i = 1, 2, 3\ldots$)
m	Ratio between AVG CP and MAX CP
SPJ	Packet-in produced by switches connected to the controller
T	Starting time of load data collection
t	Time interval
num_i	Packet-in received by each controller ($i = 1, 2, \ldots$)
CPT_i	Average rate of packet-in received by each controller in the last collection ($i = 1$, 2,..)
F	Average data fluctuation
w	Packet-in receiving rate to control the message flows by controller
M & K	Number of controllers & Number of clusters
C_j	jth controller
$l(t)_j$	Load of controller at average flow of request at the jth controller/second in t time intervals
P	Capacity of a controller to handle a number of requests/second
d_{ij}	Minimum numbers of hops/hop at a distance between C_i and C_j
S_j	Set of switches ($j = 1, 2, 3 \ldots\ldots m$)
SC_i	Set of switches to ith controller
SL_i	Load of ith switch
CL_i	Load of ith controllers
PL_i	Previous load of ith controllers
BT_i	Base threshold of an ith controller
LLC	Lightly loaded controller [C/HLC]
d_{ik}	Shortest distance between C_i and S_k
D	Maximum allowable distance between migrated switch and alternate controller

4. **Rescheduling:** This module comes into existence after collecting the details from the abovementioned three modules.

   The notations used throughout the chapter are summarized in Table 6.1.

6.4 Conclusion and Future Work

The goal of this chapter is to highlight the use of various SDN networks as per their requirements. The increase in workload is the major concern in this pandemic period. To handle this, a centralized controller is required to manage the workload in an efficient manner. The distribution of the workload on the various nodes is uneven due to lack of intelligent load distribution techniques. To balance the load on the various nodes, the different load balancing approaches are discussed in the SDN-enabled environment. The load balancing approaches are discussed with the flowchart to improve the readability. The optimal path selection methods are highlighted to avoid the congestion in the network.

References

1. Aujla, G. S., Chaudhary, R., Kaur, K., Garg, S., Kumar, N., & Ranjan, R. (2018). SAFE: SDN-assisted framework for edge-cloud interplay in secure healthcare ecosystem. *IEEE Transactions on Industrial Informatics, 15*(1), 469–480.
2. Aujla, G. S., Singh, A., & Kumar, N. (2019). Adaptflow: Adaptive flow forwarding scheme for software-defined industrial networks. *IEEE Internet of Things Journal, 4*(7):5843–5851.
3. Aujla, G. S., Chaudhary, R., Kumar, N., Rodrigues, J. J., & Vinel, A. (2017). Data offloading in 5g-enabled software-defined vehicular networks: A Stackelberg-game-based approach. *IEEE Communications Magazine, 55*(8), 100–108.
4. Aujla, G. S., Singh, M., Bose, A., Kumar, N., Han, G., & Buyya, R. (2020). BlockSDN: Blockchain-as-a-service for software defined networking in smart city applications. *IEEE Network, 34*(2), 83–91.
5. Aujla, G. S., Chaudhary, R., Kumar, N., Kumar, R., & Rodrigues, J. J. (2018). An ensembled scheme for QoS-aware traffic flow management in software defined networks. In *2018 IEEE International Conference on Communications (ICC)* (pp. 1–7). IEEE.
6. Aujla, G. S., Kumar, N., Garg, S., Kaur, K., & Ranjan, R. (2019). EDCSuS: sustainable edge data centers as a service in SDN-enabled vehicular environment. *IEEE Transactions on Sustainable Computing*.
7. Singh, A., Aujla, G. S., Garg, S., Kaddoum, G., & Singh, G. (2019). Deep-learning-based SDN model for Internet of Things: An incremental tensor train approach. *IEEE Internet of Things Journal, 7*(7), 6302–6311.
8. Aujla, G. S., Jindal, A., Kumar, N., & Singh, M. (2016). SDN-based data center energy management system using RES and electric vehicles. In *2016 IEEE Global Communications Conference (GLOBECOM)* (pp. 1-6). IEEE.
9. Latif, Z., Sharif, K., Li, F., Karim, M. M., & Wang, Y. (2019). A comprehensive survey of interface protocols for software defined networks, pp. 1–30.
10. Aujla, G. S., & Kumar, N. (2018). SDN-based energy management scheme for sustainability of data centers: An analysis on renewable energy sources and electric vehicles participation. *Journal of Parallel and Distributed Computing, 117*, 228-245.
11. Maaloul, R., Taktak, R., Chaari, L., & Cousin, B. (2018). Energy-aware routing in carrier-grade Ethernet using SDN approach. *IEEE Transactions on Green Communications and Networking, 2*(3), 844–858. https://doi.org/10.1109/TGCN.2018.2832658.
12. Aujla, G. S., Jindal, A., & Kumar, N. (2018). EVaaS: Electric vehicle-as-a-service for energy trading in SDN-enabled smart transportation system. *Computer Networks, 143*, 247–262.
13. Ren, C., Bai, S., Wang, Y., & Li, Y. (2020). Achieving near-optimal traffic engineering using a distributed algorithm in hybrid SDN. *IEEE Access, 8*, 29111–29124. https://doi.org/10.1109/ACCESS.2020.2972103.
14. Amin, R., Reisslein, M., & Shah, N. (2018). Hybrid SDN networks: A survey of existing approaches. *IEEE Communications Surveys & Tutorials, 20*(4), 3259–3306. https://doi.org/10.1109/COMST.2018.2837161.
15. Montevecchi, F. (2017). Analysis and Optimization of Hybrid Software-Defined Networks.

16. Lee, A., Wang, X., Nguyen, H., & Ra, I. (2018). A hybrid software defined networking architecture for next-generation IoTs. *KSII Transactions on Internet and Information Systems, 12*(2), 932–945. https://doi.org/10.3837/tiis.2018.02.024.
17. Gasmelseed, H., & Ramar, R. (2019). Traffic pattern-based load-balancing algorithm in software-defined network using distributed controllers. *International Journal of Communication Systems, 32*(17), 1–14. https://doi.org/10.1002/dac.3841.
18. Song, P., Liu, Y., Liu, T., & Qian, D. (2017). Flow Stealer: lightweight load balancing by stealing flows in distributed SDN controllers. *Science China Information Sciences, 60*(3), 1–16. https://doi.org/10.1007/s11432-016-0333-0.
19. Zhong, H., Sheng, J., Xu, Y., & Cui, J. (2019). SCPLBS: a smart cooperative platform for load balancing and security on SDN distributed controllers. *Peer-to-Peer Networking and Applications, 12*(2), 440–451. https://doi.org/10.1007/s12083-017-0605-1.
20. Nayyer, A., Sharma, A. K., & Awasthi, L. K. (2019). Laman: A supervisor controller based scalable framework for software defined networks. *Computer Networks, 159*, 125–134. https://doi.org/10.1016/j.comnet.2019.05.003.
21. Sufiev, H., Haddad, Y., Barenboim, L., & Soler, J. (2019). Dynamic SDN controller load balancing. *Future Internet, 11*(3), 1–21. https://doi.org/10.3390/fi11030075.
22. Priyadarsini, M., Mukherjee, J. C., Bera, P., Kumar, S., Jakaria, A. H. M., & Rahman, M. A. (2019). An adaptive load balancing scheme for software-defined network controllers. *Computer Networks, 164*, 106918. https://doi.org/10.1016/j.comnet.2019.106918.
23. Jin, H., Yang, G., Yu, B. Y., & Yoo, C. (2019). TALON: Tenant throughput allocation through traffic load-balancing in virtualized software-defined networks. In *2019 International Conference on Information Networking (ICOIN)*, (vol. 2019-Jan, pp. 233–238). https://doi.org/10.1109/ICOIN.2019.8717976.
24. Jadhav, K. A., Mulla, M. M., & Narayan, D. G. (2020). An efficient load balancing mechanism in software defined networks. In *Proc. - 2020 12th Int. Conf. Comput. Intell. Commun. Networks, CICN 2020* (pp. 116–122). https://doi.org/10.1109/CICN49253.2020.9242601.
25. Kadim, U. N., & Mohammed, I. J. (2020). A hybrid software defined networking-based load balancing and scheduling mechanism for cloud data centers. *Journal Of Southwest Jiaotong University, 55*(3), 1–8.

Chapter 7
Analysis of Energy Optimization Approaches in Internet of Everything: An SDN Prospective

Gurpinder Singh, Amritpal Singh, and Rohit Bajaj

7.1 Introduction

The Internet of Things (IoT) concept has been used from 1999 to provide efficient communication between different connected devices through the Internet connectivity. Medical, Industry, Sports and Social Networking fields are availing several benefits using IoT with the support of networking solutions. Numerous devices are connected with one another using different technologies, like, IoT (Internet of Things), Global Positioning System (GPS), etc. It has been estimated in one of the surveys that by 2025, around 75 billion devices are expected to be connected with the web services [1]. With the increase in the number of connected devices, an evolution in IoT can be observed in the form of Internet of Everything (IoE), which includes people, devices and storage. The revolutionary change from Internet of Things to Internet of Everything is shown in Fig. 7.1. The numerous sensors/devices and people are connected via Internet, generating an abundant amount of data for processing, storage or transmission using networking solutions. IoT devices forward a huge amount of data at various geographical locations for processing. To handle this elephant-like data, an abundance of energy is required to process the forwarded data from connected IoT devices. There is a need to provide the required amount of energy to the processing nodes; otherwise, the nodes can crash and the agreed service-level agreement (SLA) can be affected. To manage the workload, an energy-efficient and latency-free approach is required to meet the requirements of the end users by using the existing infrastructure. A revolutionary approach of the networking, i.e., software-defined network, is introduced to manage the network in an efficient manner to maintain the QoS and reduce the energy level as compared

G. Singh · A. Singh (✉) · R. Bajaj
Department of Computer Science and Engineering, Chandigarh University, Mohali, India
e-mail: amritpal.cse@cumail.in; rohitbajaj.cse@cumail.in

G. S. Aujla et al. (eds.), *Software Defined Internet of Everything*, Technology, Communications and Computing, https://doi.org/10.1007/978-3-030-89328-6_7

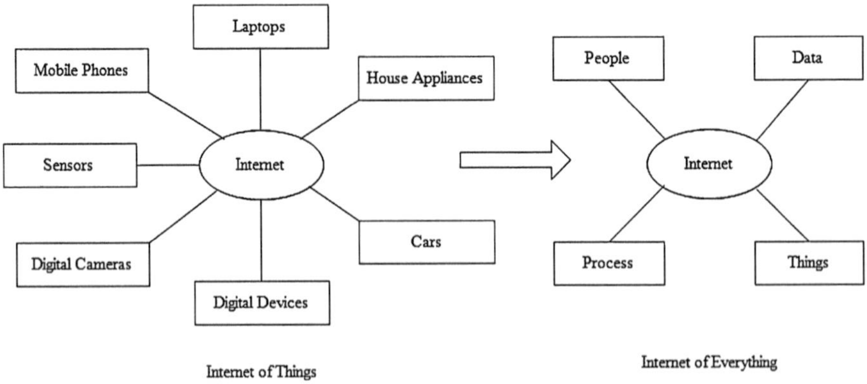

Fig. 7.1 Internet of Things to Internet of Everything

to the traditional networking approaches. The incorporation of IoE/IoT and SDN using cloud/fog computing platform provides a maximum throughput and latency-free platform for data transmission and processing. The incorporated platform is the requirement of today's scenario to connect everything via Internet to maintain the QoS [2–6].

7.1.1 Applications and Types of IoT

The concept of IoT is used in every aspect of life by connecting to the Internet using various IoT devices [1, 7]. The types of IoT are highlighted in Fig. 7.2.

1. **End User IoT:** Each digital device that is used by different users to monitor and control things in the house is called end user IoT, e.g., digital watches, health monitoring devices, smart houses, etc.
2. **Infrastructure IoT:** In organizations and educational institutions, numerous infrastructure IoT devices are configured to manage the resources efficiently, e.g., smart cities traffic management, garbage disposal, CCTV data, healthcare system, patient information, accommodation availability, environmental studies to check the level of pollutants, biodiversity in particular areas, are the various tasks managed by IoT.
3. **Industrial IoT:** The IoT devices are configured to manage the resources efficiently at various organizations and educational institutions. Traffic management in smart cities, garbage disposal, CCTV data, healthcare system, patient information, accommodation availability, environmental studies to check the level of pollutants, and biodiversity in particular areas are the various tasks managed by IoT.
4. **Emergencies and Defense Structure:** The emergency scenarios like natural disasters can be tackled by using various sensors connected to the Internet.

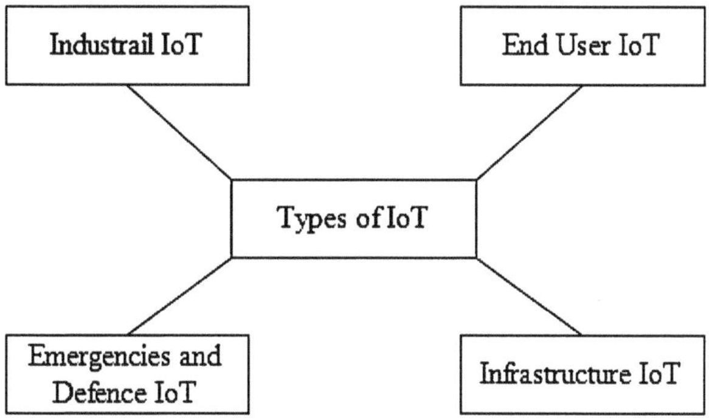

Fig. 7.2 Categories of IoT

The collected data are filtered and further used to predict any disaster in the environment. The defense sector also benefited from the IoT-enabled sensors by collecting real-time data with the help of drones, surveillance sensors, planes and battle machines for effective decision-making.

The application areas and types are not limited as any device connected to the internet is considered in the same category. The industrial IoT type requires large computational power, storage and networking capabilities to monitor and process the data generated from the various IoT sensors, automation and robotics devices and wearable devices. To handle the large amount of data, the concept of Big Data Analytics is used by following the approach of Concentric Computing Model (CCM) via cloud infrastructure as shown in Fig. 7.3. The benefits of the used approach are bandwidth utilization, security, less energy consumption, resource optimization and economical for industries. The steps used by the approach are discussed below:

1. Sensing Systems: Industrial IoT is configured with heterogeneous nature of sensors to collect the data. It also incorporated normal sensors like temperature, pressure measuring sensors. The configured devices generate a huge amount of data, and later the data are monitored and processed for effective decision-making in the industry.
2. Outer Gateway Processors: They are the servers configured for various applications, edge nodes along with the networking devices that are one hop away from the sensors. The main principle of this method is to reduce and filter the data as per the format of the input data for different processors. This method helps to control the flow of data in the network and reduce the consumption of energy by using less bandwidth.
3. Inner Gateway Processors: The servers at this level are integrated for computational power, and they are two to four hops apart from the sensors and considered as cloudlets. This method can reduce the latency and provide demanded services to the end users.

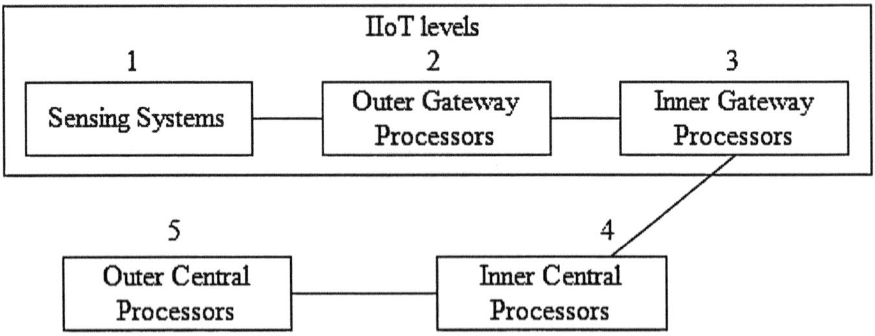

Fig. 7.3 Steps of concentric computing model

4. Outer Central Processor: To manage and process the data at large scale, powerful resources are required. To provide the limitless services to the end users, cloud platform is required. The computing nature servers are used to manage the data, route data between different cloud infrastructures and control the mechanism of services.
5. Inner Central Processor: This infrastructure is used to provide processing, storage and networking services to the large stream of data. The distance of the infrastructure is farthest from the IIoT. The data flow at this level is handled by one or more cloud platforms.

7.2 Software-Defined Networking

In this digital era, numerous service providers are trying to meet the requirements of the end users due to the configuration of a large number of IoT devices. To fulfill the requirements of the configured devices, various networking and other related services are required, e.g., storage, processing power, networking services, etc. [8, 9]. The management policies of the different services changed according to the geographical location of the end users. The service providers can meet their requirements in two manners like economically or virtually. Considering the economical factor, the service provider gets the network services on rent basis and in another way scales in and scales out the services as per the requirements of the end users. To achieve this, software-defined network (SDN) provides an interface to manage the configured devices in the network through the centralized controller [10–13].

The controller can be used as an individual unit to manage the network traffic at a specific scenario and can be used as a cluster of different controllers to provide the services in the cloud environment [14]. The detailed three-layered architecture

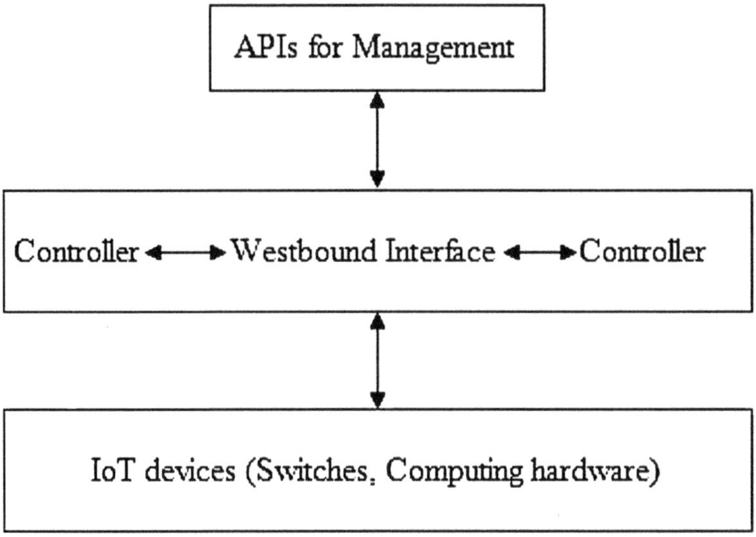

Fig. 7.4 SDN architecture for IoT devices

of the SDN approach, management, controller, and data layer and IoT devices is shown in Fig. 7.4.

All the configured IoT devices, like networking devices (switches, storage space and Virtual machines), are residing at the data layer and are monitored/managed by the management layer via APIs using hypervisor configured on the controller by using the network operating system to handle the network from single point of control. Hence, this approach can reduce the energy consumption along with economically beneficial for service provider [15–17].

The different managing and monitoring policies are directed by the controller to the connected devices. Whenever, a device is connected to the network, the health and monitoring activities are managed by the controller, e.g., energy level. The threshold value of the energy consumption of the various connected devices is adjusted by the controller. If the energy level of any device/node recorded lesser than the threshold value, then the alternate device is considered for defined role under the guidance of the controller. This approach is extended for security purpose according to the behaviour of the configured node, if the communicated node is not able to prove its authenticity, pushed to the blacklist for further actions. Overall, this approach improves the reliability of the system by avoiding the congestion of the network [18–20].

7.3 Analysis of Various Energy-Efficient Techniques in SDN-Based Environment

The energy-efficient approaches are divided into three categories as depicted in Fig. 7.5. The analysis of various energy-efficient approaches is highlighted in detail to identify the major findings of the different authors and is discussed below:

Due to an increase in the number of the IoT-based devices/sensors and an introduction of new application during this pandemic period, the workload manifold has been observed in the network. By using the standard resource allocation mechanism, the allocation of the resources to the requests generated by various end users is not justifiable. It can be seen that few nodes are above the threshold value (overloaded), and the rest of the nodes are almost free. Due to this reason, overloaded nodes require more energy to process and manage the workload. In other way, every time same path (port) will be selected to forward the incoming traffic from one node to another node, resulting into bottleneck in the network. To reconfigure/rerun the nodes, an extra amount of energy is required to restart all the nodes in the network. The load of controller is measured by the rate of packets transmitted by the switches to the controller, queries generated for flow table and round trip time between the switches and the controller. The load of each node is monitored by the centralized controller, and whenever the load of the certain node exceeds the limit of threshold set by centralized controller, immediately load balancing process comes into existence. To avoid the congestion, firstly, the overloaded path is identified on the particular switch. Onwards, alternative paths to reach the same destination are identified to reduce the load on the same port. In case the identified path goes through different controllers in the network, then, a switch is targeted whose load is less than or equal to half of the difference of the load of the overloaded controller and target controller. This mechanism is named as Dynamic Load Balancing using Alternate Path (DALBP). The approach to shift the stream of data on another path consumes less energy during processing the workload as compared to transmitting the switch [21]. Hence, this approach integrates the edge platform to handle a large amount of data to provide with Quality of Service. This approach can be enhanced by using software-defined networking by providing various benefits over traditional networking [22].

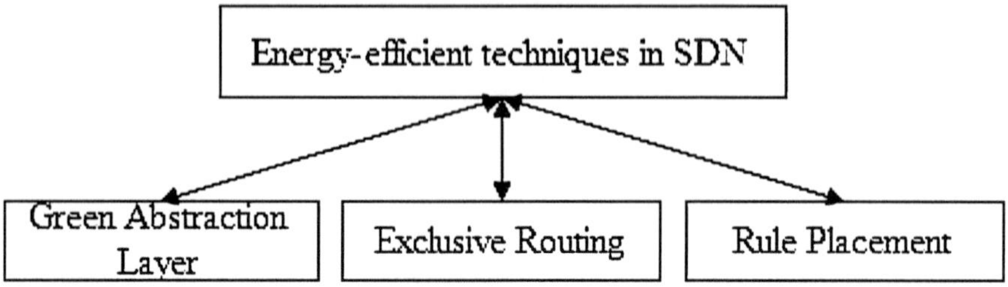

Fig. 7.5 Energy-efficient approaches

Conventional networking approaches are not able to handle the heavy workload, resulting into degradation of Quality of Service in IoT-based scenarios. Despite the traditional networking approach, flow establishment between devices can be provided by software-defined network by utilizing minimum devices and selected ports of the devices according to the rate of flow of data. In case the flow rate is minimal, use only minimal nodes and put other nodes on sleep mode. SDN-enabled network can monitor and control the devices to make the network more energy-efficient. The approach used to achieve the defined objective is ***Green Abstraction Layer (GAL)***, which is integrated with the controller for optimization of energy at different levels. Onwards, the current state of each network device is measured, and accordingly the entries are updated into the flow table. The health of the connected nodes is monitored at regular intervals of time to check the level of energy consumption. In the second approach, an extended version of the fair share route selection method, named as ***exclusive routing***, is used to manage the data flow in the network. The proposed approach is able to minimize the load on the switches, and however, the 55% resources of the switches are used. To handle the over utilization of the resources of the switches, an exclusive routing approach is introduced to manage the active and suspended flow on priority basis. For every incoming flow, the controller checks the free paths from the flow table entries to send the active flow over the network; otherwise, it checks alternate paths with suspended flow with priority value less than the current calculated value and forwards the current flow on the selected path. In case more than one condition is available to forward the flow, then, the controller checks the available of the switches to constitute a path. Though the provided solution is not suitable for the current flow, therefore, the flow is shifted to the suspended list. In another approach, the routing is processed to turn off the active devices to select the shortest path from the pool of paths except deactivated paths. The approach is extended by selecting the active paths with minimum active nodes as compared to other available paths. In a similar manner, the ***rule placement and TCAM-based energy-aware*** concept is used for optimal utilization of little memory of the switches. The heuristic method is used to select the suitable route by considering the link parameters. One port is selected to consider as a default port to handle the requests without installation of rules multiple times, and further low traffic on paths is diverted to other active ports to put the ports on sleep mode for optimal utilization of the energy [23].

Energy consumption can be reduced by incorporating the SDN platform in the existing network for intelligent decision-making. The multiple-layered architecture (control plane and data plane) of SDN helped the system to segregate the data for effective decision-making. One reason is that in the case of SDN, the control plane and data plane/forwarding plane are used to manage the network. The devices (switches and routers) considered in the category of data plane/forwarding plane and all the controlling functions (like change, update and manage the network topologies) are managed by the controller via different applications. In case of traditional networking, the decision-making is done by the single device which may be the cause of congestion in the network. It has been proved by the authors that in a topology of three routers and five switches, the total energy consumption by the network devices as compared to the traditional network is 6974 W. In SDN-enabled network, considering the same topology, the energy consumption is 1500 W. Therefore, it is clear that the difference of energy consumption in both cases is 5474 W [24].

The major contributors of the energy consumption are data processing, data forwarding, etc. The mobility issues in the mobile network and change in the services from one node to another are the major concerns of workload overloading on the nodes. To overcome the problem, SDN architecture with root switch can be used to update the location of the users by directly interacting with them. There is a mobile terminal (user) that moves in the cluster or from one cluster to another. Every cluster has its own location to manage to avoid the overloading of workload. In the SDN architecture, the main components are first hop switch (receive packets from the sender, i.e., mobile terminal), root switch (receive packet from first hop switch to establish route), controller (used to create new entries for particular sender), leaf switch (forward the packets to the receiver) and intermediate switch (send packets from source to destination). In the root-based approach, the packet is forwarded from mobile terminal to the first hop switch, onwards the packet to the root switch. The root switch matches the packet for suitable path selection with route table and forwards the packet to the next hop switch, if the path of the packet is identified successfully; otherwise, it forwards the packet to the controller for necessary action. Furthermore, the controller processes the header of packet, updates the route table by selecting a suitable path that leads to the destination and installs the new entry to intermediate switches and sends packet to the next hop switch. In this manner, the packet reaches to the destination by following the intermediate switches and leaf switch [25].

In resource sharing approaches like IoT and Big Data applications, data centre networks by using cloud infrastructure can provide virtual resources at a minimum cost. An elephant-like data is generated by various devices that affect the QoS. The SDN manages the workload in an efficient manner with the help of control plane by using either a single controller or multiple controllers as per the size of the network. To achieve the fault tolerance at various data centres, mirror controller is used. In case of scalable network, adequate numbers of switches are handled by one controller and follow the same approach for other switches to avoid overloading on different nodes. To reduce the latency in the network, the distance between the controller and switches needs to be minimum. The problem of energy consumption in data centre networks is divided into three sections as in Fig. 7.6.

In the first scenario, the cloud-based data centres required to run all the times to fulfill the requirements of the clients; however, an ample amount of energy is consumed for this purpose. In the second scenario, there is a restriction on the number of controllers to minimize the energy consumption. The numbers of controllers are decided according to the load on the switches and the load threshold value of each controller. In the third scenario, the main focus is to manage the QoS parameters (fault tolerance, route optimization throughput, energy consumption, etc.) in the network using management/control plane policies at the controller. The solution of all problems is provided in the following sections:

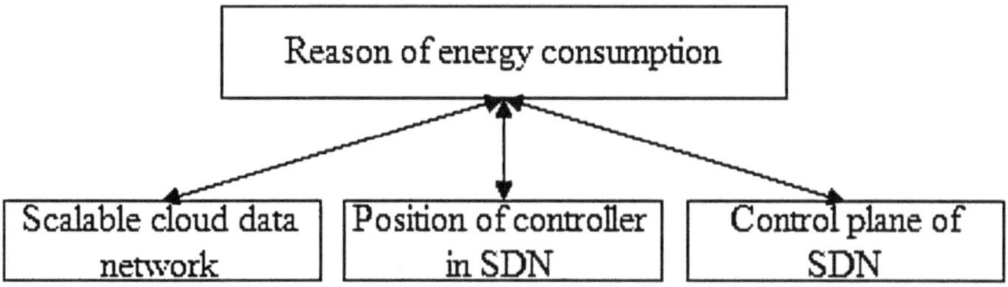

Fig. 7.6 Energy consumption problem

- Use the multi-core processors at data centre networks to process applications in a parallel manner.
- Controllers that consume lesser energy can be used frequently to process the network. The controllers with lesser frequency of processor can also save energy. Another way for the reduction of energy consumption is to relax the load of the controllers by shifting the flow to other controllers [26].

The data processing and other activities related to the management of the data are configured at virtual machines via cloud environment for various applications like IoT, Big Data and virtual resource optimization by organizations, etc. The QoS and energy optimization are the promising parameters that need to be considered to increase the performance of the network. Mostly, the overloading of the virtual machines is managed by transmitting the load of overloaded one to other machine using cold and hot data transfer approaches. The cold migration approach is one of the safest approach in terms of cost during turn off the Virtual machine then transmitting the data that causes delay in the network. The hot migration approach is complex but helps to reduce the processing time by transmitting the data in different chunks to handle the dirty pages during active and inactive mode of the machine. The dynamic flow approach can be considered for migration of data that is suitable for hot migration approach but may be used for cold migration approach. The migration time (T) for the hot approach is as follows:

$$T = M * \left[\frac{\left(1 - \frac{R}{L}\right)^n \left(1 - \frac{R}{L}\right)}{L} \right], \tag{7.1}$$

where M is the memory size of the virtual machine, L is the available bandwidth, R is the frequency of produced dirty pages and n is the number of pre-copy stages.

The dynamic flow algorithm for data migration between source and destination virtual machines (VMs) with the help of their IPs starts by retrieving the topology table from SDN controller. The extracted information from the topology table and the shortest path between the source and destination VMs are selected by using breadth-first search algorithm. In the cloud environment, the network configuration settings between the data centres do not change frequently, and the shortest path between the various VMs pair is calculated initially and use as per requirement. Furthermore, the shortest path is selected according to the maximum availability of the bandwidth during processing the workload along with the link of the path which could provide maximum byte rate of flow (frequently used). To find the shortest path according to the flow rate of the VM migration, the flow rate of the VM migration is deducted from the link flow rate where the value of the flow rate is equal to the byte rate. The calculated value is compared with the link of the maximum byte rate, and the link with more value as compared to the link's byte rate is selected. In the last stage, the shortest path with link's byte rate, i.e., less than or equal to the path's maximum residual bandwidth, is selected [27].

The wireless body area networks are frequently adopted in the area of medical, sports, industrial research, etc. Therefore, the sensors are connected to the body of the host and transmit the data via nodes. Each node is connected to the multiple sensors. The main objective of this type of network is to enable communication between the nodes by considering the QoS in terms of SNR (source-to-noise ratio), numbers of hops and energy level using infrastructure of software-defined networking. The process starts with node which is trying to send data. If the flow entries (route) are available and updated in the node memory, then data can be transferred or otherwise forwarded to the controller by route request (RREQ) message for further action and policies. The other condition concerned to the flow entries could be used when the route has one or more deactivated nodes, in that case route request (RREQ) message is being sent to the controller for alternate route. Whenever the controllers receive the RREQ, it broadcasts the topology request message (TREQ) to all the nodes of the network and collects the information related to QoS from different nodes (related to their neighbours and itself) in the form of message name topology response message (TREQ) and updates the information of the topology. In the next stage, the best shortest path is calculated along with reserve in case of failure and stores all the collected details for further use. The Dijkstra algorithm with fuzzy logic approach is considered to maintain the QoS. Then, the route reply message is generated by the controller against the route request of node to clarify the suggested route. The energy-efficient and SDN-enabled routing algorithm is used for wireless body area networks (ESR-W) to achieve the mentioned objectives [28].

Fog computing is the best platform to optimize the utilization of the resources in the network. The configured devices can be of heterogeneous nature and get the benefits of the characteristics of the fog computing platform as shown in Fig. 7.7.

A novel approach of SDN-based fog computing is another milestone to manage the network pragmatically by using the defined APIs, e.g., topology management, energy optimization, security at control level and services of fog computing are at data plane. The fog computing services are managed by cooperative and non-cooperative approaches along with the methodology of selection of appropriate servers on the platform. The name clears the functionality of the cooperative fog computing as it starts communicating the nodes as one is overloaded and the other one is underloaded. The non-coopertive fog computing servers communicate with each other, further, the devices/users wait in the queue for the servers to acquire the

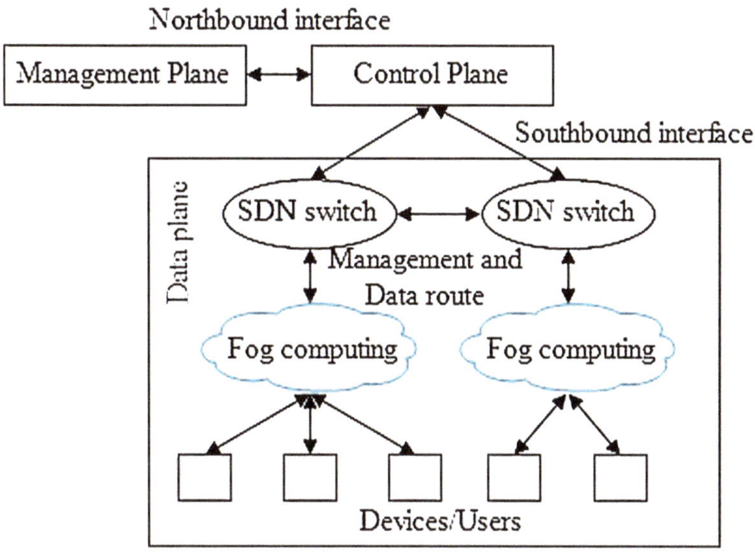

Fig. 7.7 SDN-based fog computing approach

resources. This approach reduces the complexity during load migration from one server to other in the cost of delay. At last, the cooperative approach balances the load by optimizing the resource with less power consumption as compared to the non-cooperative approach. The servers in the fog computing platform are selected on the basis of geographical distance. The server near to the user will serve the first as per the requirements. In case the resources are not available, then queuing policies are used considering different QoS parameters, e.g. FIFO, fair/weighted fair queue, priority queue, etc. [29].

The IoT-based platform incorporated with SDN provides the flexibility to manage the network centrally. The generated traffic from the various IoT devices is the major reason of network congestion. Usually, HTTP is used for transferring the messages between devices via Internet, but Message Queue Telemetry Transport (MQTT) protocol can provide better result as compared to HTTP in terms of load and energy consumption. The Raspberry pi 3 is considered as host (broker/publisher) for MQTT and connected to IoT devices/sensors, e.g., bulb to provide communication between publisher and subscriber. Despite using the client–server architecture of HTTP, the publisher–subscriber approach of MQTT provides the independent environment between the connected devices. The ON message is communicated by subscriber to Raspberry pi 3 which is directly connected to the bulb. The result of this scenario using MQTT and HTTP represents that MQTT is delivering less bytes/min as compared to HTTP by consuming less energy [30].

A comparative analysis of energy-efficient approaches in SDN-enabled environment is provided in Table 7.1.

Table 7.1 Analysis of various energy-efficient approaches in SDN-enabled environment

Authors	Approach	Description
Abbas Yazdinejad et al. [18]	Policy to send packets based on the energy level	This approach is being used to handle communication of packets by considering the energy level of each node
Suchismita Rout and S. P. Nayak [21]	This approach is being used to find the alternate path when the existing path is overloaded while using the SDN to transfer stream of data	Dynamic Load Balancing using Alternate Path (DALBP) is used to find the overloaded path by accessing the information of controller. Controller frequently uses the shortest path to transfer data, which is the reason of overload. To overcome this problem, DALBP uses alternate path to balance the network by utilizing most of the resources
Mamdouh Alenezi, 2018	Architecture of IoT network by using cloud computing resource utilization approach and using SDN	IoT devices like networking devices, storage spaces and computational resources are monitored and managed using SDN by cloud infrastructure
Muhammad Habib ur Rehman et al. [22]	Managing and processing large stream of data as Big Data of Industrial IoT (IIoT) by getting help of cloud and fog computing	Limited resource requirement is handled in IIoT by the concept of fog and for computational power at large is provided by cloud infrastructure
Péter András Agg and Z. C. Johanyák [23]	Energy saving methods of SDN	Green abstraction layer, executive routing and route placement approaches for energy optimization
Suada Hadzovic et al. [24]	Comparing traditional network approach with SDN for energy consumption	Use SDN's devices instead of traditional network's devices. Comparison shows reduction in energy consumption while using SDN
Nurul Hazrina Shahba binti Mohammad Shah-run, 2018	Location awareness of mobile nodes and route management for newly connected nodes by SDN for power efficiency	Root switch is added to SDN infrastructure for this approach
Tadeu F. Oliveira and L. F. Q. Silveria [26]	Effectively spread the controller in the network to handle load and consume less energy	Multiprocessors core for parallel processing and use controllers which need less energy along with flow migration between controllers in case of overloading
Adel Nadjaran Toosi, 2018	The method of transferring instance of virtual machine (VM) from one server to other VM using SDN network by installing different flow paths between servers and handle other network parameters	It uses dynamic data flow method for scheduling the VMs by effectively using the bandwidth available and decreasing migrate time

(continued)

Table 7.1 (continued)

Authors	Approach	Description
Murtaza Cicioğlu and A. Çalhan [28]	Effective routing approach (ESR-W) by using SDN for wireless body area network by optimizing energy consumption	Routing method using Dijkstra with fuzzy logic by following SNR (source-to-noise ratio), numbers of hops and energy level
Adnan Akhunzada et al. [29]	Fog computing using SDN for handling user's data	Cooperative and non-cooperative approaches by selecting suitable fog server for comparing energy consumption along with response and processing time where cooperative outperforms than other
Meenaxi M Raikar, 2020	IoT model using SDN by using Message Queue Telemetry Transport	Compare scenario by using HTTP and MQTT and find that MQTT is better in case of less load producing and energy consuming

7.4 Conclusion

In the IoT-based environment, numerous sensors with limited battery power are configured. Due to limited power, energy consumption is the major issue that needs to tackle. SDN-enabled platform can manage the network traffic in an efficient manner and can be the reason of reduction in energy consumption during processing the workload. The increased energy utilization can be a reason in the degradation of QoS in terms of different services provided to the end users. In this chapter, various SDN-based approaches are discussed with proper facts and figured to handle the increased rate of energy consumption in the network. In future, the analysis of various energy-efficient approaches can help identify the gaps, and a novel approach can be introduced to minimize the effect of energy utilization in the network.

References

1. Patel, K. K., & Patel, S. M. (2016). Internet of things-IOT: Definition, characteristics, architecture, enabling technologies, application and future challenges. *International Journal of Engineering Science and Computing, 6*(5), 6123–6131.
2. Peng, S.-L., Pal, S., & Huang, L. (2019). *Principles of internet of things (IoT) ecosystem: Insight paradigm* (Vol. 174).
3. Shrutika, M., Tripathi, A. R., Singh, R. S., & Priyanshu, M. (2021) *Design and implementation of internet of everything's business platform ecosystem.*
4. Kumar, R. (2021). Future of Internet of Everything (IoE). *International Research Journal of Computer Science (IRJCS), 08*(04), 84–92.
5. Hassan Shakib, K., & Faiza Neha, F. (2021). A study for taking an approach in industrial IoT based solution. *Journal of Physics: Conference Series, 1831*(1). https://doi.org/10.1088/1742-6596/1831/1/012007

6. Nitnaware, P., & Nimbarte, M. (2021). IoT Based Smart City Mangement Using IAS: A survey. *Journal of University of Shanghai for Science and Technology, 23*(4), 117–122.
7. Srinivasan, C. R., Rajesh, B., Saikalyan, P., Premsagar, K., & Yadav, E. S. (2019). A review on the different types of internet of things (IoT). *Journal of Advanced Research in Dynamical and Control Systems, 11*(1), 154–158 (2019)
8. Aujla, G. S., & Kumar, N. (2018). MEnSuS: An efficient scheme for energy management with sustainability of cloud data centers in edge–cloud environment. *Future Generation Computer Systems, 86*, 1279–1300.
9. Aujla, G. S., Garg, S., Batra, S., Kumar, N., You, I., & Sharma, V. (2019). DROpS: A demand response optimization scheme in SDN-enabled smart energy ecosystem. *Information Sciences, 476*, 453–473.
10. Singh, M., Aujla, G. S., Singh, A., Kumar, N., & Garg, S. (2020). Deep-learning-based blockchain framework for secure software-defined industrial networks. *IEEE Transactions on Industrial Informatics, 17*(1), 606–616.
11. Garg, S., Singh, A., Aujla, G. S., Kaur, S., Batra, S., & Kumar, N. (2020). A probabilistic data structures-based anomaly detection scheme for software-defined internet of vehicles. *IEEE Transactions on Intelligent Transportation Systems, 22*, 3557–3566.
12. Singh, A., Aujla, G. S., Garg, S., Kaddoum, G., & Singh, G. (2019). Deep-learning-based SDN model for internet of things: An incremental tensor train approach. *IEEE Internet of Things Journal, 7*(7), 6302–6311.
13. Ranjan, R., Thakur, I. S., Aujla, G. S., Kumar, N., & Zomaya, A. Y. (2020). Energy-efficient workflow scheduling using container-based virtualization in software-defined data centers. *IEEE Transactions on Industrial Informatics, 16*(12), 7646–7657.
14. Aujla, G. S., Kumar, N., Garg, S., Kaur, K., & Ranjan, R. (2019). EDCSuS: Sustainable edge data centers as a service in SDN-enabled vehicular environment. *IEEE Transactions on Sustainable Computing*.
15. Alenezi, M., Almustafa, K., & Meerja, K. A. (2019). Cloud based SDN and NFV architectures for IoT infrastructure. *Egyptian Informatics Journal, 20*(1), 1–10. https://doi.org/10.1016/j.eij.2018.03.004
16. Singh, A., Batra, S., Aujla, G. S., Kumar, N., & Yang, L. T. (2020). BloomStore: dynamic bloom-filter-based secure rule-space management scheme in SDN. *IEEE Transactions on Industrial Informatics, 16*(10), 6252–6262.
17. Singh, A., Aujla, G. S., & Bali, R. S. (2021). Container-based load balancing for energy efficiency in software-defined edge computing environment. *Sustainable Computing: Informatics and Systems 30*, 100463.
18. Yazdinejad, A., Parizi, R. M., Dehghantanha, A., Zhang, Q., & Choo, K. K. R. (2020). An energy-efficient SDN controller architecture for IoT networks with blockchain-based security. *IEEE Transactions on Services Computing, 13*(4), 625–638. https://doi.org/10.1109/TSC.2020.2966970
19. Wen, Z., Garg, S., Aujla, G. S., Alwasel, K., Puthal, D., Dustdar, S., Zomaya, A. Y., & Rajan, R. (2020). Running industrial workflow applications in a software-defined multi-cloud environment using green energy aware scheduling algorithm. *IEEE Transactions on Industrial Informatics, 17*, 5645–5656.
20. Aujla, G. S., Singh, A., & Kumar, N. (2019). Adaptflow: Adaptive flow forwarding scheme for software-defined industrial networks. *IEEE Internet of Things Journal, 7*(7), 5843–5851.
21. Rout, S., & Nayak, S. P. (2020). Energy minimization technique in SDN using efficient routing policy. *Journal of Xi'an University of Architecture & Technology. XII*(Iv), 3512–3519.
22. Rehman, M. H. U., Ahmed, E., Yaqoob, I., Hashem, I. A. T., Imran, M., & Ahmad, S. (2018). Big data analytics in industrial IoT using a concentric computing model. *IEEE Communications Magazine, 56*(2), 37–43. https://doi.org/10.1109/MCOM.2018.1700632
23. Agg, P. A., & Johanyák, Z. C. (2021). Energy savings in SDN networks. *Gradus, 8*(1), 205–210.

24. Hadzovic, S., Seremet, I., Mrdovic, S., & Causevic, S. (2020). Reduction of energy consumption based on replacement of routers with SDN switches. In: *2020 24th International Conference on Information Technology IT* , 2020, February (pp. 14–17). https://doi.org/10.1109/IT48810.2020.9070464
25. Shahrun, N. H. S. B. M., Mohammed, A. F. Y., Ramlie, R., Shah Newaz, S. H., & Wan, A. T. (2018). An energy efficient mobility management mechanism in software defined networking (SDN). In *2018 International Conference on Computer, Control, Electrical, and Electronics Engineering ICCCEEE* (pp. 1–6). https://doi.org/10.1109/ICCCEEE.2018.8515818
26. Oliveira, T. F., & Silveria, L. F. Q. (2019). Distributed SDN controllers optimization for energy saving. In *2019 Fourth International Conference on Fog and Mobile Edge Computing (FMEC)* (pp. 86–89). https://doi.org/10.1109/FMEC.2019.8795343
27. Nadjaran Toosi, A., & Buyya, R. (2019). Acinonyx: Dynamic flow scheduling for virtual machine migration in SDN-enabled clouds. In *2018 IEEE International Conference on Parallel & Distributed Processing with Applications, Ubiquitous Computing & Communications, Big Data & Cloud Computing, Social Computing & Networking, Sustainable Computing & Communications* (pp. 886–894). https://doi.org/10.1109/BDCloud.2018.00131
28. Cicioğlu, M., & Çalhan, A. (2020). Energy-efficient and SDN-enabled routing algorithm for wireless body area network. *Computer and Communications, 160*, 228–239. https://doi.org/10.1016/j.comcom.2020.06.003
29. Akhunzada, A., & Sherali, Z. (2021). Power and performance efficient SDN-enabled fog architecture. *arXiv preprint arXiv:2105*.14607
30. Raikar, M. M., Meena, S. M., & Mulla, M. M. (2020). Software defined internet of things using lightweight protocol. *Procedia Computer Science, 171*(2019), 1409–1418. https://doi.org/10.1016/j.procs.2020.04.151

Chapter 8
Network Function Virtualization

Haotong Cao

8.1 What Is Network Function Virtualization

Network function virtualization, abbreviated as NFV [1], is an emerging network technology for the next generation network. NFV enables network functions, usually implemented in software, to run on top of the general-purpose hardware. In addition, NFV can build many types of network equipment, such as servers, switches, and storage, into a data center (DC) network [2]. That is to say, dedicated hardware, such as the middleboxes, can be virtualized into virtual network functions (VNFs). These VNFs can be managed and operated to implement a specific service, in the way of software.

8.2 NFV Architecture and Model

The high-level overview of the NFV architecture [3] is shown in Fig. 8.1. Generally speaking, the NFV architecture is made up of three main components: NFV infrastructure (NFVI), virtual network functions (VNFs), and NFV management and orchestration (MANO).

With respect to the NFVI, it includes the virtualization layer (hypervisor, container management system, such as Docker, and vSwitch) and physical resources, such as COTS servers, switches, storage devices, etc. NFVI can be seen as a general

H. Cao (✉)
Jiangsu Key Laboratory of Wireless Communications, Nanjing University of Posts and Telecommunications Nanjing, Nanjing, China

Department of Computing, The Hong Kong Polytechnic University, Hong Kong SAR, China
e-mail: haotong.cao@polyu.edu.hk

G. S. Aujla et al. (eds.), *Software Defined Internet of Everything*, Technology, Communications and Computing, https://doi.org/10.1007/978-3-030-89328-6_8

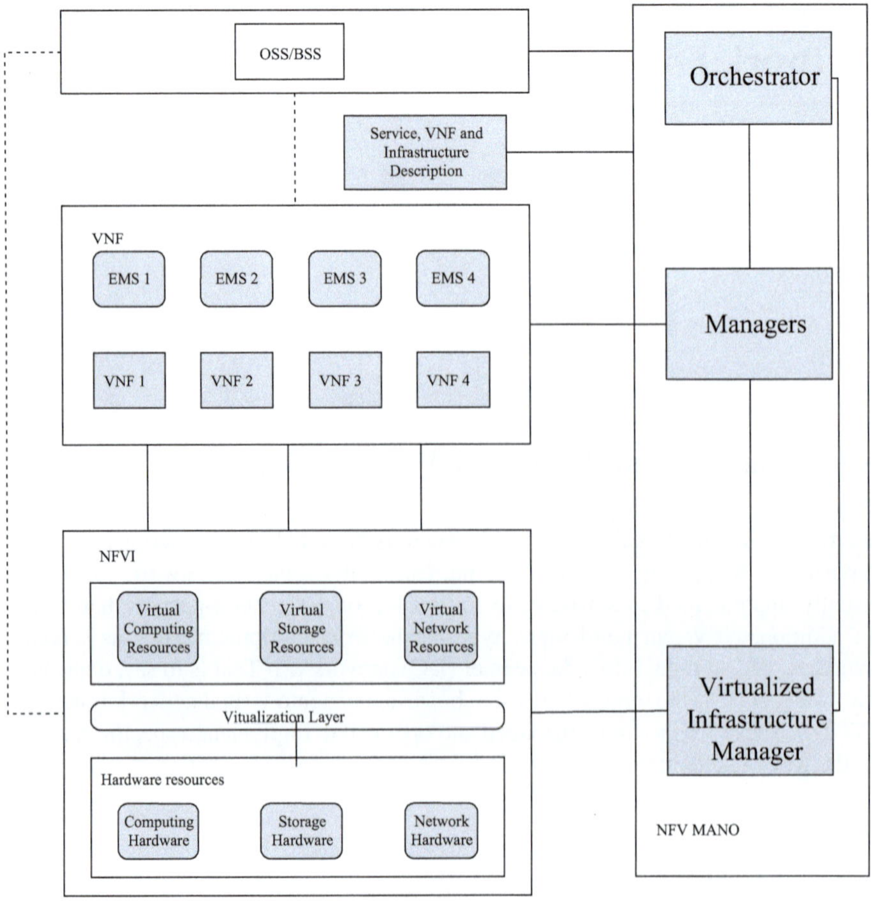

Fig. 8.1 High-level overview of NFV architecture

virtualization layer. All virtual resources should be in a unified and shared resource pool. In the resource pool, certain VNFs will run on top of it.

With respect to the VNFs, they refer to specific virtual network functions, such as the firewall function. These specific functions can constitute the network service, such as the remote video. VNFs are software that is usually deployed in virtual machines, containers, or bare-metal physical machines using the infrastructure provided by NFVI. Compared with VNF, traditional hardware-based network elements can be called as PNF. VNF and PNF can be networked separately or combined to form another so-called service chain to provide end-to-end (E2E) network services required in specific scenarios [4].

With respect to the NFV MANO, it aims at providing the overall management and orchestration of NFV. MANO is also connected to OSS/BSS upward. MANO is usually composed of NFVO (NFV orchestrator), VNFM (VNF manager), and

VIM (virtualized infrastructure manager). The original meaning of orchestration is an orchestra. While in the NFV architecture, all components must have a certain choreography effect. All VNFs, PNFs, and other various resources can only be properly choreographed so as to do the right thing at the right time. NFVI is managed by VIM. VIM controls the allocation of virtual resources of VNF, such as virtual computing, virtual storage, and virtual network. Both OpenStack and virtual machine can be used as VIM. The OpenStack is open source while the latter is commercial. VNFM manages the life cycle of VNF, such as online and offline, status monitoring, image onboard. VNFM is based on VNFD (VNF description) to manage VNF. NFVO has the function of managing the life cycle of NS (network service, network business) and coordinating the management of NS life cycle. In addition, VNFM can coordinate the management of VNF life cycle. The life circle needs to be supported by the VNF manager VNFM. VNFM can coordinate the management of various resources of NFVI so as to ensure the optimal configuration of various resources and connections required. Onboard new network business, VNF forwarding table, VNF package, and NFVO run based on the NSD (network service description). NSD usually includes the service chaining, NFV description, and performance goal.

The future evolution of NFV will undergo two main stages [2, 3]: Initial Stage and Advanced Stage. In the Initial Stage, NFV will be used as a new method of implementing traditional services. NFV mainly aims at completing the one-to-one conversion of the traditional software execution environment based on dedicated hardware into a dedicated virtualization environment based on general-purpose hardware VMs. In the Advance Stage, NFV will be used as a new method to implement new services. It includes decomposing VNFs into micro-services and even single-function VNFs, then recombining them, using container technology to slice a single VM into smaller containers, and implementing software-programmable data models.

8.3 NFV Applications and Implementations

This part firstly talks about the benefits of NFV applications. With adopting NFV technology, service providers can deploy network functions on top of standard hardware instead of deploying dedicated hardware. In addition, the network functions can be fully virtualized. That is to say, multiple functions can run on the corresponding servers. This means that the required physical hardware is reduced as much as possible. More required resources can be integrated so as to reduce physical space consumption and overall costs [3].

NFV allows service providers to run VNFs on different servers flexibly or move VNFs as needed when service requirements change. This approach can speed up the delivery of services and applications by service providers. For example, if a customer requests a new network function, he (she) can start a new virtual machine to handle the request. When the function is no longer needed, this virtual machine

[4] can be replaced. This is also a low-risk way to test the value of potential new services.

As discussed above, service providers can earn benefits by doing NFV applications. However, some key flaws exist in NFV, limiting NFV implementations. The first key flaw is the high software maintenance cost. The second flaw is that the integration of NFV is not streamlined enough. Although NFV can eliminate the need of proprietary hardware, it will only transfer proprietary hardware from different vendors to proprietary software from different vendors. Therefore, the deployment of NFV is difficult to manage and very complicated. Due to the complex types of equipment [4], the chaotic combination and the lack of general and conventional ways are impossible to manage them on a large scale. NFV cannot provide scalable and automated capabilities. In order to make matters worse, many software licenses are attached to the mixed dedicated software package, which causes the above-mentioned cost problem.

In general, the lack of standardization, complex and incoherent deployment models, and high deployment cost restrict the NFV applications and implementations.

8.4 Resource Allocation in NFV-Enabled Networks

Multiple technical issues exist in NFV. One key technical issue is the resource allocation problem in NFV (RA-NFV) [5–7]. In NFV, the virtual network service (NS) is modeled as an ordered set of chained virtual network functions (VNFs), also called as service function chain (SFC). The SFC is required to be deployed into the physical network efficiently. During the deployment, the specific function orders and resource demands (e.g. computing, storage resource) of the SFC must be fulfilled. In the NFV-enabled networks, the RA-NFV is called as SFC [6, 7] problem.

Researchers from the academia and industry have paid much effort so as to push SFC research ahead [7–9]. According to [5], SFC problem is usually conducted in four stages: SFC description (SFC-D), SFC composition (SFC-C), SFC embedding (SFC-E), and SFC scheduling (SFC-S). With respect to the first two stages, the industry is mainly responsible for developing NS description languages and VNF concatenating methods. With respect to the last two stages, the academia mainly focuses on proposing and developing SFC embedding and scheduling algorithms. The SFC embedding and scheduling issues are researched in this chapter.

Since the optimal SFC embedding and scheduling solution [10–19] are required to be achieved, the chapter concentrates on researching this optimal solution.

In the first place, it is the introduction of model for SFC problem. The model for SFC embedding and scheduling research consists of two sub-models: physical network model and SFC model.

In NFV-enabled networks, the underlying physical network must support the virtualization scheme. That is to say, the network functions and resources of networks are fully virtualized and isolated. The physical network is modeled by undirected weighted graph [20] $PNetwork = (PNode, PLink)$. $PNode$

represents the physical nodes set, while $PLink$ represents the physical links set. With respect to certain physical node M^P in $PNode$, specific functions that run on it can be represented by $Func1(M^P)$, $Func2(M^P)$, and so on. Take note that $Func1(\,)$ refers to one specific network function, such as firewall. NAT network function can be represented by $Func2(\,)$. With respect to resources of M^P, the computing $Computing(M^P)$ and storage resources $Storage(M^P)$ are considered in this chapter. In addition, the node deployment (deployment) time of M^P is represented by $Deploy(M^P)$, revealing the scheduling delay of embedding one virtual node onto M^P. With respect to certain one physical link MN^P, its bandwidth is represented by $Band(MN^P)$. In this chapter, the link deployment time that can serve as the SFC scheduling delay is not considered. In addition, the spectrum and backhaul resource are not taken into account in this chapter.

In NFV research, multiple NSs are usually requested and arrive independently. Hence, there will exist multiple NSs. Each NS is represented by SFC. With respect to ith SFC, it can be modeled by using directed weighted graph $SFC_i^V = (VNFs_i^V, VLinks_i^V)$. $VNFs_i^V$ and $|VNFs_i^V|$ represent the VNF set and a number of VNFs, respectively. With respect to jth VNF in SFC_i^V, it is labeled by VNF_{ij}^V. As usual, each VNF in its SFC just has one specific function demand. The VNF_{ij}^V is selected as the example. VNF_{ij}^V needs the $Func1(\,)$ function, thus labeling as $Func1(VNF_{ij}^V)$. With respect to the resource demands of VNF_{ij}^V, the computing $Computing(VNF_{ij}^V)$ and storage resources $Storage(VNF_{ij}^V)$ are considered in this chapter. In addition, the maximum allowed node deployment (scheduling) time of VNF_{ij}^V is represented by $MaxDeploy(VNF_{ij}^V)$. In this chapter, the total deployment time of SFC to represent its QoS performance is considered. When a service is requested, implementing and deploying the service quickly will enhance the user experience. With respect to certain one virtual link $VNF_{ij}^V VNF_{i(j+1)}^V$, its required bandwidth is represented by $Band(VNF_{ij}^V VNF_{i(j+1)}^V)$. In addition, ith SFC has its maximum allowed waiting time $MaxWait(SFC_i^V)$. That means the SFC_i^V must be embedded and scheduled successfully within $MaxWait(SFC_i^V)$.

In order to assist readers to understand model for SFC embedding and scheduling, Fig. 8.2 is plotted. Figure 8.2 consists of one physical network and one SFC. Physical attributes (functions and resources) and virtual attributes (functions and resources) are highlighted. The deployment time attributes that play a vital role in SFC rescheduling example description are plotted, as well. The network scale of physical work is set to be small in Fig. 8.2, having six physical nodes. The initial SFC deployment results are plotted, as well: $VNF1$ embedded onto B, $VNF2$ embedded onto C, and $VNF3$ embedded onto D. $VNF1VNF2$ is embedded onto BC, while $VNF2VNF3$ is embedded onto CD. In Fig. 8.2, the function and resources demands of the SFC are fulfilled successfully.

In the second place, it is the formulation details of exact and optimal SFC embedding and scheduling algorithm (*SFC-Optimal*). The *SFC-Optimal* is based on the known integer linear programming (ILP) [21, 22] method, which is widely accepted as the optimal approach in virtual resource allocation era [5].

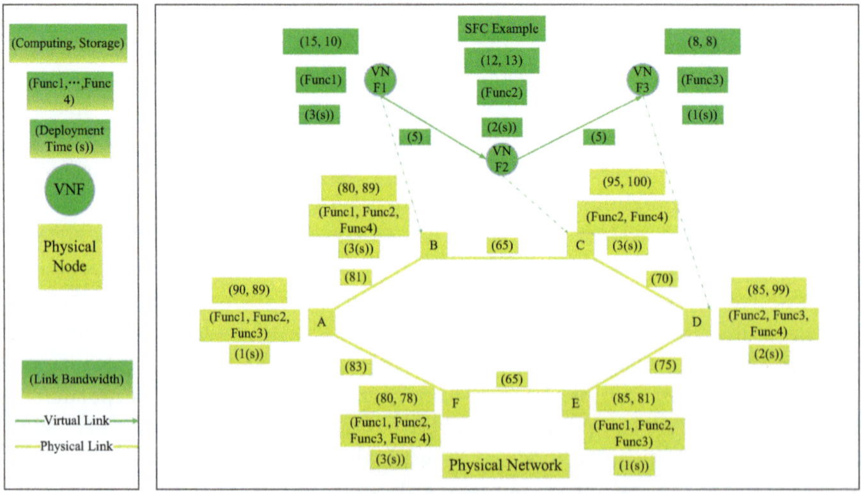

Fig. 8.2 Resource allocation model for NFV-enabled networks

Firstly, two kinds of binary variables are defined. The first type of binary variables is X, indicating the relationship between virtual nodes and physical nodes. The other type of binary variables is Y, indicating the relationship between virtual links between physical paths (links).

Secondly, it is the objective function. In this chapter, the aim is minimizing the consumed physical resources for accommodating one SFC (e.g., ith SFC SFC_i^V). This is in accordance with the goal of TSP. Minimizing the consumed resources enables to leave more space for accommodating more virtual networks, thus improving the profit in the long run. The objective function of *SFC-Optimal* is formulated in Expression 8.1. With respect to other objects, such as balancing the network loading, it can be realized by setting the proper objective functions.

$$
\begin{aligned}
&Obj : Cost\left(SFC_i^V\right) = \\
&\alpha(cost) \cdot \sum_{VNF_{ij}^V \in VNFs_i^V} \mathbf{Computing}\left(VNF_{ij}^V\right) \\
&\quad + \beta(cost) \cdot \sum_{VNF_{ij}^V \in VNFs_i^V} \mathbf{Storage}\left(VNF_{ij}^V\right) \\
&\quad + \gamma(cost) \cdot \sum_{VNF_{ij}^V VNF_{i(j+1)}^V \in VLinks_i^V} \sum_{path^P \in Path^P} \\
&Num_{VNF_{ij}^V VNF_{i(j+1)}^V}^{path^P} \cdot \mathbf{Band}\left(VNF_{ij}^V VNF_{i(j+1)}^V\right)
\end{aligned}
\tag{8.1}
$$

where $Num^{path^P}_{VNF^V_{ij}VNF^V_{i(j+1)}}$ records the number of physical links in the selected physical path that accommodates the virtual link $VNF^V_{ij}VNF^V_{i(j+1)}$. $\alpha(cost)$, $\beta(cost)$, and $\gamma(cost)$ are three different kinds of weighting factors. This weighting method aims at balancing different types of virtual resources.

Thirdly, it is the constraints that must be fulfilled, while the service provider tries to accommodate the ith SFC^V_i, ranging from Expressions 8.2 to 8.8.

$$\forall VNF^V_{ij} \in VNFs^V_i, \quad \sum_{PNode} X^{VNF^V_{ij}}_{M^P} = 1 \tag{8.2}$$

where Expression 8.2 aims at ensuring each virtual NF in the virtual SFC embedded onto only one physical node.

$$\forall M^P \in PNode, \quad \sum_{VNFs^V_i} X^{VNF^V_{ij}}_{M^P} \leq 1 \tag{8.3}$$

where Expression 8.3 aims at ensuring that at least one virtual NF is embedded onto one corresponding physical node while doing the SFC embedding and scheduling. In both Expressions 8.22 and 8.3, the relationship between each virtual NF and each physical node can be achieved.

$$\forall VNF^V_{ij}VNF^V_{i(j+1)}, \exists path^P \in Path^P,$$
$$\sum_{path^P} \left(Y^{VNF^V_{ij}VNF^V_{i(j+1)}}_{path^P} \right) = 1 \tag{8.4}$$

where Expression 8.4 indicates that there must exist one isolated physical path $path^P$ for accommodating the virtual link $VNF^V_{ij}VNF^V_{i(j+1)}$. Take note that two end nodes (NFs) of virtual link $VNF^V_{ij}VNF^V_{i(j+1)}$ are embedded onto corresponding physical nodes ahead.

$$\forall VNF^V_{ij} \in VNFs^V_i, \quad \sum_{M^P} X^{VNF^V_{ij}}_{M^P} \cdot Computing(VNF^V_{ij})$$
$$\leq Computing(M^P) \tag{8.5}$$

$$\forall VNF^V_{ij} \in VNFs^V_i, \quad \sum_{M^P} X^{VNF^V_{ij}}_{M^P} \cdot Storage(VNF^V_{ij})$$
$$\leq Storage(M^P) \tag{8.6}$$

$$\forall VNF_{ij}^{V} \in VNFs_{i}^{V}, \quad \sum_{M^{P}} X_{M^{P}}^{VNF_{ij}^{V}} \cdot Deploy\left(M^{P}\right)$$

$$\leq MaxDeploy\left(VNF_{ij}^{V}\right) \tag{8.7}$$

where Expressions 8.5, 8.6, and 8.7 indicate the VNF constraint (e.g., VNF_{ij}^{V}). When certain physical node accommodates the VNF, the physical node must reserve enough computing and storage resources for accommodating the VNF. In addition, its deployment time must be within the maximum allowed deployment time of VNF_{ij}^{V}.

$$\forall VNF_{ij}^{V} VNF_{i(j+1)}^{V} \in VLinks_{i}^{V}, \exists path^{P} \in Path^{P},$$

$$\sum_{path^{P}} Y_{path^{P}}^{VNF_{ij}^{V} VNF_{i(j+1)}^{V}} \cdot Band\left(VNF_{ij}^{V} VNF_{i(j+1)}^{V}\right)$$

$$\leq Band\left(path^{P}\right) \tag{8.8}$$

where Expression 8.8 indicates that the accommodated physical path $path^{P}$ must have available bandwidth resource for accommodating the embedded virtual link $VNF_{ij}^{V} VNF_{i(j+1)}^{V}$.

After formulating the integer linear programming model of *SFC-Optimal*, service provider can use the professional optimization software tool, such as GPLK [23], to achieve the optimal SFC embedding and scheduling of SFC_{i}^{V}. However, the fact that the programming model method has the flaw of high time complexity. The complexity of directly solving ILP model is usually approaching the exponential level [24]. When the network scale is large, no matter physical or virtual, the number of binary variables increases, thus further increasing the time complexity. That is why the heuristics and the meta-heuristics [25, 26] are fully developed in SFC research.

References

1. Giordani, M., Polese, M., Mezzavilla, M., Rangan, S., & Zorzi, M. (2020). Toward 6G networks: Use cases and technologies. *IEEE Communications Magazine, 58*(3), 55–61 (2020).
2. *ETSI GS NFV 002 V1.2.1:Network functions virtualisation (NFV)*; Architectural framework (2014).
3. *Cisco Visual Networking Index: Global Mobile Data Traffic Forecast Update, 2015–2020, White Paper*, Cisco (2016).
4. You, X., Wang, C., Huang, J., Gao, X., Zhang, Z., Wang, M., Huang, Y., Zhang, C., Jiang, Y., Wang, J., Zhu, M., Sheng, B., Wang, D., Pan, Z., Zhu, P., Yang, Y., Liu, Z., Zhang, P., Tao, X.,

et al. (2021). Towards 6G wireless communication networks: vision, enabling technologies, and new paradigm shifts. *Science China Information Sciences, 64*(1), 110301 (2021).
5. Mijumbi, R., Serrat, J., Gorricho, J., Bouten, N., Turck, F., & Boutaba, R. (2016). Network function virtualization: State-of-the-art and research challenges. *IEEE Communications Surveys and Tutorials, 18*(1), 236–262.
6. Herrera, J. G., & Botero, J. F. (2016). Resource allocation in NFV: A comprehensive survey. *IEEE Transactions on Network and Service Management, 13*(3), 518–532.
7. Mirjalily, G., & Luo, Z. (2018). Optimal network function virtualization and service function chaining: A survey. *Chinese Journal of Electronics, 27*(4), 704–717.
8. Cao, H., Wu, S., Hu, Y., Li, Y., & Yang, L. (2019). A survey of embedding algorithms for virtual network embedding. *China Communications, 16*(12), 1–33 (2019).
9. Davalos, E. J., & Baran, B. (2018). A survey on algorithmic aspects of virtual optical network embedding for cloud networks. *IEEE Access, 6*(1), 20893–20906.
10. Yu, M., Yi, Y., Rexford, J., & Chiang, M. (2008). Rethinking virtual network embedding: Substrate support for path splitting and migration. *SIGCOMM Computer Communication Review, 38*(2), 17–29.
11. Cao, H., Yang, L., & Zhu, H. (2018). Novel node-ranking approach and multiple topology attributes-based embedding algorithm for single-domain virtual network embedding. *IEEE Internet of Things Journal, 5*(1), 108–120.
12. Cao, H., Wu, S., Hu, Y., Mann, R., Liu, Y., Yang, L., & Zhu, H. (2020). An efficient energy cost and mapping revenue strategy for inter-domain NFV-enabled networks. *IEEE Internet of Things Journal, 7*(7), 5723–5736.
13. Huu, T., Mohan, P., & Gurusamy, M. (2019). Service chain embedding for diversified 5G slices with virtual network fu6ction slicing. *IEEE Communications Letters, 23*(5), 826–829.
14. Zhong, X., Wang, Y., & Qiu, X. (2018). Service function chain orchestration across multiple clouds. *China Communications, 15*(10), 99–116.
15. Liu, J., Lu, W., Zhou, F., Lu, P., & Zhu, Z. (2017). On dynamic service function chain deployment and readjustment. *IEEE Transactions on Network and Service Management, 14*(3), 543–553.
16. Fu, X., Yu, F., Wang, J., Qi, Q., & Liao, J. (2019). Service function chain embedding for NFV-enabled IoT based on deep reinforcement learning. *IEEE Communications Magazine, 57*(9), 102–108.
17. Fu, X., Yu, F. R., Wang, J., Qi, Q., & Liao, J. (2020). Dynamic service function chain embedding for NFV-enabled IoT: A deep reinforcement learning approach. *IEEE Transactions on Wireless Communications, 19*(1), 507–519.
18. Liu, S., Cai, Z., Xu, H., & Xu, M. (2015). Towards security-aware virtual network embedding. *Computer Networks, 91*(11), 151–163.
19. Besiktas, C., Gozupek, D., Ulas, A., & Lokman, E. (2017). Secure virtual network embedding with flexible bandwidth-based revenue maximization. *Computer Networks, 93*(1), 89–98.
20. Newman, M. (2010). *Networks: An introduction*. Oxford, UK: Oxford University Press.
21. Cormen, T. H., Stein, C., Rivest, R., & Leiserson, C. (2001). *Introduction to algorithms* (2nd ed). McGraw-Hill Higher Education.
22. Cao, H., Hu, Y., & Yang, L. (2021). Towards intelligent virtual resource allocation in UAVs-assisted 5G networks. *Computer Networks, 185*, 107660.
23. *GLPK [EB/OL]*. [2021-03-08]. Available: http://www.gnu.org/software/glpk/.
24. Cao, H., Hu, S., & Yang, L. (2016). New functions added to ALEVIN for evaluating virtual network embedding. In *2016 IEEE International Conference on Computer and Communications* (pp. 2411–2414).
25. *IBM ILOG Optimization Products [EB/OL]*. [2021-03-08]. Available: www=01.ibm.com/software/websphere/products/optimization
26. Cao, H., Du, J., Zhao, H., Luo, D., Kumar, N., Yang, L., & Yu, F. (2021). Resource-ability assisted service function chain embedding and scheduling for 6G networks with virtualization. *IEEE Transactions on Vehicular Technology, 70*(99), 1–14.

Part III
Application of Software-Defined Networking in Cloud Computing

Chapter 9
Prospective on Technical Considerations for Edge–Cloud Cooperation Using Software-Defined Networking

Amritpal Singh, Rasmeet Singh Bali, and Gagangeet Singh Aujla

9.1 Introduction

In our daily life, many devices are connected wirelessly with the other devices using radio frequency identification (RFID), Electronic Product Code (EPC), Global Positioning System (GPS) and mobile devices using sensors, embedded devices, and cloud. It has been estimated that by 2025, 75 billion devices will be connected to the web to access the services [64]. This inflation in connected devices introduced one technology, known as Internet of Things (IoT) [48]. According to a survey, 93% of organizations adopted IoT platform, 80% of companies moved toward IoT technology, and 90% of vehicles connected with the IoT platform in 2020 [64]. IoT can be integrated with a different application as shown in Fig. 9.1. The different versions of IoT are discussed below:

- ***Consumer IoT:*** In this, numerous devices are connected to the Internet, like home assistant appliances and voice recognition devices.
- ***Commercial IoT:*** Different applications including healthcare services and vehicle-to-vehicle communication services are considered under this category.
- ***Industrial Internet of Things (IIoT):*** It considered smart agriculture appliances and industrial big data services.
- ***Infrastructure IoT:*** It included smart cities sensors and user-friendly applications.

A. Singh · R. S. Bali
Department of Computer Science and Engineering, Chandigarh University, Mohali, India
e-mail: amritpal.cse@cumail.in

G. S. Aujla (✉)
Department of Computer Science, Durham University, Durham, UK

G. S. Aujla et al. (eds.), *Software Defined Internet of Everything*, Technology, Communications and Computing, https://doi.org/10.1007/978-3-030-89328-6_9

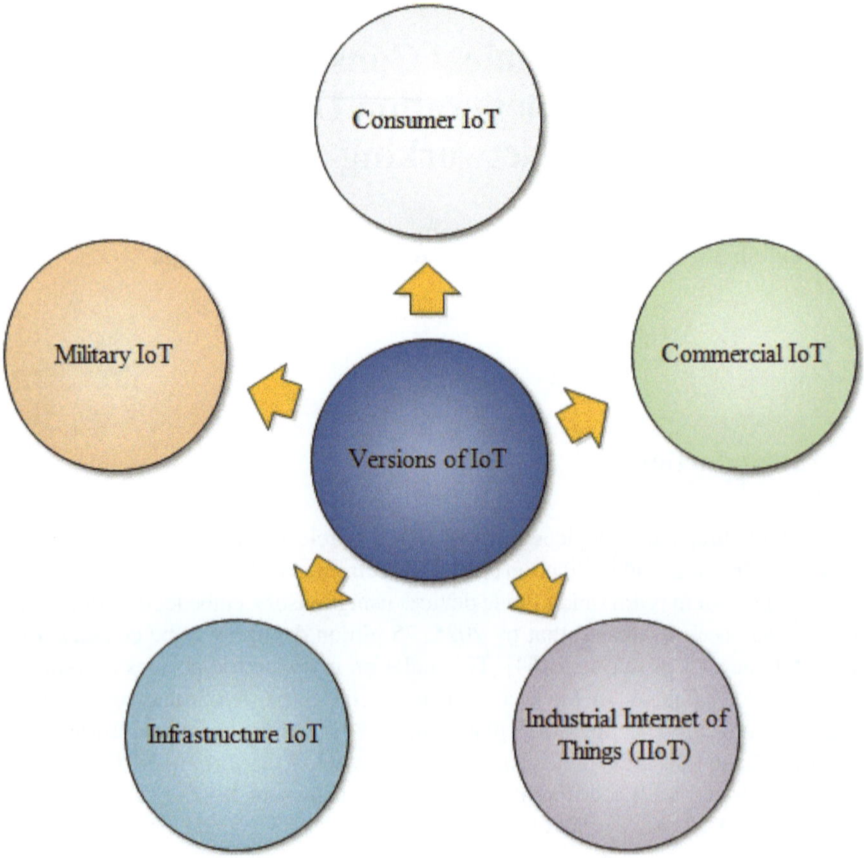

Fig. 9.1 Different versions of IoT

IoT contribution can be analyzed in different eras in daily routine, like ambient intelligence, smart homes, and smart cites. The major issue during configuration of IoT devices is data privacy and security. In this concern, many infrastructure organizations, like CISCO, AT&T, IBM, Intel, etc., introduced new intelligent devices to handle the data. The most commonly used architectures based on different operations in IoT are mentioned below [53]:

- *Event-Based Architecture:* The operational data are collected and analyzed for particular applications in this category.
- *Time-Based Architecture:* In this category, the data are collected in a specified interval of time from certain devices.

In our daily routine, IoT platform is used for regularly used applications for specific tasks. A scenario integrating various applications is shown in Fig. 9.2, where different sensors are configured in the smart city to collect the data. The collected

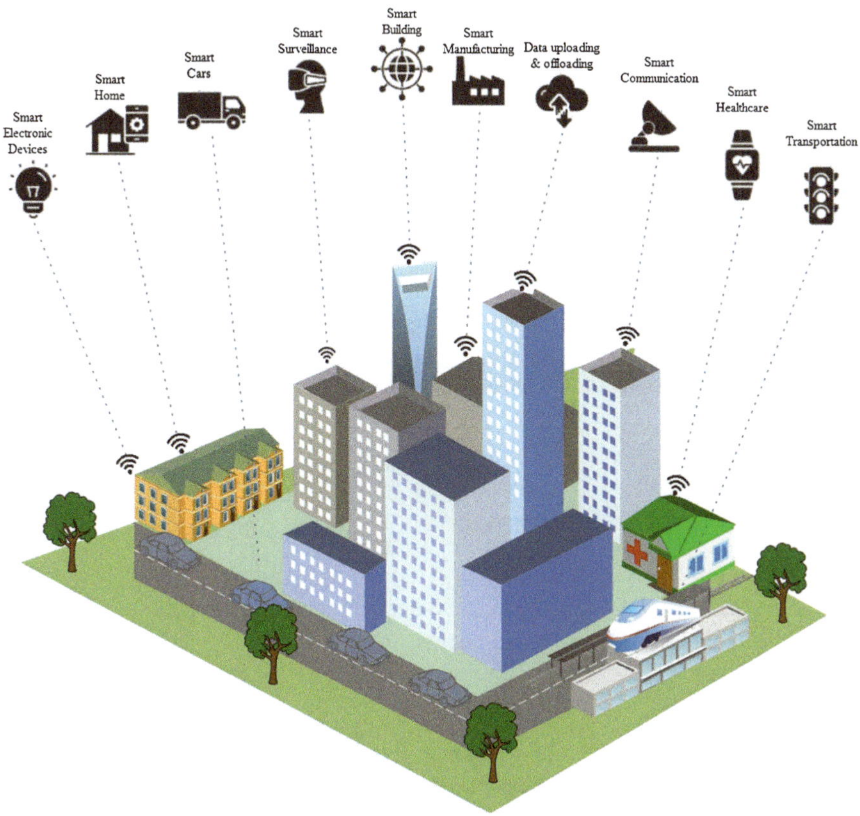

Fig. 9.2 Smart city

data are analyzed and further used to administer the city services and resources [9]. In the similar manner, the data from people, buildings, and other devices are analyzed to manage the transportation system, hospitals, smart organizations, agriculture, banking, libraries, and other services. The defined concept helps to improve the efficiency of the operations related to the citizens of the city. The generated information is used by the administrator officials to communicate with the citizens of the smart city to manage the effective usage of the infrastructure and services.

According to a survey [22], 90% of organizations are using cloud services for various routine activities. Cloud computing provides hardware- and software-based services to the end users through Internet. The services offered by the cloud service provider are highlighted below:

- *Software-as-a-Service (SAAS):* Instead to install software on your personal systems, the required software can be used by the client on the basis of pay-per-use module.

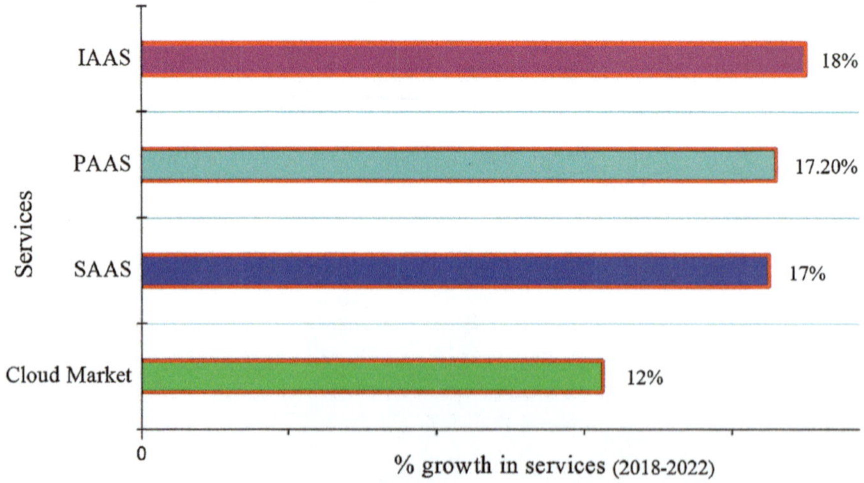

Fig. 9.3 Percentage increase in cloud services

- *Infrastructure-as-a-Service (IAAS):* On-demand storage and network usage are provided in this service to the end clients.
- *Platform-as-a-Service (PAAS):* Provide platform to run the applications as per requirement.

During this pandemic period, the increase in cloud services by various organizations is shown in Fig. 9.3. The improved flexibility and reliability in the cloud services increase the popularity in organizations and academia. However, with the increase in demand of IoT sensors, there is an abrupt inflation in the usage of cloud services. With the increase in the workload on the cloud, the different challenges that need to be addressed are mentioned below:

- *Lack of resources:* Organizations rely on cloud resources for storage and even computational power to process their workloads. This leads to an increase in the load on cloud resources that may further lead to a scarcity of the required resources.
- *Latency:* Cloud servers are deployed at a remote location at far-off locations across the globe. So, the round trip delay between the end users and the remote cloud services often leads to degradation in the performance.
- *Degradation in QoS:* With the increase in the latency, the agreement of the service provider with the user to provide continuous service hampers.
- *Interoperability:* In case of cloud computing, the application running on one platform can incorporate the services from another platform using web services only. However, developing such type of platform is very complex and cannot handle the interoperability of the services.

- *Computing performance:* Very high bandwidth is required to process the requests of the users. However, with an increase in the workload, the bandwidth is divided resulting into degradation in the computing performance.

9.2 Edge Computing

The IoT devices collect the data at local networks, and after monitoring the data, again, it is required at the same local network to make further decisions. However, it is not efficient to send the data to the centralized servers to process the data, as it increases the latency in the network. A solution is provided in the form of "*Edge Computing*" to process the data near the ground rather to forward it to the located servers [6]. Edge devices are located closer to the location of the IoT devices in contrast to the cloud data centers (located at remote locations) in order to provide improved availability and scalability [13, 55]. The evolution of the computing architecture is shown in Fig. 9.4.

There are applications in the IoT platform that require quick response from the data centers, and the other application may require to upload large workload on the network to process the task. Cloud computing is not a compatible solution for certain scenarios. Thus, real-time applications (like smart vehicles, intelligent transportation systems, etc.) require a suitable platform (like edge computing) that is delay sensitive in nature. Therefore, the cloud computing environment does not provide required QoS to such applications. The basic architecture of

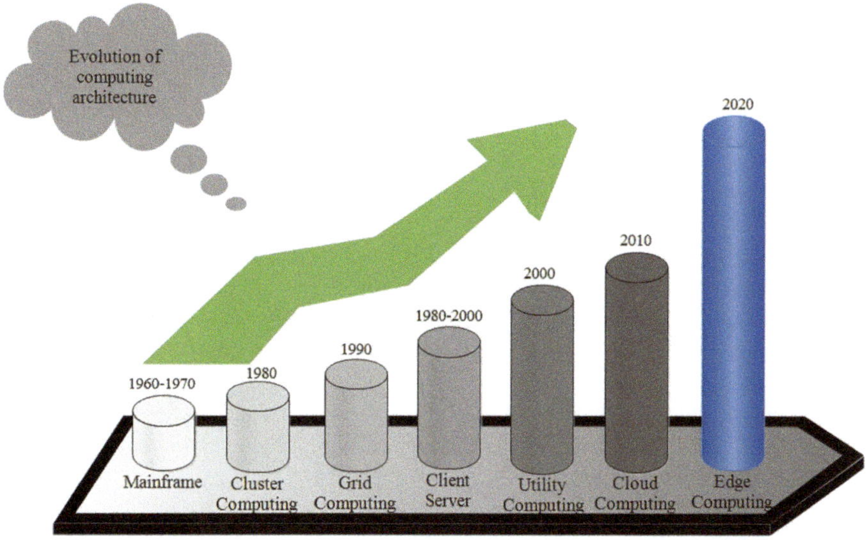

Fig. 9.4 Evolution of computing architecture

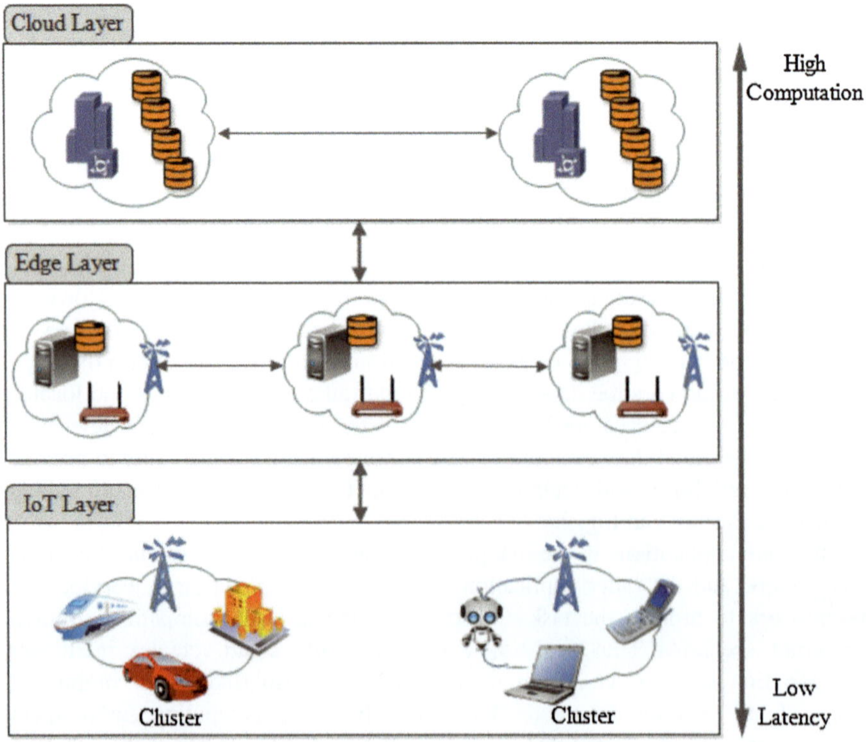

Fig. 9.5 Edge computing architecture in IoT

the edge computing considering IoT platform is shown in Fig. 9.5. The device having a sufficient computational power and required storage space, like mobiles, routers, gateway, etc., can be considered as edge device. The benefits of using edge computing platform as compared to cloud computing are mentioned below:

- *Better computing and network communication:* As the computation devices are near to the IoT devices, improve the response time and minimize the latency in the network. The approach is more effective in a scenario considering large workload to process.
- *Improved QoS:* As per the priority or sensitivity of the task, it can be processed either on edge devices or on cloud platform to maintain the QoS between service provider and end user.
- *Interoperability between modern devices and end users:* It is easier to switch the application from one platform to another using edge devices or merge another platform to the executing task to maintain the service-level agreement (SLA).
- *More reliable approach:* The edge computing platform is more reliable even with low bandwidth and during migration of services from one edge device to another.

Cloud computing is an efficient approach to process the data having huge computing processing power. However, the QoS is not maintained by the cloud data centers and does not meet the required standards in terms of latency and breakage of Internet connection due to requirement of high bandwidth. Edge computing is the solution of all the challenges, as it processes the task near to the end users [11]. Due to close in the proximity of the edge devices, it reduces the level of bandwidth requirement to process the task. Firstly, a task is recommended to the edge devices for processing; if the required resources are not available, then it is forwarded to the cloud platform. Before forwarding the workload to the cloud, a preprocessing is done at the edge devices to improve the quality and reduce the overall response time of the task. It is recommended to compress the workload on the edge platform before forwarding to the cloud to reduce the size of the workload, resulting into reduction in the level of bandwidth to upload the data to the destination servers. If the data are analyzed at the edge level, only the required information is further passed to the cloud to improve the efficiency of task. The architecture of edge computing considering IoT devices and cloud platform is shown in Fig. 9.5. It is clear from the figure that latency is reduced with the closeness of the edge devices to the IoT devices and computation power is increased as the request is forwarded to the cloud platform. For the clarity of the concept, the intercommunication between different layers is highlighted in the figure.

9.2.1 Technical Challenges for Edge Computing

In this platform, data from various end users are forwarded to the layer with availability of the resources for processing. The generic scheme is not efficient to handle all the heterogeneous requests from the end users, and it is very difficult to design an adoptable approach for efficient data handling. Following are the various challenges of the edge computing:

- ***Generic purpose computing:*** As the resources at the edge level are very restricted, therefore, computing is very generic at the platform. The platform is not efficient to handle heavy workload and only specific applications are processed.
- ***Discovery of edge nodes:*** The location of the edge devices is not static, every time there is a need to discover the edge devices for processing the tasks. To discover the edge devices, additional energy and time are required for the same.
- ***Partitioning and task offloading:*** The resources on the edge devices are restricted, and therefore, it is required to partition the incoming task into different available edge nodes for processing. Data offloading from one edge node to another is not an energy-efficient approach. Additional time is required during migrating the data from one edge node to another.
- ***Security measure:*** The selected edge nodes are selected randomly in the generic scenario, and security is the major concern in this approach. As authentication of

the edge nodes still requires more energy and time, therefore, it is not a viable approach to implement in the proposed platform.

- ***Network congestion:*** The escalation in the number of IoT devices deployed across different applications overloads the network resources and thus leads towards the congestion in the network. A controller is required to handle the congestion in the network for smooth processing of the allocated tasks.

9.3 Software-Defined Networking

In order to handle the defined computing scenario, a solution is required to dominate all the mentioned challenges of the edge computing. *Software-defined networking (SDN)* is a required solution for programmable network deployment and an efficient network management [7, 12]. Due to an increase in the user demand and use of IoT devices (in smart cities, smart transportation system, smart medical system, smart buildings, etc.), the need of an intelligent network management approach becomes essential to maintain the QoS committed to the end users. SDN provides a programmable interface to manage the communication between the deployed controller and configured devices on data plane. By using this interface, the available resources are proactively managed and can be extended by controller as per the requirement [70]. The SDN platform comprises different planes, which helps the service provider to handle and upgrade the configuration of the resources dynamically as per the requirement. The SDN controllers are configured centrally, all the available resources at the specific network are accessed, and accordingly, the network traffic can be managed. The centralized management approach of SDN can decrease the implementation complexities of the edge computing platform and improve the resource management scheme.

If at a certain instance, the requested resources are not available at the edge nodes, then, the workload needs to be shifted toward cloud platform for processing. The edge device can forward the workload from the same port, resulting into an increase in congestion in the network. Therefore, an intelligent device is required to control the traffic on the network. SDN directs the traffic to choose the suitable port for workload forwarding to reach the destination device. To make communication with the available devices in the network, OpenFlow switch protocol is used by the SDN controller [47].

9.3.1 Architecture

The standard architecture of the SDN platform includes three basic layers, i.e., data, control, and application layers. The detailed discussion on each layer depicted in Fig. 9.6 is mentioned below:

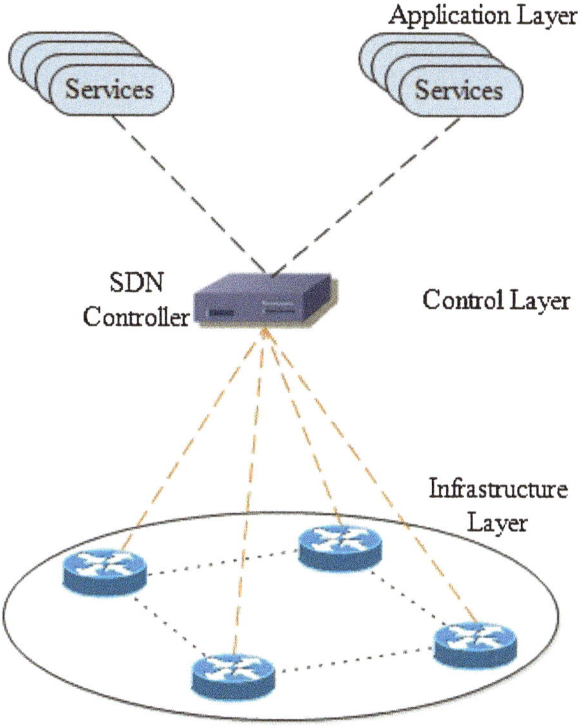

Fig. 9.6 SDN layer-wise architecture

- ***Infrastructure Layer:*** The lowest layer in the architecture of SDN contains various infrastructural devices to build a network, mainly switching devices (like routers and switches). To connect the various infrastructural devices, different transmission media are used as per the requirement of the organizations.
- ***Control Layer:*** This is the middle layer that provides communication between the infrastructure layer and the application layer. The centralized controller is considered as the brain of the SDN-enabled network that forwards the traffic intelligently in the network.
- ***Application layer:*** This layer is the upper most layer, and the applications are privileged to access global network view of the network and direct the changes in the policies in the network to the SDN-enabled layer.

The incoming traffic is managed intelligently by the centralized controller by configuring defined policies as per the end user. The following installation modes are used in the controller:

- ***Proactive Mode:*** In this mode, the flow policies are configured on the available switches in the data plane layer to manage the incoming traffic/packets in the network. It makes the network to direct the incoming traffic without the involvement of the controller.

- ***Reactive Mode:*** In this mode, if a new packet arrives in the network, firstly, it matches with the flow tables to forward it in a proper channel; otherwise, *PACKET-IN* message is forwarded to the controller to make a new flow entry for a particular packet, and *PACKET-OUT* message is updated on each data plane switch flow table for further use.
- ***Hybrid Mode:*** This mode is the combination of the Proactive mode and Reactive mode to make the network more intelligent and flexible.

In case of SDN, OpenFlow switches are used for forwarding the traffic from source to destination. It comprises three main components, including flow tables, OpenFlow protocol to set policies, and secure communication channel. Flow tables keep the policies, and accordingly the incoming traffic is directed. They are updated on each OpenFlow switch to improve the efficiency of the configured topology. In flow table, three attribute values are stored, like "*Rule*," "*Action*," and "*Status*." The header information of each packet is defined in the *Rule* attribute, the directions to forward the certain traffic are defined in the *Action* attribute, and the current status of the traffic is defined in the *Status* attribute.

9.4 SDN-Edge Cooperation

The problems identified in the cloud platform related to the response time and degradation in QoS, and the problems in the edge computing platform can be solved by using a centralized controlling mechanism in the distributed environment. The barriers in the performance of the edge computing platform can be removed by using the programmable interface provided by the SDN in the network. The SDN interface is transparent to the end user and can manage the data flow, service orchestration, and other mandatory services.

9.4.1 Architecture

The architecture of the SDN-IoT-Edge-enabled is highlighted in Fig. 9.7. The architecture comprises of different layers, like *Infrastructure Layer*, *Edge Layer*, *Control Layer*, and finally *Application Layer*. At *Infrastructure Layer*, various IoT devices request for the different services as per the behavior of the task. The requested services list is forwarded to the edge layer for offloading, and further this layer is managed by the controller configured at control layer to handle the network traffic efficiently. The motivation of the SDN-IoT-Edge-enabled architecture is from the limited resource constraints at the IoT devices. An inflation in the number of the IoT devices has been observed, and therefore, an intelligent and manageable platform is required to handle the network traffic more efficiently from source to destination.

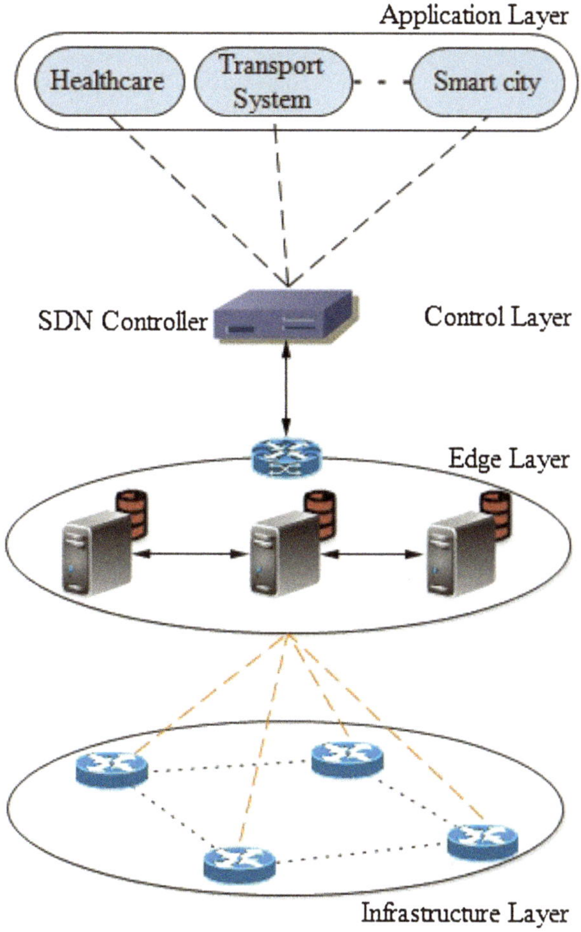

Fig. 9.7 SDN-IoT-Edge architecture

9.4.2 A Real Scenario of SDN-IoT-Edge Architecture

Let us consider an example of SDN-enabled edge computing system in daily life scenario. The use of SDN-enabled network in various fields is discussed with prospective to edge computing platform. A complete model with all the basic components of the SDN-enabled network is highlighted in Fig. 9.8. In the defined model, cloudlets are used in the edge computing technology. Cloudlets are defined as small-scale mobility-enhanced data centers available at the edge devices. These cloudlets are able to process the services requested by the end users that are close to the ground as compared to the cloud data centers. The cloudlets are distributed at various locations and are connected through the OpenFlow-enabled switches to provide services. The configured switches are controlled by the centralized controller (SDN) to provide communication in the network protocols like OF-

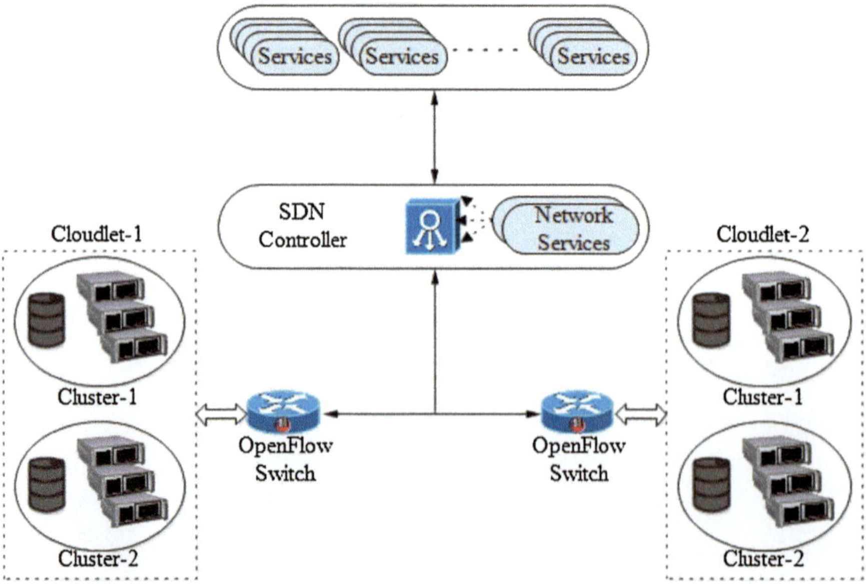

Fig. 9.8 SDN-IoT-Edge example

Config [52]. The controller receives the requests from the end users, and according to the defined policies the traffic is managed. The services at the northbound provide the feedback periodically to the controller after analyzing the behavior of the incoming traffic.

9.4.3 Benefits of Using SDN-IoT-Edge Platform

An inflation in the network traffic can be observed with the increase in the number of IoT devices and requests from the end users. The challenge of constraint resources can be managed by using the edge-enabled platform to provide the services at the edge of the Internet and by integrating the centralized SDN controller to manage the network more intelligently. Various components in the SDN-IoT-Edge platform have their own benefits in the network. A proper discussion on the requirements and benefits is discussed below:

- ***Management of IoT-Edge platform using SDN:*** The management of network traffic is one of the core responsibilities of the service provider to maintain the QoS [15]. The intelligent devices are required to forward, manage, and control the incoming traffic in the network to overcome the network delay. There might be a case where load balancing in the network is at higher stake, and therefore, an intelligent load balancing device is required. The solution of the abovementioned issues is provided by the SDN controller to control the IoT-Edge platform.

- ***IoT-Edge authentication:*** With the increase of workload, a concept is required to segregate the workload on different machines, and inter-operability is required to make communication among all the allocated machines. While allocating the workload on various machines, it is required to authenticate the systems in the edge environment. An authenticated and trusted approach is the demand of the network to guarantee the genuineness of the connected machines to provide services to the end user. SDN provides an interface to manage the network and the connected devices to ensure the authenticity centrally.
- ***Inter-operability among heterogeneous platform:*** With the potential increase in the size of the organizations, the infrastructure cost also inflated. To optimize the cost, inter-operability among the configured systems also escalated. While managing the dependency among various systems, it raised the complexity in the infrastructure. SDN is the one solution to reduce the complexity in the infrastructure by providing the centralized infrastructure with defined policies as per the end user requirements.
- ***Workload dispersion:*** In standard networks, the requests of the end users are forwarded to the cloud data centers to process the tasks, and the result is returned to the respective user. This mentioned scenario, resulting into the elephant-like traffic in the network. To reduce the network congestion, initially workload is preprocessed at the edge devices. There is a requirement of data dissemination to reduce the energy consumption and bandwidth consumption to process the workload. SDN has the report of all connected devices in the network, and by using this, the raised workload can be dispersed among the available devices in the network [7].
- ***Latency minimization:*** A different nature of workload is generated from the data plane (like real-time data, video streaming, etc.), and they need response without any delay in the response time. Again, security is the major issue during transmission of data from source to destination, and it increases the level of complexity in the infrastructure, resulting into an increase in latency in the network. Applications, like self-driven cars, smart organizations, and real-time applications, require services with no delay. Therefore, a fault-tolerant infrastructure is required to reduce the overhead in the network.
- ***Flexibility in SDN-IoT-Edge platform:*** The standard networks are not able to configure virtualization, and therefore, integration of new technologies is not possible. The network programmable property nature of the SDN enhances the dynamic configuration of the services on the distributed devices.
- ***Preprocessing on edge devices:*** The workload is increasing with the configuration of smart devices in different fields. This bulk amount of workload creates overhead and congestion in the network. Therefore, a layer is required to filter the workload to reduce the level of energy utilization and latency in the network. The workload generated at data plane is preprocessed at the edge layer before forwarding to the cloud data centers.

9.5 Technical Requirements of SDN-Edge vs Energy Efficiency

A constant communication is required between the IoT devices and edge platform to manage the incoming traffic in the network [9, 10]. Following are the services that are required by the SDN-IoT-edge architecture to maintain the QoS:

- ***Discovery of services:*** The requested services have different nature, and accordingly there is a need to discover the data centers at different locations to provide the required resources. SDN handles the network centrally, and it has the details of all the available resources in the network. Therefore, SDN can help to identify the required resources to process the service by the end users.
- ***Service and data migration:*** There might be limited resources at various cloudlets, and therefore, there is a need to migrate the services or data to the different cloudlets to maintain the QoS. The defined model decides the migrated cloudlets as per the requested services and locations.
- ***Resource provision:*** With the enhancement in the technology and resource constraint at IoT devices, a platform is required to provide the resources at all times. Edge devices provide the service near to the end users to improve the efficiency of the configured network. Cloud platform is an abundance of resources, and applications that can tolerate latency can request the resources from the cloud directly.
- ***Lightweight algorithms for workload processing:*** The lightweight algorithms are required to process the incoming workload to improve the throughput of the platform with a limited number of resources. As it is already clear that IoT devices and edge platform are short of resources, an approach is required to optimize the utilization of the resources to provide QoS to the end user.
- ***Energy-efficient platform:*** With the inflation in workload, more filtration and preprocessing are required at the edge level to improve the efficiency of the task. The level of energy consumption is also increased at a great extent, and therefore an energy-efficient approach is required to optimize the use of resources.
- ***Pay-per-use module for edge devices:*** As cloud services work with pay-per-use module, similarly, standards are required to build a model for edge devices. Applications that are sensitive with the latency can avail the services from the edge devices; however, no policies are defined for the payment of the used resources at edge nodes.
- ***One-point failure:*** The benefits of the SDN-enabled network are already discussed. However, the key challenge of the SDN-IoT-Edge platform is one-point failure, which can degrade the performance of the configured network. To overcome this challenge, distributed controllers are required rather than centralized ones to build fault-tolerant networks.

9.6 Stat of the Art

In the last few years, the authors proposed numerous load balancing approaches to minimize the overhead on the DCs. In the current scenario, the major focus of the service providers, like Facebook, Google, Amazon, Apple, and Microsoft, concentrates toward the proposal of energy efficient DCs [1]. As per global survey report [25], the major contribution for energy consumption in the DCs is cooling and power provision servers. It can be noticed from the energy consumption breakup that 43% is the energy consumption during power supply to the servers and cooling the servers as shown in Fig. 9.9 [20]. The consumption of energy by the servers is minimal when compared with cooling and powering the systems, i.e., 11%, and a minute energy is used by the network for communication among the end devices.

In the same prospective, Fig. 9.10 highlights the classification of various existing approaches for energy management of DCs. On the basis of the abovementioned taxonomy, the reviews and discussions based on the proposed research are illustrated in the below sections.

9.6.1 Cloud Computing

The main contribution to design an optimal solution for energy management of DCs is to minimize the energy consumption. The key factors to reduce the energy consumption by distributed DCs are analyzed first. In this direction, Dayarathna et al. [20] recommended that energy modeling is a major parameter to develop energy-efficient DCs. The authors studied numerous approaches related to the energy efficiency of DCs. When a thorough study was done on energy consumption, it has been observed that consumption is at different levels (hardware, server, and system). There is a need to explore the levels of the energy consumption to propose

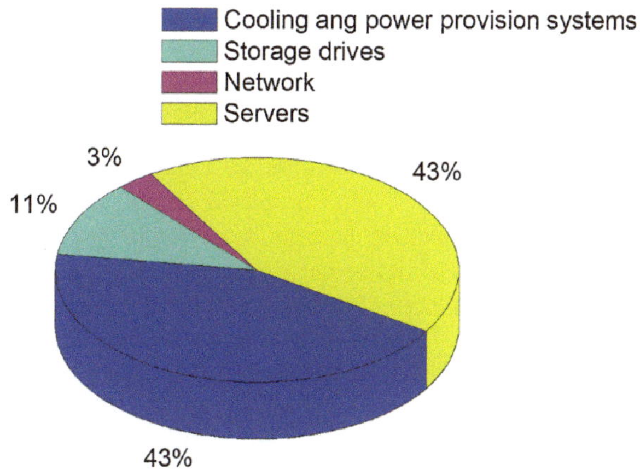

Fig. 9.9 Energy consumption breakup for a particular DC

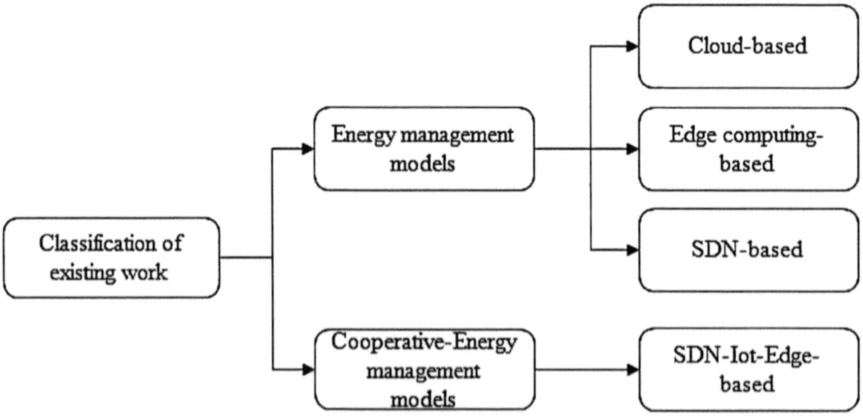

Fig. 9.10 Classification of existing proposals

an optimal solution. The consumption of energy at various levels, like servers, processors, memory, storage, network, and software, is considered below.

The productive work is done by the servers available at various DCs. To process the various tasks, a large amount of energy is consumed by the servers. For example, Roy et al. [57] suggested an efficient model which focused on the components of the server as central processing unit (CPU) and memory. Other authors, Jain et al. [31], discussed a model by bifurcating the total energy consumption into data and instruction of CPU and memory. In a similar manner, Tudor et al. [65] suggested a conflux approach considering the above discussed power models with input/output metrics. In a similar direction, Ge et al. [23] segregated the power model into different components, like CPU, memory, etc., of the system. In the same direction, Warkozek et al. [71] divided the operation of the CPU into two parts, the energy consumption by the CPU for self-running application (virtual machine manager) and the other part to consume the energy for providing services to the end users. The power profile of different tasks was analyzed by Jaintilal et al. [30] to check the dependency of the processor and consumption of energy.

Along with the CPU, the *Memory* is a major participant toward energy consumption from the last few years. According to a survey, *Memory* stands second in terms of energy consumption in different servers [19]. In a similar way, Malladi et al. [46] proposed a power model considering the energy consumption by the memory in a server. In the proposed model, the requested memory stream was considered as a Poison process, and the energy consumption during power on and power down of the server was depleted. Including CPU and memory, *storage systems* contribute a hefty consumption of energy for permanent storage of the data. In storage system, hard disk drive (HDD) is the major contributor of energy consumer [33]. In a similar direction [74], the power consumption during storage on HDD is divided into two classes: (1) static and (2) dynamic. Considering the infrastructure of the network, optical fiber is installed for communication among

Table 9.1 Energy models considering CPU, servers, memory, and storage systems

Authors	Considered area	Description
Ahn et al. [3]	Routers	Energy utilization based on the packet size and type of the link used
Daya et al. [20]	Energy model	Discussed that energy is the major consideration to develop energy-efficient DCs
Heller et al. [24]	Network links	Focused on the rate of energy consumption by a link
Lewis et al. [37]	Memory utilization	The rate of energy consumption is proportional to the number of read/write operations
Mohan et al. [49]	Solid state drive (SSD)	Introduced an analytical energy model, i.e., FlashPower used for two types of memory chip variants named as single-level cell and two-bit multi-level cell
Roy et al. [57]	Server	Highlighted the consumption of energy including CPU and memory
Heddeghem et al. [66]	Optical type network	Considered static and dynamic energy consumption in the network
Vishwanath et al. [68]	Entire network	Energy consumption by all devices configured in the network
Yao et al. [72]	CPU utilization	Energy utilization during performing various operations
Zhang et al. [74]	Hard disk (HDD)	Energy consumption is divided into two classes, i.e., dynamic and static

the servers. The energy consumption by the optical based network is minimal and provides high throughput. In this direction, Heddeghem et al. [66] discussed the energy consumption at different layers of an optical network. Table 9.1 highlights the power models for CPU, servers, memory, storage systems, and network.

9.6.2 *Edge Computing*

With an increase in the number of IoT devices, the workload generated by these devices also increased, thereby increasing the load on the existing resources and infrastructure to a great extent. To forward the generated workload on the designated data centers, the end users have to spend more in terms of energy and latency in the network. Therefore, a terminology is required to optimize the energy consumption and reduce the level of latency in the network. "Edge computing," a new paradigm, is introduced to overcome the said issues to maintain the QoS between the service provider and the end users. A number of researchers are focusing on the collaboration of IoT with edge platform to reduce the level of energy consumption. Figure 9.11 shows the number of publications considering IoT and IoT cooperating with the edge platform.

It can be clearly observed from the figure that the publications cooperating IoT and edge platform are more in trend in the recent years. In a similar direction, a unique bio-inspired clustering algorithm was introduced by Agbehadji et al. [2]

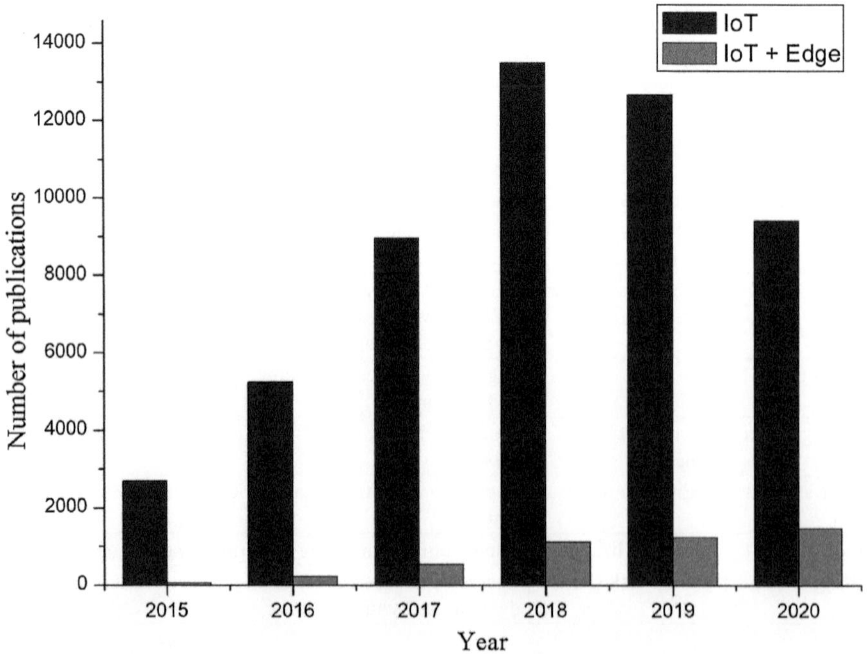

Fig. 9.11 Year-wise publications of IoT and IoT cooperating with edge platform [27]

to reduce the energy consumption in IoT-Edge envisioned platform. A trade-off between the network latency and energy consumption to process the workload was defined by Cui et al. [18]. The authors considered it into a constrained multi-objective optimization problem and introduced an identical solution by using Nondominated Sorting Genetic Algorithm (NSGA-II). The performance of the proposed scheme was enhanced by integrating problem-specific encoding scheme. In the same manner, Zhan et al. [73] studied unmanned aerial vehicle (UAV) energy optimization issue without considering the pre-determined completion time. The authors decoupled the optimization algorithm into two portions using successive convex approximation (SCA)-based algorithms. Another energy saving approach was proposed by Sodhro et al. [62]. The authors considered the execution time of various sensors and time taken to process in IoT devices using forward central dynamic and available approach (FCDAA). A system model was designed to evaluate the energy consumption in IoT devices, and an AI-based edge-IoT platform was introduced to reduce the energy consumption during processing the tasks.

The authors, named Zhang et al. [75], discussed different computation strategies for terminal devices. In the first approach, the authors defined a scenario where a terminal device can compute the tasks by itself. In the second approach, terminal devices can forward the task to the unmanned aerial vehicle for processing, and in the final case, the tasks can be forwarded to the access points using unmanned aerial vehicle by terminal devices. The authors proposed an optimized solution to reduce the overall energy consumption in all the defined scenarios. The window-based rate control algorithm was proposed by Sodhro et al. [61] considering the required network parameters to enhance the medical Quality of Service. The authors considered peak-to-mean ratio, standard deviation, delay, and jitter to check and improve the efficiency of the network. In a similar manner, Sitton et al. [60] discussed the benefits of edge computing in terms of real-time data processing and energy efficiency. The authors introduced Edge-IoT and social computing platform to optimize the energy usage in smart cities and smart devices.

The authors, Li et al. [38], integrated deep learning approach with the IoT-Edge platform to reduce the gap between the multilayered architecture. In the same manner, Liu et al. [42] highlighted the long-term energy efficiency problem using IoT-Edge platform. The authors incorporated reinforcement method to make the IoT-Edge platform more intelligent to manage the incoming workload in the network. The authors, Li et al. [40], discussed unmanned aerial vehicle used as cloudlets to store and process the tasks forwarded by the end users. The energy efficiency was improved by using nonconvex fractional programming, the Dinkelbach algorithm, and the successive convex approximation (SCA) approach. Liu et al. [41] proposed a triple-layer architecture, named edge device plane, edge server plane, and cloud server plane, to optimize the energy consumption in the network. To reduce the complexity due to heterogeneity nature of different devices, tensor-based model was composed. In the similar direction, Chen et al. [17] studied energy-efficient workload offloading scheme in mobile edge computing. The authors introduced stochastic optimization approach and proposed a novel stochastic optimization techniques (EEDOA) to optimize the energy consumption.

Table 9.2 summarizes the existing approaches for IoT-Edge platform.

9.6.3 SDN

The benefits of using IoT-Edge platform are highlighted in the abovementioned section, including reduction in latency, improvement in QoS, etc. The IoT-Edge platform is suitable for small networks, and however, with the increase in the number of devices, the traffic level is also increased, resulting into congestion in the network. Therefore, it becomes difficult to manage the configured infrastructure with the standard policies. Therefore, a platform is required to manage the network traffic intelligently for smooth processing of the incoming workload. To handle all the abovementioned issues, researchers start proposing new techniques using software-defined networking platform. The year-wise consideration of the SDN platform is also highlighted in Fig. 9.12.

In the similar direction, Sezer et al. [58] discussed the importance of energy efficiency and security in the daily networks. The authors used the dynamic network functionality of SDN approach to improve to handle the elephant-like traffic. The authors, named Rawat et al. [56], surveyed the various techniques used to propose an energy-efficient platform in heterogeneous network using

Table 9.2 Comparison analysis of existing approaches for IoT-Edge platform

Authors	1	2	3	4	5	6	7	8	9	10
Agbehadji et al. [2]	Clustering approach was used to optimize the energy consumption	✓	✓	×	×	×	✓	✓	×	×
Cui et al. [18]	Introduced an identical solution for energy optimization using Nondominated Sorting Genetic Algorithm (NSGA-II)	✓	✓	×	×	×	✓	✓	×	×
Zhan et al. [73]	Decoupled the optimization algorithm into two portions using Successive Convex Approximation (SCA)-based algorithms	✓	✓	×	×	✓	✓	×	×	×
Sodhro et al. [62]	AI-based edge-IoT platform was introduced to reduce the energy consumption during processing the tasks	✓	✓	×	×	✓	×	×	×	×
Li et al. [38]	Integrated deep learning approach with the IoT-Edge platform to reduce the gap between the multilayered architecture	✓	✓	×	×	×	×	×	×	✓
Zhang et al. [75]	An optimized solution to reduce the overall energy consumption in all the available scenarios	✓	✓	×	×	✓	✓	×	×	×
Liu et al. [42]	Incorporated reinforcement method to make the IoT-Edge platform more intelligent to manage the incoming workload in the network	✓	✓	×	✓	✓	×	×	×	✓
Sodhro et al. [61]	Considering the required network parameters to enhance the medical Quality of Service	✓	✓	×	✓	×	✓	×	×	×
Sitton et al. [60]	Introduced edge-IoT and social computing platform to optimize the energy usage in smart cities and smart devices	✓	✓	×	×	✓	×	×	×	×
Li et al. [40]	Using nonconvex fractional programming, the Dinkelbach algorithm, and successive convex approximation (SCA) approach	✓	✓	×	×	✓	×	×	×	×
Liu et al. [41]	Proposed a triple-layer architecture, named edge device plane, edge server plane, and cloud server plane, to optimize the energy consumption in the network	✓	✓	×	✓	✓	×	×	×	×

1: Description, 2: IoT + Edge, 3: Energy optimization, 4: Storage management, 5: Network management, 6: Workload management, 7: QoS, 8: Latency , 9: Virtualization, 10: Deep learning approach

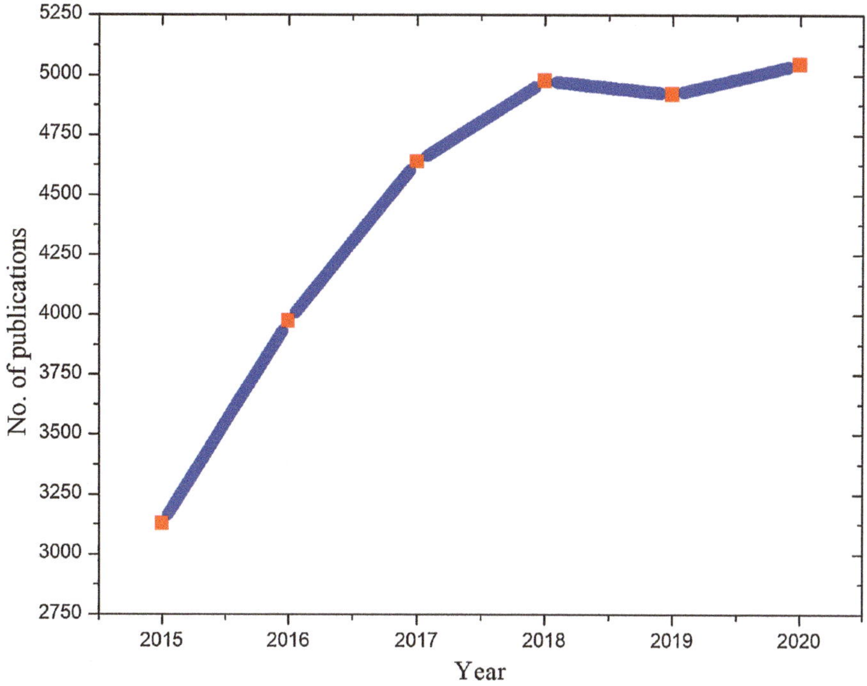

Fig. 9.12 Year-wise publications considering SDN platform [29]

SDN architecture. The authors focused on network security, energy efficiency, and network virtualization concept. In the similar way, Jammal et al. [32] highlighted the benefits of SDN platform in different environments, like data centers, enabled networks, and Network-as-a-Service. The authors also focused on the various challenges using the SDN platform in terms of scalability, reliability, and security. Focusing on the same concept, Zhu et al. [77] highlighted the overflow of energy consumption even when there is no traffic in the network. Therefore, the authors discussed the routing strategies considering energy efficiency as one of the major parameters during processing the tasks. Considering SDN architecture, Hsieh et al. [26] proposed mobile edge computing environment using Container-as-a-Service platform. The proposed IoT gateway reduces the network latency and improved the overall network efficiency. In the similar way, Dorsch et al. [21] combined the SDN platform with the Multi-Agent System to control the energy consumption and handle the overloaded network and voltage stability in the network. The authors configured a control agent to make direct connection with the SDN controller to define the overall demand of the end users. The authors, Ranjan et al. [54], proposed container-based virtualization to develop an energy-efficient scheme for scheduling the resources in data centers. Container-based platform provides an interface to allocate the resources as per the requirement. To access the available servers in the developed infrastructure, doubly linked list was used to enhance the scalability.

In the similar direction, Son et al. [63] studied various techniques of energy optimization in distributed data centers, traffic system, virtualized networks, and security. Ma et al. [45], proposed a load balancing approach in the SDN-enabled environment to control the energy flow in the platform using various planes. The lower plane collects the details of all the available resources and upper layer to optimize the resources, resulting into avoidance of network bottleneck. Moving ahead in the same direction, Zinner et al. [78] introduced a dynamic and requirement-based resource allocation to various applications controlled by SDN-enabled network. Kobo et al. [36] discussed the importance of software-defined wireless sensor network (SDWSN) approach in wireless sensor networks. In this work, the authors highlighted the issues of wireless network with prospective to energy utilization and memory usage to process the workload. The authors proved that the SDN-enabled network can control the wireless network in an efficient manner to reduce the overall cost of the network.

To proceed in similar direction, Buyya et al. [14] defined architectural framework and benefits of programmable behavior of the SDN-enabled network in the distributed environment. In this work, the authors focused on the resource scheduling and provisioning to reduce the overall cost and maintain the service level agreement between service provider and end users. In another work, Morabito et al. [50] integrated two technologies to make the platform more intelligent. The authors considered container-based virtualization and SDN-enabled controller to configure an energy-efficient network. Cardoso et al. [16] proposed a scheme to activate the on-demand resources for end users using containerization controlled by SDN-enabled network. In another work, Zhao et al. [76] proposed a scheme to separate the coupling property in the control plane and data plane layer to control and increase the security during usage of the network resources. In a similar way, Kobo et al. [35] focused on software-defined wireless sensor networks and provide a fragmentation-based control system in distributed environment to control the energy utilization. The authors configured two level controllers, first architecture contained local controller and the other considered global controller to minimize the gap between the resource elements and the controllers. Violettas et al. [67] achieved Routing over Low Power and Lossy Networks (RPL) by using software-defined networking (SDN) to improve the efficiency of the network. The authors focused on routing control strategies for dynamic re-configuration of the network and link-coloring technique for point-to-point communication in mobile networks.

Table 9.3 summarizes the existing approaches for SDN-enabled networks.

9.6.4 SDN-Edge Cooperation

The edge devices are close to the end devices and provide services with an efficient manner and without any delay. According to the available resources at edge platform, the specific application can be processed there only, and there is no need to forward the requests to the centralized static data centers at cloud

Table 9.3 Comparative analysis of existing approaches for SDN-enabled networks

Authors	1	2	3	4	5	6	7	8	9	10	11
Sezer et al. [58]	Discussed the dynamic network functionality of SDN approach to improve the handle the elephant-like traffic	✓	✓	×	×	✓	✓	×	×	×	×
Rawat et al. [56]	Surveyed the various techniques used to propose an energy-efficient platform in heterogeneous network using SDN architecture	✓	✓	✓	✓	×	×	✓	×	×	×
Jammal et al. [32]	Highlighted the benefits of SDN platform in different environments, like data centers, enabled networks, and Network-as-a-Service	✓	✓	✓	✓	✓	×	×	×	×	×
Zhu et al. [77]	Discussed the routing strategies considering energy-efficiency as one of the major parameters during processing the tasks	✓	✓	✓	✓	✓	×	×	×	×	×
Hsieh et al. [26]	Proposed mobile edge computing environment using Container-as-a-Service platform	✓	✓	✓	×	✓	✓	×	×	×	✓
Dorsch et al. [21]	Configured a control agent to make direct connection with the SDN controller to define the overall demand of the end users	✓	✓	✓	✓	✓	×	×	×	✓	×
Ranjan et al. [54]	Proposed container-based virtualization to develop an energy-efficient scheme for scheduling the resources in data centers	✓	✓	✓	×	✓	×	✓	×	×	✓
Son et al. [63]	Studied various techniques of energy optimization in distributed data centers, traffic system, virtualized networks, and security	✓	✓	✓	×	✓	✓	✓	×	×	×
Ma et al. [45]	Proposed a load balancing approach in the SDN-enabled environment to control the energy flow in the platform using various planes	✓	✓	✓	×	✓	×	×	×	×	×

1: Description, 2: SDN, 3: Energy efficiency, 4: Resource utilization, 5: Cost saving, 6: Workload management, 7: QoS, 8: Virtualization, 9: Deep learning approach, 10: Game model, 11: Container-as-a-Service (CoaaS)

layer, resulting into reduction in latency and energy consumption during pushing the workload onto the cloud layer. However, with the scalability of IoT devices, there is an abrupt increase in the traffic and can create congestion in the network. Therefore, a platform is required to control the traffic in an efficient manner to make an acceptable platform for the end users. By using the programmed and centrally controlled interface of the *SDN*, the configured network can work more intelligently and efficiently by optimizing the available resources. To increase the efficiency

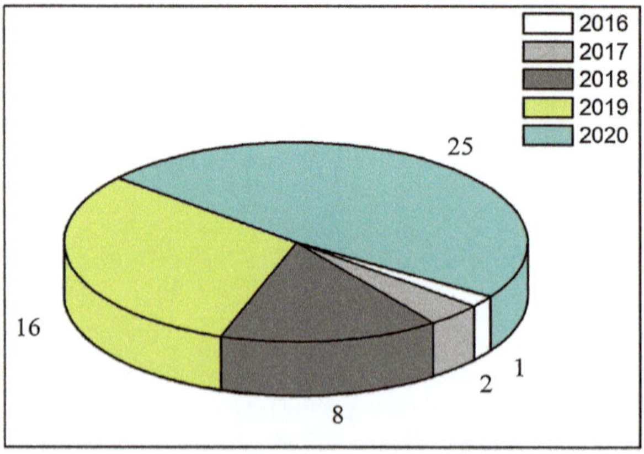

Fig. 9.13 Year-wise publications of SDN-Edge cooperation [28]

and manage the incoming traffic of the network, SDN can be cooperated with edge platform, known as *"SDN-Edge cooperation"* platform. Many researchers start proposing various energy-efficient schemes considering "SDN-Edge" platform, and the count is highlighted in Fig. 9.13.

In the similar direction, Aujla et al. [8] discussed that in edge–cloud environment, there was a massive data migration, resulting into increase in computation cost. To overcome the highlighted issue, the authors proposed an efficient workload slicing scheme to manage the elephant-like data in the network using centralized controller, i.e., *"SDN"*. In this work, the authors handled inter-datacenter data migration using SDN to develop an energy-efficient platform. The authors, Sharma et al. [59], highlighted the issue of intensive real-time data analysis at various data centers. To handle this issue, the authors introduced a SoftEdgeNet model using SDN-enabled network integrated with blockchain concept to improve the security concerns. The authors also proposed an algorithm for data flow and resource allocation for resource management. In another work, Liu et al. [43] introduced orchestrate data as a service approach and to eliminate the data redundancy, and a data aggregation scheme was integrated with the standard methods. To make the network more responsive, the architecture was divided into three layers, namely data center layer, middle routing layer, and vehicle network layer. In the same way, Munoz et al. [51] discussed that collection and storage of large scale of information on the cloud was not a feasible solution to maintain the integrity of the data. Therefore, the authors introduced a scalable and energy-efficient solution by disseminate the data into edge layer and cloud layer. In the work, the authors made efficient and dynamically configuration of the available resources. The authors, Kaur et al. [34], highlighted the congestion problem in the network due to a large number of migrations in the edge and cloud devices. To manage the network traffic, authors introduced SDN-enabled programmable and scalable paradigm. In the work, the authors proposed multi-objective evolutionary algorithm based on Tchebycheff decomposition to manage the data flow in the network. Assefa et al. [5] introduced a novel classification and a comprehensive solution using SDN-enabled network to categorize the incoming

Table 9.4 Comparative analysis of existing approaches for SDN-Edge cooperative network

Authors	Description	1	2	3	4	5	6	7	8	9	10
Aujla et al. [8]	Proposed an efficient workload slicing scheme to manage the elephant-like data in the network using centralized controller	✓	✓	✓	✓	×	×	×	×	✓	✓
Sharma et al. [59]	Introduced a SoftEdgeNet model using SDN-enabled network integrated with blockchain concept to improve the security concerns	✓	✓	✓	✓	×	×	×	×	✓	✓
Liu et al. [43]	Introduced orchestrate data as a service approach and to eliminate the data redundancy	✓	✓	✓	✓	×	×	×	×	×	×
Munoz et al. [51]	Introduced a scalable and energy-efficient solution by disseminate the data into edge layer and cloud layer	✓	✓	✓	✓	×	×	×	×	×	×
Kaur et al. [34]	Highlighted the congestion problem in the network due to a large number of migrations in the edge and cloud devices	✓	✓	✓	✓	×	×	×	×	✓	✓
Assefa et al. [5]	Introduced a novel classification and comprehensive solution using SDN-enabled network to categorize the incoming traffic into different classes	✓	✓	✓	✓	×	✓	×	×	×	✓
Aujla et al. [13]	Proposed EDCSuS: Sustainable EDC as a service framework in SDN-enabled network		✓	✓	✓	✓	×	×	×	✓	✓
Li et al. [39]	Introduced an adaptive transmission model by cooperating SDN with edge envisioned devices in Industrial Internet of Things (IIoT)	✓	✓	✓	✓	×	✓	×	×	×	✓
Alnoman et al. [4]	Introduced an energy-efficient approach by configuring SDN-enabled network to switch ON/OFF the edge nodes as per the requirement of the end users	✓	✓	✓	✓	×	×	×	×	✓	✓
Lv et al. [44]	Proposed mobile edge computing (MEC) framework integrating with SDN and network function virtualization (NFV)	✓	✓	✓	✓	×	×	×	×	✓	✓

1: SDN, 2: Edge computing, 3:Energy efficiency, 4: Resource utilization, 5: Game model, 6: Deep learning approach, 7: Container-as-a-Service (CoaaS), 8: Caching model, 9: SLA, 10: Latency consideration

traffic into different classes. The authors provided an energy-efficient optimization model by using objective function, sensitive parameters for defined models.

In a similar manner, Aujla et al. [13] highlighted the sensitive data requirements like low latency and higher bandwidth to maintain the QoS. To handle the issue, the authors proposed EDCSuS: Sustainable EDC as a service framework in SDN-enabled network. In this work, the authors configured SDN platform for intelligently handling the traffic flow and direct the optimal path for the same. The authors integrated Stackelberg game model for efficient resource allocation to the end users, and finally, a cooperative model was used for resource utilization to improve the efficiency of the proposed scheme. The authors, Li et al. [39], introduced an adaptive transmission model by cooperating SDN with edge envisioned devices in Industrial Internet of Things (IIoT). The authors classified the incoming requests into two categories as per the priority of the task, i.e., ordinary and emergent stream. Furthermore, the authors proposed an efficient approach to select an optimal path for the incoming traffic to avoid the overhead in the network. Alnoman et al. [4] introduced an energy-efficient approach by configuring SDN-enabled network to switch ON/OFF the edge nodes as per the requirement of the end users. The authors designed the topology by using the M/M/k queuing model, and a load balancing approach was used for optimal utilization of the available resources in the network. In the same direction, Lv et al. [44] discussed the service migration platform to balance the load among various devices to provide the QoS to the end users. Furthermore, the authors proposed mobile edge computing (MEC) framework integrating with SDN and network function virtualization (NFV). Wang et al. [69] proposed an energy-efficient routing approach on control plane for optimal resource utilization. The authors used heuristic algorithm at control plane to select the optimal route for incoming traffic. Furthermore, the authors integrated multi-objective evolutionary approach for best route selection. Table 9.4 summarizes various SDN-Edge cooperative energy management schemes.

9.7 Conclusion

In this chapter, the benefits and challenges of the IoT-Cloud platform have been discussed with required facts. Afterward, the layered architecture of the edge platform has been highlighted with pros and cons of the same. The challenges of the edge platform are handled by the SDN to avoid the congestion in the network. In the end, to define the benefits of the SDN envisioned platform, edge–cloud cooperation platform has been discussed with layered architecture, and the complete analysis has been highlighted considering the standard techniques.

References

1. *12 Green Data Centers Worth Emulating*. Accessed February 2016.
2. Agbehadji, I. E., Frimpong, S. O., Millham, R. C., Fong, S. J., & Jung, J. J. (2020). Intelligent energy optimization for advanced IoT analytics edge computing on wireless sensor networks. *International Journal of Distributed Sensor Networks, 16*(7), 1550147720908772.
3. Ahn, J., & Park, H.-S. (2014). Measurement and modeling the power consumption of router interface. In *16th International Conference on Advanced Communication Technology* (pp. 860–863). IEEE.

4. Alnoman, A., & Anpalagan, A. (2019). A SDN-assisted energy saving scheme for cooperative edge computing networks. In *2019 IEEE Global Communications Conference (GLOBECOM)* (pp. 1–6). IEEE.
5. Assefa, B. G., & Özkasap, Ö. (2019). A survey of energy efficiency in SDN: Software-based methods and optimization models. *Journal of Network and Computer Applications, 137*, 127–143.
6. Aujla, G. S., & Kumar, N. (2018). MEnSuS: An efficient scheme for energy management with sustainability of cloud data centers in edge–cloud environment. *Future Generation Computer Systems, 86*, 1279–1300.
7. Aujla, G. S., Jindal, A., Kumar, N., & Singh, M. (2016). SDN-based data center energy management system using RES and electric vehicles. In *2016 IEEE Global Communications Conference (GLOBECOM)* (pp. 1–6). IEEE.
8. Aujla, G. S., Kumar, N., Zomaya, A. Y., & Ranjan, R. (2017). Optimal decision making for big data processing at edge-cloud environment: An SDN perspective. *IEEE Transactions on Industrial Informatics, 14*(2), 778–789.
9. Aujla, G. S., Jindal, A., & Kumar, N. (2018, October 9). EVaaS: Electric Vehicle-as-a-Service for energy trading in SDN-enabled smart transportation system. *Computer Networks, 143*, 247–262.
10. Aujla, G. S., Chaudhary, R., Kaur, K., Garg, S., Kumar, N., & Ranjan, R. (2018). SAFE: SDN-assisted framework for edge–cloud interplay in secure healthcare ecosystem. *IEEE Transactions on Industrial Informatics, 15*(1), 469–480.
11. Aujla, G. S., Chaudhary, R., Kumar, N., Kumar, R., & Rodrigues, J. J. P. C. (2018). An ensembled scheme for QoS-aware traffic flow management in software defined networks. In *2018 IEEE International Conference on Communications (ICC)* (pp. 1–7). IEEE.
12. Aujla, G. S., Singh, A., & Kumar, N. (2019). Adaptflow: Adaptive flow forwarding scheme for software-defined industrial networks. *IEEE Internet of Things Journal, 7*(7), 5843–5851.
13. Aujla, G. S. S., Kumar, N., Garg, S., Kaur, K., & Ranjan, R. (2019) EDCSuS: Sustainable edge data centers as a service in SDN-enabled vehicular environment. *IEEE Transactions on Sustainable Computing*. https://doi.org/10.1109/TSUSC.2019.2907110
14. Buyya, R., & Son, J. (2018). Software-defined multi-cloud computing: A vision, architectural elements, and future directions. In *International Conference on Computational Science and Its Applications* (pp. 3–18). Springer.
15. Cao, H., Wu, S., Aujla, G. S., Wang, Q., Yang, L., & Zhu, H. (2019). Dynamic embedding and quality of service-driven adjustment for cloud networks. *IEEE Transactions on Industrial Informatics, 16*(2), 1406–1416.
16. Cardoso, P., Moura, J., & Marinheiro, R. (2020). A software-defined solution for managing fog computing resources in sensor networks. Preprint, arXiv:2003.11999.
17. Chen, Y., Zhang, N., Zhang, Y., Chen, X., Wu, W., & Shen, X. S. (2019). Energy efficient dynamic offloading in mobile edge computing for internet of things. *IEEE Transactions on Cloud Computing*. https://doi.org/10.1109/TCC.2019.2898657
18. Cui, L., Xu, C., Yang, S., Huang, J. Z., Li, J., Wang, X., Ming, Z., & Lu, N. (2018). Joint optimization of energy consumption and latency in mobile edge computing for internet of things. *IEEE Internet of Things Journal, 6*(3), 4791–4803.
19. Yuventi, J., & Roshan M. (2013). A critical analysis of Power Usage Effectiveness and its use in communicating data center energy consumption. *Energy and Buildings 64*, 90–94.
20. Dayarathna, M., Wen, Y., & Fan, R. (2016). Data center energy consumption modeling: A survey. *IEEE Communications Surveys Tutorials, 18*(1), 732–794.
21. Dorsch, N., Kurtz, F., Dalhues, S., Robitzky, L., Häger, U., & Wietfeld, C. (2016). Intertwined: Software-defined communication networks for multi-agent system-based smart grid control. In *2016 IEEE international conference on smart grid communications (SmartGridComm)* (pp. 254–259). IEEE.
22. Galov, N. (2020, November 24). 25 must-know cloud computing statistics in 2020.
23. Ge, R., Feng, X., & Cameron, K. W. (2009). Modeling and evaluating energy-performance efficiency of parallel processing on multicore based power aware systems. In *2009 IEEE International Symposium on Parallel & Distributed Processing* (pp. 1–8). IEEE.

24. Heller, B., Seetharaman, S., Mahadevan, P., Yiakoumis, Y., Sharma, P., Banerjee, S., & McKeown, N. (2010). Elastictree: Saving energy in data center networks. In *Nsdi* (Vol. 10, pp. 249–264).
25. How much energy do data centers really use?, March 17, 2020.
26. Hsieh, H.-C., Lee, C.-S., & Chen, J.-L. (2018). Mobile edge computing platform with container-based virtualization technology for IoT applications. *Wireless Personal Communications, 102*(1), 527–542.
27. IEEE Xplore. IoT and IoT-edge.
28. IEEE Xplore. SDN-edge cooperation.
29. IEEE Xplore. Software defined networking.
30. Jaiantilal, A., Jiang, Y., & Mishra, S. (2010). Modeling CPU energy consumption for energy efficient scheduling. In *Proceedings of the 1st Workshop on Green Computing* (pp. 10–15).
31. Jain, R., Molnar, D., & Ramzan, Z. (2005). Towards understanding algorithmic factors affecting energy consumption: Switching complexity, randomness, and preliminary experiments. In *Proceedings of the 2005 Joint Workshop on Foundations of Mobile Computing* (pp. 70–79).
32. Jammal, M., Singh, T., Shami, A., Asal, R., & Li, Y. (2014). Software defined networking: State of the art and research challenges. *Computer Networks, 72*, 74–98.
33. Kansal, A., Zhao, F., Liu, J., Kothari, N., & Bhattacharya, A. A. (2010). Virtual machine power metering and provisioning. In *Proceedings of the 1st ACM Symposium on Cloud Computing* (pp. 39–50).
34. Kaur, K., Garg, S., Aujla, G. S., Kumar, N., Rodrigues, J. J. P. C., & Guizani, M. (2018). Edge computing in the industrial internet of things environment: Software-defined-networks-based edge-cloud interplay. *IEEE Communications Magazine, 56*(2), 44–51.
35. Kobo, H. I., & Abu-Mahfouz, A. M. (2019). A distributed control system for software defined wireless sensor networks through containerisation. In *2019 International Multidisciplinary Information Technology and Engineering Conference (IMITEC)* (pp. 1–6). IEEE.
36. Kobo, H. I., Hancke, G. P., & Abu-Mahfouz, A. M. (2017). Towards a distributed control system for software defined wireless sensor networks. In *IECON 2017-43rd Annual Conference of the IEEE Industrial Electronics Society* (pp. 6125–6130). IEEE.
37. Lewis, A. W., Tzeng, N.-F., & Ghosh, S. (2012). Runtime energy consumption estimation for server workloads based on chaotic time-series approximation. *ACM Transactions on Architecture and Code Optimization (TACO), 9*(3), 1–26.
38. Li, H., Ota, K., & Dong, M. (2018). Learning IoT in edge: Deep learning for the internet of things with edge computing. *IEEE Network, 32*(1), 96–101.
39. Li, X., Li, D., Wan, J., Liu, C., & Imran, M. (2018). Adaptive transmission optimization in SDN-based industrial internet of things with edge computing. *IEEE Internet of Things Journal, 5*(3), 1351–1360.
40. Li, M., Cheng, N., Gao, J., Wang, Y., Zhao, L., & Shen, X. (2020). Energy-efficient UAV-assisted mobile edge computing: Resource allocation and trajectory optimization. *IEEE Transactions on Vehicular Technology, 69*(3), 3424–3438.
41. Liu, H., Yang, L. T., Lin, M., Yin, D., & Guo, Y. (2018). A tensor-based holistic edge computing optimization framework for internet of things. *IEEE Network, 32*(1), 88–95.
42. Liu, Y., Yang, C., Jiang, L., Xie, S., & Zhang, Y. (2019). Intelligent edge computing for IoT-based energy management in smart cities. *IEEE Network, 33*(2), 111–117.
43. Liu, Y., Zeng, Z., Liu, X., Zhu, X., & Bhuiyan, Md. Z. A. (2019). A novel load balancing and low response delay framework for edge-cloud network based on SDN. *IEEE Internet of Things Journal, 7*, 5922–5933.
44. Lv, Z., & Xiu, W. (2019). Interaction of edge-cloud computing based on SDN and NFV for next generation IoT. *IEEE Internet of Things Journal, 7*(7), 5706–5712.
45. Ma, Y.-W., Chen, J.-L., Tsai, Y.-H., Cheng, K.-H., & Hung, W.-C. (2017). Load-balancing multiple controllers mechanism for software-defined networking. *Wireless Personal Communications, 94*(4), 3549–3574.

46. Malladi, K. T., Shaeffer, I., Gopalakrishnan, L., Lo, D., Lee, B. C., & Horowitz, M. (2012). Rethinking dram power modes for energy proportionality. In *2012 45th Annual IEEE/ACM International Symposium on Microarchitecture* (pp. 131–142). IEEE.
47. McKeown, N., Anderson, T., Balakrishnan, H., Parulkar, G., Peterson, L., Rexford, J., Shenker, S., & Turner, J. (2008). Openflow: Enabling innovation in campus networks. *ACM SIGCOMM Computer Communication Review, 38*(2), 69–74.
48. Miazi, Md. N. S., Erasmus, Z., Razzaque, Md. A., Zennaro, M., & Bagula, A. (2016). Enabling the internet of things in developing countries: Opportunities and challenges. In *2016 5th International Conference on Informatics, Electronics and Vision (ICIEV)* (pp. 564–569). IEEE.
49. Mohan, V., Bunker, T., Grupp, L., Gurumurthi, S., Stan, M. R., & Swanson, S. (2013). Modeling power consumption of NAND flash memories using flashpower. *IEEE Transactions on Computer-Aided Design of Integrated Circuits and Systems, 32*(7), 1031–1044.
50. Morabito, R., & Beijar, N. (2017). A framework based on SDN and containers for dynamic service chains on IoT gateways. In *Proceedings of the Workshop on Hot Topics in Container Networking and Networked Systems* (pp. 42–47).
51. Muñoz, R., Vilalta, R., Yoshikane, N., Casellas, R., Martínez, R., Tsuritani, T., & Morita, I. (2018). Integration of IoT, transport SDN, and edge/cloud computing for dynamic distribution of IoT analytics and efficient use of network resources. *Journal of Lightwave Technology, 36*(7), 1420–1428.
52. Open Networking Foundation (2016, October). Of-Config.
53. Rafique, W., Qi, L., Yaqoob, I., Imran, M., Rasool, R. ur., & Dou, W. (2020). Complementing IoT services through software defined networking and edge computing: A comprehensive survey. *IEEE Communications Surveys & Tutorials, 22*(3), 1761–1804.
54. Ranjan, R., Thakur, I., Aujla, G. S., Kumar, N., & Zomaya, A. Y. (2020). Energy-efficient workflow scheduling using container based virtualization in software defined data centers. *IEEE Transactions on Industrial Informatics, 16*(12), 7646–7657.
55. Rao, T. V. N., Khan, A., Maschendra, M., & Kumar, M. K. (2015). A paradigm shift from cloud to fog computing. *International Journal of Science, Engineering and Computer Technology, 5*(11), 385.
56. Rawat, D. B., & Reddy, S. R. (2016). Software defined networking architecture, security and energy efficiency: A survey. *IEEE Communications Surveys & Tutorials, 19*(1), 325–346.
57. Roy, S., Rudra, A., & Verma, A. (2013). An energy complexity model for algorithms. In *4th Conference on Innovations in Theoretical Computer Science* (pp. 283–304). ACM.
58. Sezer, S., Scott-Hayward, S, Chouhan, P. K., Fraser, B., Lake, D., Finnegan, J., Viljoen, N., Miller, M., & Rao, N. (2013). Are we ready for SDN? Implementation challenges for software-defined networks. *IEEE Communications Magazine, 51*(7), 36–43.
59. Sharma, P. K., Rathore, S., Jeong, Y.-S., & Park, J. H. (2018). SoftEdgeNet: SDN based energy-efficient distributed network architecture for edge computing. *IEEE Communications Magazine, 56*(12), 104–111.
60. Sittón-Candanedo, I., Alonso, R. S., García, Ó., Muñoz, L., & Rodríguez-González, S. (2019). Edge computing, IoT and social computing in smart energy scenarios. *Sensors, 19*(15), 3353.
61. Sodhro, A. H., Luo, Z., Sangaiah, A. K., & Baik, S. W. (2019). Mobile edge computing based QoS optimization in medical healthcare applications. *International Journal of Information Management, 45*, 308–318.
62. Sodhro, A. H., Pirbhulal, S., & de Albuquerque, V. H. C. (2019). Artificial intelligence-driven mechanism for edge computing-based industrial applications. *IEEE Transactions on Industrial Informatics, 15*(7), 4235–4243.
63. Son, J., & Buyya, R. (2018). A taxonomy of software-defined networking (SDN)-enabled cloud computing. *ACM Computing Surveys (CSUR), 51*(3), 1–36.
64. The IoT rundown for 2020: Stats, risks, and solutions, January 13, 2020.
65. Tudor, B. M., & Teo, Y. M. (2013). On understanding the energy consumption of arm-based multicore servers. In *Proceedings of the ACM SIGMETRICS/International Conference on Measurement and Modeling of Computer Systems* (pp. 267–278).

66. Van Heddeghem, W., Idzikowski, F., Vereecken, W., Colle, D., Pickavet, M., & Demeester, P. (2012). Power consumption modeling in optical multilayer networks. *Photonic Network Communications, 24*(2), 86–102.
67. Violettas, G., Petridou, S., & Mamatas, L. (2019). Evolutionary software defined networking-inspired routing control strategies for the internet of things. *IEEE Access, 7*, 132173–132192.
68. Vishwanath, A., Hinton, K., Ayre, R. W. A., & Tucker, R. S. (2014). Modeling energy consumption in high-capacity routers and switches. *IEEE Journal on Selected Areas in Communications, 32*(8), 1524–1532.
69. Wang, J., Chen, X., Phillips, C., & Yan, Y. (2015). Energy efficiency with QoS control in dynamic optical networks with SDN enabled integrated control plane. *Computer Networks, 78*, 57–67.
70. Wang, L., Li, Q., Sinnott, R., Jiang, Y., & Wu, J. (2018). An intelligent rule management scheme for software defined networking. *Computer Networks, 144*, 77–88.
71. Warkozek, G., Drayer, E., Debusschere, V., & Bacha, S. (2012). A new approach to model energy consumption of servers in data centers. In *2012 IEEE International Conference on Industrial Technology* (pp. 211–216). IEEE.
72. Yao, Y., Huang, L., Sharma, A. B., Golubchik, L., & Neely, M. J. (2012). Power cost reduction in distributed data centers: A two-time-scale approach for delay tolerant workloads. *IEEE Transactions on Parallel and Distributed Systems, 25*(1), 200–211.
73. Zhan, C., Hu, H., Sui, X., Liu, Z., & Niyato, D. (2020). Completion time and energy optimization in UAV-enabled mobile edge computing system. *IEEE Internet of Things Journal, 7*(8), 7808–7822.
74. Zhang, Y., Gurumurthi, S., & Stan, M. R. (2007). Soda: Sensitivity based optimization of disk architecture. In *Proceedings of the 44th Annual Design Automation Conference* (pp. 865–870).
75. Zhang, T., Xu, Y., Loo, J., Yang, D., & Xiao, L. (2019). Joint computation and communication design for UAV-assisted mobile edge computing in IoT. *IEEE Transactions on Industrial Informatics, 16*(8), 5505–5516.
76. Zhao, Y., Li, Y., Zhang, X., Geng, G., Zhang, W., & Sun, Y. (2019). A survey of networking applications applying the software defined networking concept based on machine learning. *IEEE Access, 7*, 95385–95405.
77. Zhu, H., Liao, X., de Laat, C., & Grosso, P. (2016). Joint flow routing-scheduling for energy efficient software defined data center networks: A prototype of energy-aware network management platform. *Journal of Network and Computer Applications, 63*, 110–124.
78. Zinner, T., Jarschel, M., Blenk, A., Wamser, F., & Kellerer, W. (2014). Dynamic application-aware resource management using software-defined networking: Implementation prospects and challenges. In *2014 IEEE Network Operations and Management Symposium (NOMS)* (pp. 1–6). IEEE.

Chapter 10
Software-Defined Networking in Data Centers

Priyanka Kamboj and Sujata Pal

10.1 Introduction

In the past years, cloud computing has attained huge attention as it processes a large volume of data by using various computing cluster servers. Earlier web servers were maintained by organizations at their place, but today many organizations are hosting their web services on the cloud, as their infrastructure is not adequate to meet the growing application's needs. The major reasons for enterprises to shift toward cloud computing because of its various characteristics that include economic factors, scalability, security, rapid elasticity, and manageability. Further, the different interaction mediums like social networking sites (such as Facebook, Instagram, Twitter), the Internet of things, and research are generating a tremendous amount of data each day [19, 20]. Due to a large-scale increase in cloud services and to support scalability, the need for data centers has emerged [5, 17].

With the increase in real-time streaming applications, it demands high-speed access networks with fast computation and storage. Traditional networking of its complex architecture is inadequate to find optimal routing paths for these applications. However, the data center network (DCN) does not comply with real-time application demands and needs traffic monitoring to measure traffic loads. Therefore, cloud service providers are adopting SDN technology in data centers for effective traffic management.

SDN has gained attention because of the network programming paradigm [6, 8, 35]. SDN detaches the data from the control plane to build and program network architecture flexibly. Due to centralized control in SDN, it easily facilitates transmission, processing, and storage of cloud applications [20]. Thus, we can say

P. Kamboj · S. Pal (✉)
Department of Computer Science and Engineering, Indian Institute of Technology Ropar, Punjab, India
e-mail: 2018csz0003@iitrpr.ac.in; sujata@iitrpr.ac.in

G. S. Aujla et al. (eds.), *Software Defined Internet of Everything*, Technology, Communications and Computing, https://doi.org/10.1007/978-3-030-89328-6_10

that combining cloud data centers and SDN enables efficient, scalable, dynamic, and cost-effective platforms to support application deployments.

In this chapter, we highlight the challenges of data centers and the importance of SDN. The chapter discusses the different routing and traffic engineering schemes in data centers and how they manage network traffic for better resource utilization. Section 10.2 presents the building blocks of SDN and its applications. Section 10.3 introduces cloud computing, its different service models, the importance of DCNs, and its various challenges. Section 10.4 shows SDN in cloud data centers for flow management, resource management, and energy management. This chapter emphasizes by presenting a unified view by combining technologies: cloud computing and SDN. We wind up the chapter in Sect. 10.5 with a short description of SDN that performs traffic engineering and energy-aware routing. SDN provides efficient utilization of network resources and minimizes power consumption in DCNs. At last, we have given references for various research articles and papers on SDN and cloud data centers.

10.1.1 Software-Defined Networking: An Overview

SDN is a rising concept that has lighten the interest of researchers toward programming network devices. It has gathered the interest of network researchers to build and manage the network more innovatively and flexibly. SDN improves the network performance and avoids the pitfalls of present Internet architecture [45].

With the growing number of mobile devices and advancements, in-network service's trends become complex to reconsider the current Internet architecture. It has become difficult to effectively design and administer the present network architecture because of its static nature. The coupling of the control layer (takes decisions of traffic management) and the data layer (consists of network elements for packet forwarding) [39]. It becomes a tedious, complex, and error-prone task for network administrators to manually configure the network and apply access control policies to the network devices. As the world is becoming network centric, organizations need modern ways to add more flexibility to the network architecture. SDN achieves the above functionality by decoupling the control and data functions due to centralized network architecture [7, 35, 45].

SDN is a new technology that manages and controls the complete network with the centralized logical entity [35, 45]. Due to its agile nature, it allows network administrators to adjust traffic flows in the network. It also introduces the concept of virtualization of network functions to perform load balancing and traffic forwarding using standard architectures having generalized hardware in place of proprietary hardware and software [39]. SDN enables enterprises to simplify the design and management of network infrastructure. SDN also eliminates the need to understand different protocol standards as it uses open standards or APIs.

The Open Networking Foundation (ONF) gives network operator's privilege to build and expand SDN in dynamic, secure, flexible, and cost-effective networks.

It uses open APIs that comply with changing business objectives or landscape [42]. Besides, SDN abstraction effectively implements applications such as routing, bandwidth management, security, quality of service, access control policies, and traffic engineering to meet the end-user requirements. Additionally, SDN leverages IT to monitor and alter network functions in real time and scales down its burden to deploy new applications in a few hours or days, while weeks or months are required today.

OpenFlow is the primitive interface used for communications between the control and the data plane in SDN [42]. OpenFlow identifies network traffic flows using the flow table entries that are configured either statically or dynamically at the SDN controller. The traffic flows are configured on a per-flow basis. It also enables the network to adapt to the real-time fluctuations in the traffic patterns of end-user applications. The SDN architecture addresses the change in the business needs and reduces the operational costs of enterprises. The SDN architecture is shown in Fig. 10.1.

1. **Data layer:** The forwarding layer or the data layer comprises network elements like the OpenFlow switches for packet forwarding. The data layer communicates with the controller layer using Southbound APIs such as OpenFlow, Open Virtual Switch Database (OVSDB).
2. **Controller layer:** The controller layer consists of one or more controllers for forwarding the data traffic. The controller receives and sends OpenFlow messages between the network switches and the controller. This layer performs communication with the application layer using Northbound APIs.
3. **Application layer:** This layer performs control functions according to the end-user requirements. Applications such as routing, security, monitoring, and topology discovery are deployed on the controller to monitor network resources. The topology discovery module discovers the network topology using the Link Layer Discovery Protocol (LLDP) [2]. This module provides the information to the monitoring module. The monitoring module periodically monitors the throughput, link delay of the links to gain the network status.

10.1.2 SDN Building Blocks

As mentioned earlier, the SDN provides an abstraction to the network by separating the control from the data plane, thus simplifies the network management. In this section, we will have an in-depth discussion of the building blocks of the SDN architecture.

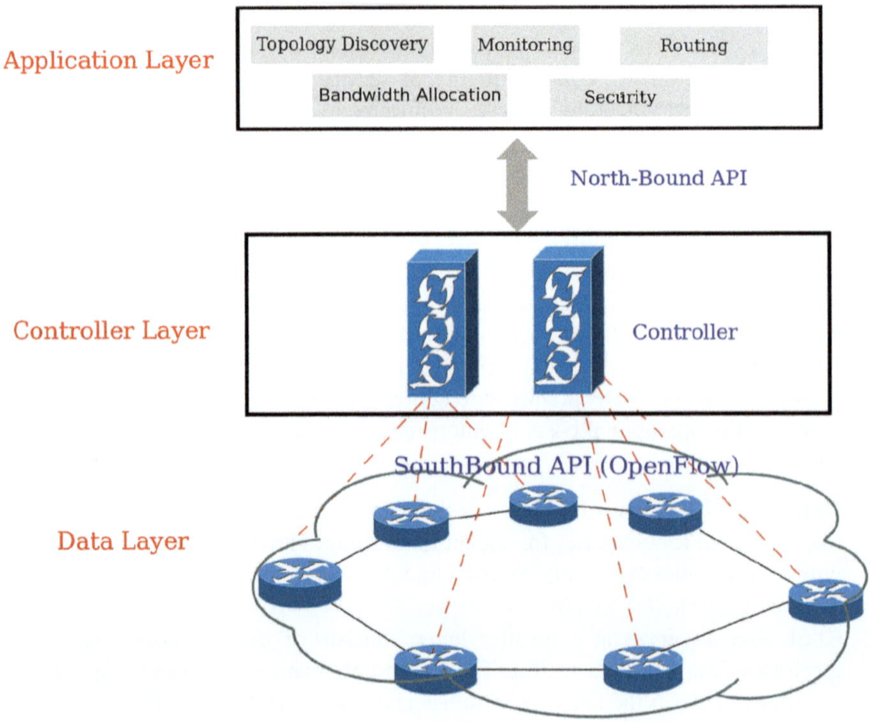

Fig. 10.1 System architecture

10.1.2.1 SDN Switches

The network infrastructure is an essential part of the network architecture. The data layer or bottom layer consists of the network devices that encapsulate all the functionalities to operate the network. These devices are known as network *switches* or routers, which constitute the forwarding tables to route the traffic flows. These devices store forwarding rules in their ternary content-addressable memory (TCAM) that is a costly hardware, size is limited, and also it requires high energy consumption [55]. Each network device has limited TCAM and thus can only store hundreds or thousands of flow rule entries. However, the changing traffic demands due to the increasing number of devices generates tremendous forwarding rules. Therefore, a shortage of TCAM affects the rule placement in SDN.

The OpenFlow switches comprise two main components: flow and group table. We will discuss the flow table in detail where each of its entry has three main parts: (1) matching rule details in a packet, (2) actions set that matches on packet header, and (3) counter for packet statistics update. Further, the OpenFlow-based forwarding devices maintain a pipeline of flow tables as shown in Fig. 10.2. A path through a pipeline of flow tables inside the forwarding device defines how the packet will be handled [55]. The lookup process begins as a new packet arrives with the first flow

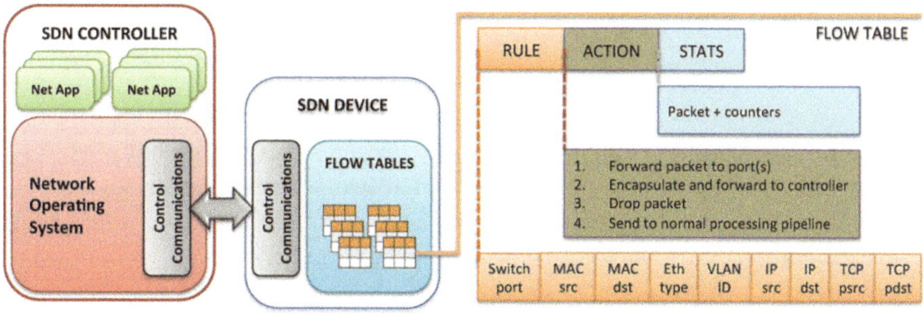

Fig. 10.2 OpenFlow-enabled devices and flow table in OpenFlow switches [35]

Table 10.1 Flow table components in OpenFlow

Match fields	Priority	Counters	Instructions	Timeouts	Cookies

table in the forwarding device. The header fields of the incoming packet match the entries in the flow table of the switches. The controller takes action to forward or discard packets on the desired port. If the header fields do not match the entries in the flow table, it results in a table miss. Thus, the switch forwards the incoming packet to the controller. Now, we will briefly discuss the flow table and group table entries.

Flow Table The entries in the flow table consist of the matched entities (or fields), priority, counter, instructions, timeouts, and cookies as shown in Table 10.1.

- **Match fields:** This field comprises ingress port, Ethernet source, a destination address, packet headers, and metadata laid out by preceding flow tables.
- **Priority:** This field indicates the matching precedence of the header fields of the flow entries.
- **Instructions:** This field contains actions set for packet forwarding.
- **Timeout:** It represents the time-to-live for a packet before its expiry.
- **Cookie:** It represents an identifier defined by the controller for flow entry.

The instructions in a flow table contain actions that include packet forwarding, modification, and pipeline processing. Finally, the controller decides to add, drop, and delete the flow table's flow entries. There exist various methods to delete the entries from the flow table. First, the controller can explicitly request to remove the entry. Second, a flow expiry approach exists at the switch that deletes entries from the table after either the hard timeout expires or the entry does not match within a certain period.

Group Table A group table comprises four group entries: group identifier, group type, counter, and action buckets, which is shown in Table 10.2.

- **Group identifier:** It is an unsigned integer of 32 bits that uniquely provides the identification to the group entry.

Table 10.2 Group entry components present in the group table

Group identifier	Group type	Counter	Action buckets

- **Group type:** This field facilitates to signify the semantics of a group.
- **Counters:** The value of counters is updated as the group handles the packet.
- **Action buckets:** It represents the ordered list of actions in the bucket where each bucket comprises actions to be taken along with its parameters.

10.1.2.2 SDN Controllers

The network operating system has a responsibility to manage the network resources and communicates with the applications deployed on the controller. The controller acts as a logical entity and is known as brain of the network. It manages traffic flows to enhance network management and simplifies to administer application performance. The central entity controls architecture, and policy enforcement at the network devices has become easy for the network administrator.

The controller receives the Packet-In messages from the switches and takes actions such as add, modify, or drop as Packet-Out messages. The logical entity controller regulates all the switches. There are several open-source options for SDN controllers such as NOX, POX, Floodlight, OpenDaylight, OpenContrail, etc. Although with advantages, the centralized approach suffers from the single point of failure. This drawback gives rise to the attacks that are the Distributed Denial of Service (DDoS) attacks. Therefore, to overcome these issues, multiple controllers are connected to a switch, and the controller is used to handle such failures through backup paths. Hence, all the controllers need to maintain consistency to avoid discrepancies for the proper functioning of applications.

Another significant issue with SDN controllers is managing the incoming traffic in the network. The decisions of routing the traffic have an impact on network performance. Traffic management is an essential subject that dynamically monitors network performance to analyze and regulate data transmission. Policy enforcement by the controller on the network devices has affected traffic management [45]. It has found that the number of devices over the Internet using various applications is thriving each day at an alarming rate.

The increase in applications such as online gaming, surveillance, and video conferencing results in abundant data generation. Today, these multimedia applications demand an underlying architecture to give responses to user requests in real time. However, existing Internet architecture is not flexible and even scalable to adapt to changes in traffic patterns. It has become a constraint for the network administrator to provide QoS that ensures performance guarantees by ensuring bandwidth, delay, and packet loss to applications. Therefore, highly scalable and efficient network management enhances resource utilization based on the end-user application demand.

To develop QoS policy management, the Internet Engineering Task Force (IETF) has come up with several QoS models—Integrated Services (IntServ) [15] and Differentiated Services (DiffServ) [14], but neither was successful and globally established. In DiffServ, the "class" of a packet is directly marked in the packet. It is a scalable mechanism that classifies and manages network traffic to provide QoS. However, IntServ uses "resource reservations" to maintain QoS for real-time traffic applications. Both IntServ and DiffServ fail to provide a global view of resources in the current Internet architecture.

10.2 SDN Applications

SDN builds the groundwork for scalable, flexible, and programmable networks. The controller defines new ways to handle the traffic flows in the network. The flow rules are installed in the flow table of the switches. The centralized network architecture motivates the novel applications to program network functions. Thus, SDN has become applicable in various networking domains such as cloud computing, the Internet of things, data centers, cellular networks, wide area networks (WANs), optical networks, etc. In this section, we will briefly discuss the major application domains.

10.2.1 Internet of Things

In simple terms, in the Internet of things (IoT) technology, people, devices (such as mobile phones, laptops, cars, sensors) are connected to the Internet to form the network. With the rapid increment in the devices, an enormous amount of data is getting generated. The data set tends to grow as more information is being collected and gathered from sensors, mobile devices, microphones, and other devices. The management and control of billions of connected objects is a complex task in the traditional Internet architecture.

SDN supports vendor independence due to the separation of the control from the data plane. Therefore, IoT leverages the benefits of SDN for supporting multiple technologies. SDN introduces programmability to the network devices to forward and control traffic flows in IoT architecture. SDN facilitates data transmission, resource allocation, energy management, and mobility management, which meets the growing user's needs in IoT.

10.2.2 Home Networks

Despite advancements in the transmission medium to offer high-speed services, network bandwidth always remains a limited resource. Internet service provider (ISP) serves users with limited bandwidth using a best-effort approach. We know that the best-effort service does not give any assurance of data delivery and data quality. Further, ISP provides services to several users in the local vicinity to simplify bandwidth allocation. Consequently, the users do not get the desired service, and their data quality is affected as they get the shared bandwidth from the resource pool.

Nowadays, SDN can be of great use in home networks. With SDN, ISP can enable dynamic bandwidth allocation by acknowledging the users to control their bandwidth consumption and generate revenue [46]. ISP can monitor the bandwidth usage of the users and assign or reassign the bandwidth as per the user demand in a local area network. For example, the user can request ISP for additional bandwidth for a finite time and pay according to the consumed bandwidth.

10.2.3 Cellular Networks

With the onset of mobile devices, cellular networks have become vital communication systems. The cellular networks support numerous applications with the advancement in the wide range of technologies like the Internet of things (IoT), self-driving cars, and Industry 4.0. The rapid upsurge in mobile devices with different application requirements has driven cellular networks to their limit. The increasing demands to improve the network performance of cellular devices have forced the operators to think about the current network architecture.

SDN plays an important role in satisfying the application requirements to solve the issues in cellular networks [38]. Initially, SDN decouples control functions from the data plane in the network architecture. The controller acts as a centralized logical entity for managing network resources and thus reduces infrastructure and operational costs. Additionally, SDN provides key functionalities needed in the core network (CN) of the cellular networks. Using SDN, it has become easy and flexible to manage routing, mobility, policy enforcement, resource allocation, and real-time monitoring in cellular networks. However, the SDN controller instructs the forwarding layer (comprising base stations) for traffic routing and simplifies its operation. It reduces load and interference during the coordination of the base stations.

10.2.4 Optical Networks

Optical networks play a significant role in today's network as it provides fast and high transmission capability. It is a form of communication that uses signals in the form of light to exchange information. It uses optical fiber cable to communicate from one end point to another in various telecommunications networks. The communication mainly depends on optical amplifiers, LEDs, and lasers to transmit the information across metropolitan, regional areas through the optical fiber cable.

High bandwidth during data transmission achieved using packet switching characteristics through wavelength channels has posed various challenges in optical networks [59]. SDN provides network programmability to monitor and control network infrastructure. Additionally, SDN controls physical layer components of optical communication. The optical transmitters or receivers (also known as transponders or transceivers) transmit or receive the optical signals using SDN.

10.3 Cloud Computing and Challenges

Since the past decade, cloud services have become a famous computing model that processes a large volume of data by utilizing computing server clusters. Conventionally, every organization used to maintain web servers and email servers at their site. But, on a large scale, they do not meet the growing needs of applications. One of the most popular services provided by the cloud is web hosting, as it helps the small enterprises who cannot maintain their servers because of cost factors. Furthermore, most enterprises shift toward cloud services due to scalability, economic factors, manageability, and security [49]. Cloud also stores documents such as images, videos, and files, and the user can share them with another end user. Therefore, the cloud helps to provide storage, computation, and infrastructure based on the application's needs. The users find cloud services to be reliable, efficient, and secure in their use.

10.3.1 Cloud Computing and Service Models

Cloud computing is known today for its five most essential characteristics. These are as follows:

- **On-demand self-service:** The organizations have the provision to use the resources such as computation, storage space, and virtual machines as per the application's needs. The organizations can use web interfaces to interact to have provision or de-provision of the services as per their requirements.
- **Resource pooling:** By resource pooling, it represents that the resources are shared within the customers using multi-tenancy. The resources—physical or

virtual—are assigned or released dynamically based on the customer's demands. The resources can be storage space, memory, computation, and network bandwidth. The customer will not know the exact location of the provision of resources.

- **Broad network access:** The cloud provides network access capability to users to connect and use the cloud services from anywhere and any time. The user can access the data from or upload data to the cloud using any thick or thin client mediums (such as mobile phones, laptops, tablets) and Internet connection.
- **Rapid elasticity:** It allows the scalable provisioning of resources automatically to the users. The cloud providers can allocate or deallocate the resources based on user requests. The user can very easily and quickly scale or descale the resources as per their demands. It helps the users in their cost savings.
- **Measured service:** Cloud providers optimize the use of resources by using metering capabilities based on the type of service. The cloud service providers monitor resource usage and provide transparency to the customers. The organizations use Pay As You Go model that states that the users have to pay based on the actual consumption of the services.

Cloud computing offers various services using the different models: Infrastructure as a Service (IaaS), Platform as a Service (PaaS), and Software as a Service (SaaS). Each higher-level model provides an abstraction to the lower-level models. The users use different client mediums (such as mobile phones, laptops, workstations), and the enterprises can access the cloud services via the Internet as shown in Fig. 10.3. The architecture represents the high-level abstraction of interaction between the users and enterprises using the cloud data center. Now, we will briefly discuss the above models as follows:

- **Infrastructure as a Service (IaaS):** The cloud provider provides computation, storage, and network to the enterprises. IaaS provides physical machines or workstations, virtual machines (VMs), and resources to the users from data centers. The user either owns or manages the applications running on the infrastructure and pays according to its usage.
- **Platform as a Service (PaaS):** It provides components like middleware, the operating system to develop and test customer applications. This service provides the environment for customers without buying or maintaining the hardware or software to run their business applications. One of the famous examples of this service is web hosting. The customers pay for the service based on the usage of the platform.
- **Software as a Service (SaaS):** It provides software services that are hosted by the cloud provider as per the need of the customers. The customers do not have to buy the licenses for the software to use it. The users can very easily access the software over the Internet based on its subscription.

Nowadays, cloud providers find it challenging to meet the proliferation of web services for different business customers. The cloud providers offer various service classes to the users based on their needs. Furthermore, the infrastructure of cloud

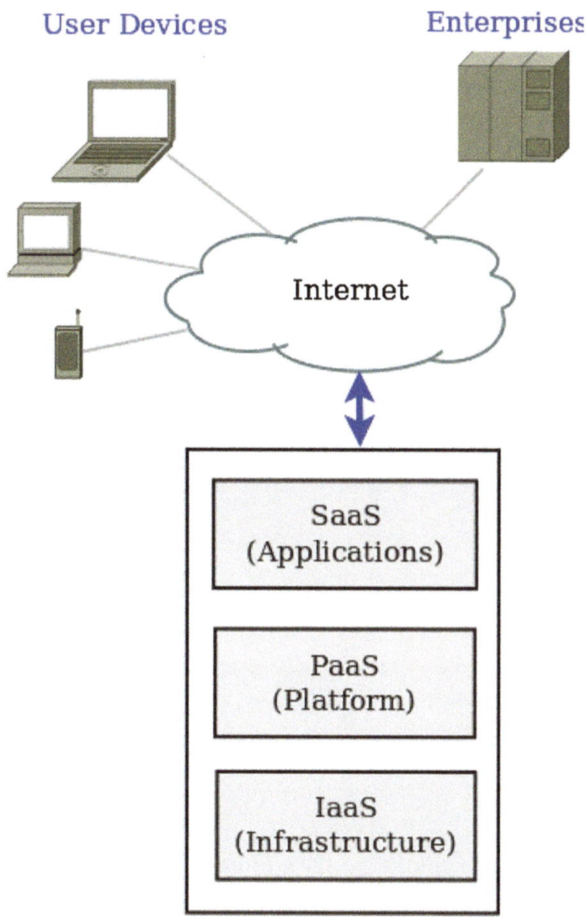

Fig. 10.3 Architecture of cloud computing

providers may not always be sufficient for all types of application needs. However, it is to note that there arises the need for data centers to offer flexibility and scalability to provide cloud services on a large scale. Thus, data center networks should have ease of management with the scalability issues and also be fault tolerant.

10.3.2 Data Center Networks

The data center provides a better understanding of its requirements to address the increase in cloud demands. It acts as a physical entity used by the organizations to house their applications and critical data. The data center design is based on a densely packed frame of workstations to provide computation and storage facilities to host and share the business-critical applications. The data center's main

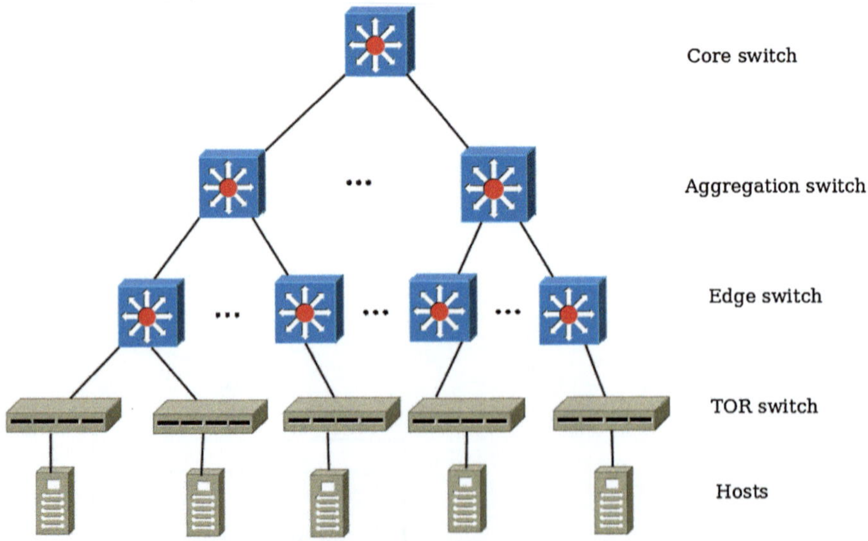

Fig. 10.4 A tree structure-based data center network's topology

components comprise routers, switches, storage systems, servers, firewalls, and delivery controllers. Over time, these data centers have served to host a large amount of computation power to the applications in just a single room. The data centers offer web hosting, SaaS, PaaS, and social networking services to form the network backbone. Virtualization is the key factor that is broadly used in the cloud data centers to provide different services and efficiently achieve scalability, flexibility, and resource allocation [34]. Moreover, virtualization effectively manages proper resource utilization and results in cost savings of power consumption and hardware. The several reasons that lead to the increase in usage of virtualization are:

- scalability and flexibility
- load balancing
- hardware coupling
- to ease the backup, recovery, and migration tasks
- to run and operate legacy applications on new operating systems
- large cost savings

The data center networks (DCNs) play a vital role as it connects resources of data centers. The DCN architecture is based on the *layered* approach that aims to improve flexibility, scalability, resiliency, and performance [63]. The DCNs have a tree structure as shown in Fig. 10.4. The fat-tree DCN architecture comprises three tiers—core, edge, and aggregation switches. In the bottom layer, end or physical hosts connect with the switches at the Top-of-Rack (TOR). The switches in the TOR layer connect with the edge switches. Further, the edge switches on every row interact with the aggregation switches. The aggregation switches perform the traffic aggregation of the data received from the edge switches. At last, the aggregation switches communicate with single or multiple switches at the core layer. The core-tier switches have a responsibility to interconnect intranetworks.

10.3.3 Challenges in Data Center Networks

With the increase in data center applications, it results in massive data sets. Real-time streaming of data creates high demands on the physical infrastructure for its computation and storage. Thus, a large amount of data streaming requires high-speed access to low-latency networks. With the expansion in operating systems, servers, and applications, data center management in real time has become complicated.

Today, the deployment of hundreds of virtual machines (VMs) in data centers has introduced new scalability challenges [12]. The DCNs find ways to how millions of VMs connect to thousands of servers. It is through the sharing of computing resources among multiple tenants, and hence it becomes uncertain to attain security. Further, the migration of applications between the VMs imposes novel mobility threats.

The tremendous increase in the devices puts pressure on the servers for delivery in service for the user applications. However, traffic management has become a critical concern in DCNs for better performance of network functions. In traditional data center architecture, with the expansion in the network, the manual configuration of the DCNs is a challenging task for the network administrators. Additionally, with the upsurge in several devices, it becomes extensively difficult to operate DCNs properly and hence cannot adapt to the dynamic end-user application requirements.

Today, modern data centers are facing challenges due to the scale of user requests and the deployment of thousands of VMs. It includes recovery from failure, data security, multi-tenant environment, traffic management, and energy management. Now we will give an overview of some of the challenges of the data centers. These are as follows:

- **Failure Recovery:**

  The data centers have a fundamental role in the economic and operational impact of cloud computing. With the virtualization in the cloud, the resource pool sharing among multiple clients is prone to failures and faults. Today, scale in data centers has become a critical task to recover from a failure and also leads to further ramifications of recovery decisions as the size grows. A small fault or dispossession in services in the cloud environment leads to severe economic and functional impacts.

  Google announced a financial loss of 20% to get the response time with an additional delay of 500 ms in an experiment [24]. Similarly, Amazon mentioned a 10% reduction in sales because of an additional delay of 100 ms in their search result [24]. In another incident, a minor network failure in O2, a well-known cellular service provider, affected seven million customers in only 3 days [13]. Besides, due to core switch failure in the network of Blackberry troubled millions of people to access the Internet as it lost its connectivity for 3 days [13]. Additionally, due to the distributed protocols in the current network architecture, it turns difficult to predict network behavior. The cloud environment should be robust to deliver QoS to meet the service-level agreement (SLA) requirements signed between the user and cloud provider despite any software or hardware failure. Any violation in SLA by the cloud may result in enormous financial and reputation loss. Therefore to ensure failure resiliency in a cloud environment is of vital importance. Moreover, the cloud data centers need proper functioning to deliver QoS even in the existence of failures.

- **Multi-tenancy:**

  Previously, we discussed that multiple individuals share resources like storage space, computation, and network in the data centers. The challenge arises how to isolate and separate the individuals from one another. It is necessary for the organizations that provide a multi-tenant environment to differentiate the resources that have been assigned and belong to an individual client. Even in network traffic, the data packets should be segregated and insulated for different clients. This requirement is necessary to provide security as well as to guarantee QoS for the applications.

- **Traffic management:**

  Traffic management in today's DCNs is a vital area of concern. In a multi-tenant environment, resources such as network, computation, and storage space are shared among multiple clients to run the applications. It enforces the cloud service provider to monitor network traffic to optimize resource utilization. Therefore, it has become essential to measure traffic loads and take suitable actions to route traffic flows.

  Now coming toward the state-of-the-art, link-state technology is used for route calculation in the traditional networks. Each switch builds up a forwarding table to direct traffic between the sources and the destinations in the network. The paths taken by packets in a flow are determined by numerous protocols like shortest path routing algorithm, spanning tree, and multipath routing. The shortest path routing algorithm is used for path computation of the packets, not always finding the optimal path since it does not consider vital factors like traffic load.

  We can say that the increase in real-time traffic has gained a lot of attention to traffic management in data centers. Therefore, cloud service providers need to find novel ways to monitor and control traffic flows in the network.

10.3.4 SDN in Data Center Networks

We can say that SDN offers many advantages to fill the previously mentioned gaps in the data centers. SDN controller makes optimal routing decisions for traffic forwarding in the network. The controller regulates the respective forwarding tables

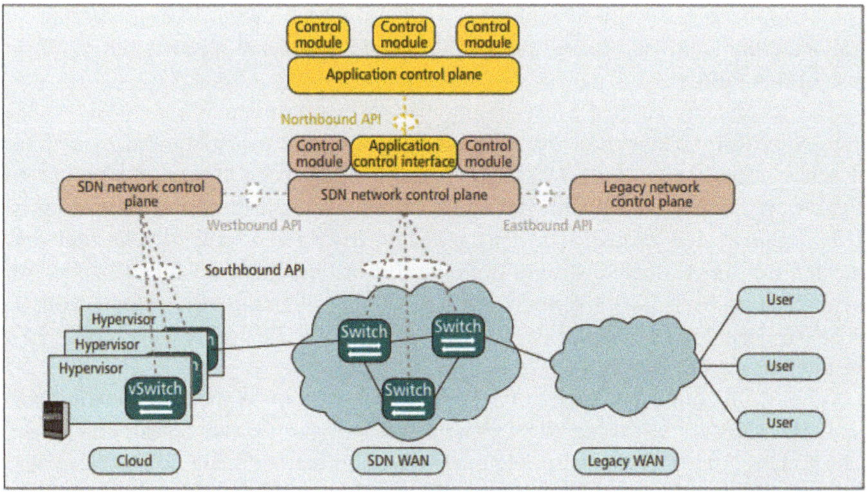

Fig. 10.5 Example: interfaces of a software-defined network [31]

in network devices to have control over routing. The SouthBound Interface is used for interaction between the control and data layer in SDN as shown in Fig. 10.5. The traditional network architecture suffers from interoperability issue, whereas, on the other hand, open interfaces in SDN allow the network to reach its full functionality [31].

The controller acts as a centralized logical entity, which helps to simplify the traffic management in the DCNs. SDN dynamically manages traffic flows and performs traffic load balancing, resource allocation, and bandwidth provisioning in data centers. It also improves network performance by adjusting the resources according to the application needs in the data centers [3].

The various cloud providers such as Apple, Google, and Microsoft serve their services to the customers using data centers that are distributed worldwide. Even with the rise in demand for the services, the data centers achieve traffic exchange between the interdata centers. Thus, SDN's global network view leverages to carry out centralized traffic and optimizes the use of network resources. Both Microsoft and Google have developed SDN in their data centers and also published their systems' technicalities.

SDN offers many advantages as compared to the traditional networks that create interests in the industry. SDN is a way to simplify network management or to develop commercial solutions. One of the most popular use cases of SDN adoption is in Google production networks. In Google networks, SDN is used to connect its data centers over the wide-area network (WAN) and is Known as B4 [30]. It is one of the prime and largest developments of SDN and OpenFlow. B4 carries a large amount of traffic than Google's WAN. In particular, the custom switches are managed and controlled by OpenFlow in B4. Further, B4 can scale to fit application demands efficiently and adapts network behavior to respond to failures.

The centralized traffic engineering (TE) solution enables the SDN controller to reallocate the bandwidth based on application demands and reroutes the traffic in case of link failures.

The second use case of SDN adoption is software-driven WAN (SWAN) [28] from Microsoft, which is a TE solution proposed to carry a significant amount of traffic. The objective of SWAN is to meet policy rules to give preference to high-priority traffic along with providing fairness among the same service class. SDN's global view helps to assign bandwidth to different paths in the network. Further, the fine-grained control of traffic in SDN enforces bandwidth utilization on an application basis [43]. Consequently, the overprovisioning of resources reduced as SWAN decides the amount of traffic a service class can forward and configures the data plane accordingly.

Other giant companies such as Facebook and Amazon are planning to build their network infrastructure using SDN principles. The various networking companies such as Cisco, Hewlett Packard (HP), VMware, Juniper, and Big Switch have also shown keen interest in SDN to provide commercial solutions for their enterprises.

10.4 Routing and Traffic Engineering in Data Center Networks

As per the discussion, we know that SDN uses a centralized approach that simplifies traffic management in the network. The flow table at switches allows us to have fine-grain level granularity. The granularity level depends on the size of the switch flow table and how the controller wishes to enforce its control on the traffic flows. The flow entries in a flow table are placed either in *reactive* or *proactive* manner. In the *reactive* approach, the incoming first packet of traffic flows comes and the switch sends it to the controller. The controller communicates with the switches to enter the flow entries of the incoming packets in the flow table. However, in the *proactive* approach, the flow entries are computed by the controller and entered in switches timely. In addition, it ignores the flow setup time, and flow entries are inserted based on wildcard rules. Further, in case of connection loss between switch and controller, traffic moves independently in the network.

In this context, OpenFlow adopts two main routing options: flow based and aggregated. In *flow-based routing*, the controller sets up during *every* flow individually packet header details that exactly match with the flow table entries. If any flow rule entries do not match the forwarding device's flow table, the packet is forwarded to the controller and the flow entries are installed in the flow table of the routed path. This method fits for *fine-grain* control only if switches have adequate capacity to store all entries. However, ingress traffic at OpenFlow switches is increasing and occurring frequently. It creates the overhead at the controller side and becomes difficult to process all the requests. This routing approach does not fit for increasing traffic flows.

Table 10.3 Example of flow entries in the switch flow table

Entry	Source	Destination	Action
1	0111	0000	To forward on Port 1
2	0111	0110	To forward on Port 2
3	1111	0110	To forward on Port 2
4	1100	1010	To forward on Port 1
5	$$$$	$$$$	Drop entry

Table 10.4 Example of aggregated flow entries

Entry	Source	Destination	Action
1	$111	0000	To forward on Port 1
2	0111	0110	To forward on Port 2
3	1111	0110	To forward on Port 2
5	$$$$	$$$$	Drop entry

The flow table implemented using TCAM memory is expensive and has high energy consumption. A large number of flow rules need to be placed in the flow table with the increase in traffic flows. As the network devices fit with small TCAM, they can store only limited flow entries. However, we can say that the flow entries get reduced in the switch flow table using an *aggregated* routing approach [68]. For example, there can be a single entry for different traffic flows in OpenFlow switches belonging to the same or particular IP prefix destination. The flow entries aggregate at OpenFlow switches and associate a single path for a set of flows. For instance, Table 10.3 shows the switch flow table entries. Similarly, in Table 10.4 flow entries 1 and 4 are aggregated in the flow table. The "$" in Table 10.3 represents a single digit (0 or 1) in a specific position. The "$$$$" in Table 10.4 represents any four digits (0 or 1) in the table.

10.4.1 Flow Management in Data Center Networks

SDN controller has control over all the routing and traffic forwarding decisions. The network should prioritize business traffic over other applications. For example, traffic from applications such as online gaming, v2x, virtual reality, and audio streaming has higher priority than the best-effort traffic. These applications have stringent requirements for latency, bandwidth, and QoS. A study [33] determines that congestion was observed in 86% of data center links due to immense requests arrived for large flows. Nowadays, the classification and scheduling of flows have become substantial to utilize the available bandwidth in DCNs.

Several studies have categorized the traffic flows in data centers into two types: elephant (long-lived) and mice flow [3, 18, 40]. With SDN-based data centers, the controller has to choose an optimal path for individual traffic flow. Generally, the controller computes the shortest path for forwarding traffic flows. But the end-to-end path may have congestion in the links that result in delay, jitter, and packet loss.

Thus, a solution is required to distribute and route the traffic flows on different paths based on their bandwidth requirements [18]. Therefore, bandwidth requirement has become a constraint to satisfy the QoS guarantee in the network. Further, the SDN controller has to take into account various network parameters for routing decisions.

Some of the applications in DCNs demand higher throughput in elephant flows, while delay-sensitive applications require lower latency [73]. The DCNs experience congestion due to an imbalance in the distribution of the traffic flows in the network. Therefore, some of the links are under-utilized, which leads to an upsurge in link latency and low resource utilization. Thus, reducing link congestion of DCNs and ensuring QoS guarantee are critical issues for cloud service providers. Therefore, we can say that dynamic flow management has vital importance in data centers to enhance QoS for the user applications [70].

10.4.2 Traffic Engineering in Data Center Networks

Traffic engineering (TE) has shown an advanced development to measure and manage network traffic. TE states that network operators handle numerous data flows in the network [26]. TE regulates network traffic to have better utilization of network resources [53]. The objectives of TE include traffic load balancing, control congestion, and minimizing network utilization.

The traditional TE approaches include IP-based TE and multi-protocol label switching (MPLS). The *IP-based* technology optimizes the routing algorithm to avoid network congestion by adjusting the traffic flows on multiple paths [25]. For instance, this approach uses the Open Shortest Path First (OSPF) routing algorithm and concept of link weights to compute multiple shortest routing paths to balance the traffic load [22]. This technology suffers from many drawbacks: (1) while using the concept of link weights in OSPF, it fails to split the traffic in a suitable proportion, so network resources are not utilized; and (2) as the link weights in topology change, the routing protocol takes a lot of time to converge to a novel state, which causes congestion, packet loss, and delay in the network.

To avoid the issues in IP-based TE, different researchers proposed a new method for sending the packets using *Multi-Protocol Label Switching (MPLS)* [56]. However, the MPLS routing technique is considered very complicated and creates difficulty for DCNs to satisfy the growing application demands. We have discussed that in the traditional networks, fine-grain control over traffic is difficult to achieve. Therefore, there is a need to develop a network architecture to solve the above problems. Many organizations have shown keen interest in SDN as it decouples network functions to introduce flexibility in architecture. TE can be applied easily to SDN switches as it modifies flow tables in switches [1]. With SDN, traffic flows dynamically change in data centers for easy management of workload.

The traffic engineering system architecture shown in Fig. 10.6 comprises three main components: data center network (DCN), controller, and traffic engineering manager. The DCN's architecture uses *layered* approach and consists of the core,

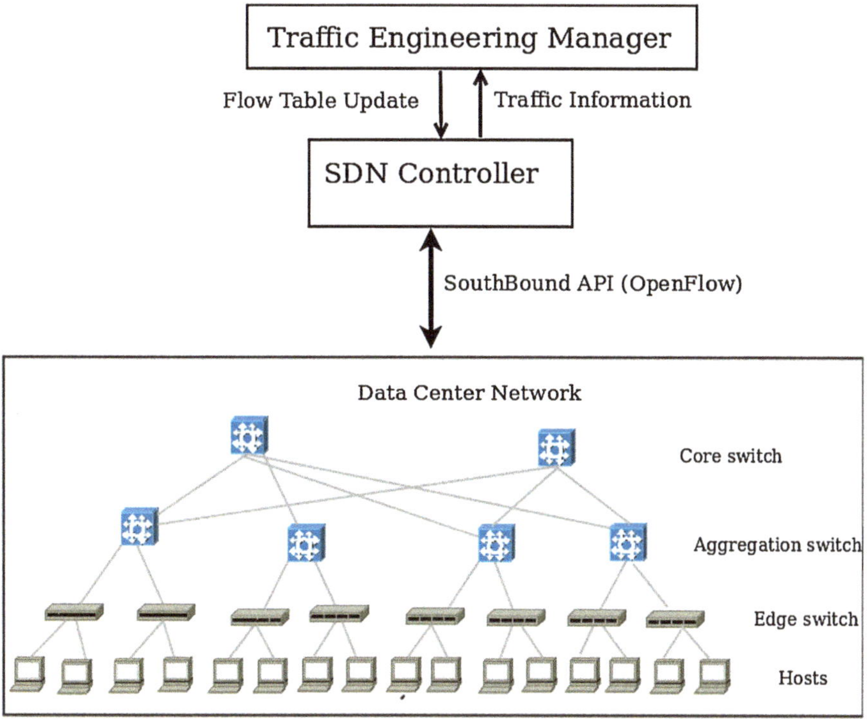

Fig. 10.6 Traffic engineering system architecture

aggregation, and edge switches. The DCN's switches send the status information to the controller using the SouthBound interface such as OpenFlow. The controller collects and aggregates all the information received from the switches. The traffic engineering manager takes all the aggregated information to make TE decisions and sends the notification to the controller [26]. Therefore, traffic engineering plays an essential role in reducing latency and balancing the traffic load in the network.

10.4.3 Load Balancing in Data Center Networks

The load balancing at the switch uses a routing scheme known as *Equal-Cost Multipath (ECMP)*, which adopts hashing techniques to split the traffic flows onto multiple paths [47]. ECMP moves the traffic flow along the paths based on a hash value computed from the packet headers [29]. In the example of *per-Flow* technique, it balances the traffic load among multiple paths with an equal cost. ECMP routing scheme faces congestion if a collision occurs at hash value, which results in forwarding flows to the same port.

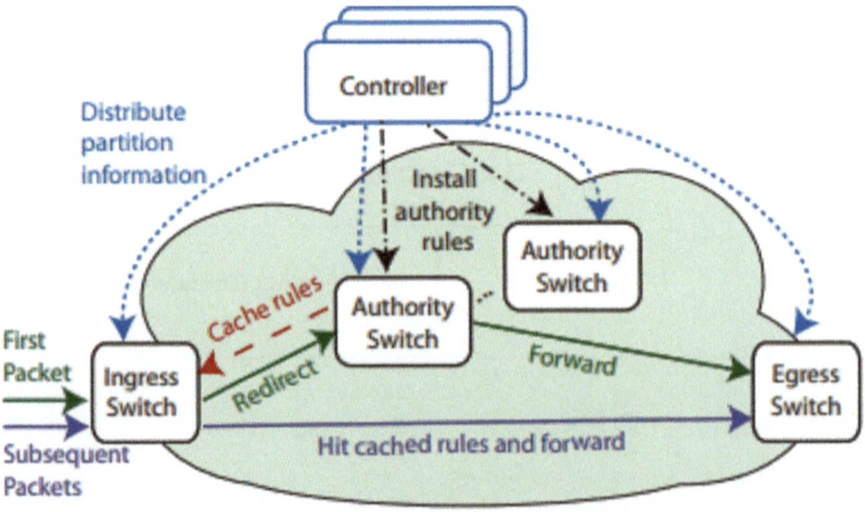

Fig. 10.7 DIFANE flow management architecture [69]

To address the bottleneck of the ECMP routing scheme, the researchers have proposed various methods. Hedera [3] used flow scheduling scheme in data centers for multi-stage switch topology. It gathers flow information from the network and computes nonconflicting paths for the traffic flows. By the global view, the scheduling system can see the bottleneck in paths and instruct the switches to reroute the traffic flows accordingly. However, Mahout [21] is a traffic management system that identifies the incoming flow as *elephant flow*. It deploys an in-band mechanism to manage traffic flows between end hosts and the controller. Thus, it notifies the controller about *elephant flows* and computes the routing paths.

Another approach used in switch load balancing is known as *wildcard rule flow forwarding*. This approach suffers from latency issues due to the presence of a single centralized controller. We will highlight some other proposed approaches to address these latency issues. DevoFlow [44] scheme was designed to place the flow rules at the OpenFlow switch to minimize the interaction with the controller. With this approach, the controller can easily monitor and detect the *elephant flows*.

Furthermore, ReWiFlow [52] restricts the class of Openflow wildcard rules to make it simple to use and overcome the previously defined issues. It reduces programming complexity and manages the group of flows without loss in performance. However, in the DIFANE [69] architecture the data plane switches use wildcard rules. In this architecture, the controller allocates the switch rules known as "authority switches" when rule-matching does not occur at ingress switches as shown in Fig. 10.7. The controller divides the rule by the use of a partitioning algorithm across authority switches. Further, the switches take action based on the packet-matching occurs in the flow table.

We have already stated that SDN simplifies traffic management by separating control functions in the network. Thus, we can say that SDN-based load balancing helps to deal with congestion and optimize link utilization in DCNs. SDN also helps to distribute traffic load on multiple paths and efficiently handles the workload in data centers.

10.4.4 Resource Management in Data Center Networks

The resource management in DCNs is found to be a crucial factor in cloud computing. The resource requirements of cloud applications are drastically changing in cloud data centers. Therefore, it is essential to maintain data center resources to meet the SLA of different business applications. Some of the SLAs for the desired cloud applications are response time, failure recovery, security, maintenance time, and data loss [20]. Moreover, the technology drift toward cloud and big data applications has created a lot of pressure on DCNs to enhance cloud services for the users by improving flexibility, performance, and security. SDN greatly benefits big data applications in different aspects that involve data delivery, scheduling, data processing, and resource utilization.

Today, SDN-enabled data centers are widely applicable in big data applications. An SDN-based OpenFlow bandwidth provisioning method for big data applications is proposed [32]. Bandwidth provisioning is a necessary component to isolate and separate the traffic flows of different users or service classes. The controller aggregates information from the network switches to maintain and update the flow table to allocate bandwidth for traffic flows. The switches receive the updates to run the scheduling algorithm to allocate the bandwidth for big data applications. Hence, resource allocation is done efficiently and reduces the power consumption in data centers.

We can state that bandwidth has become one of the crucial resources as shared between multiple network applications. Therefore, ensuring fairness among different traffic flows along with QoS is also an important criterion for a network manager [11]. The absence of a fairness policy leads to unfair allocation of resources and traffic distribution in a network. Substantial work has been proposed for optimal resource allocation [16] and to support different kinds of fairness in resource allocation that mainly includes max–min fairness [41] and proportional fairness [36].

The aim of resource allocation is to attain better SLAs between cloud service providers and users. The resource allocation strategy plays an important role that motivates clients to access the cloud services or make them reluctant to use their services [51]. Thus, the SDN controller efficiently handles resource assignments in virtual machines as per the user's requests between the data centers. Further, SDN also minimizes the cost incurred by the service provider while satisfying the user's requests. Hence, resource management is necessary to improve bandwidth utilization and to guarantee QoS in a network.

10.4.5 Energy Management in Data Center Networks

The DCNs allow enterprises to interact with the outside world and is called the "backbone" for an enterprise. In data centers, requests can arrive at any period, so it becomes necessary for the devices to function 24*7 to provide services to the customers. The devices need a large amount of energy to function, thus lead to an increase in the total power consumption cost in the data center. Additionally, in data centers resources are found to be underutilized about 30–40% of the time [54, 71].

Energy efficiency is the primary area of interest in modern data centers because of environmental factors. The inflation in energy cost is an immense threat to cloud providers as it leads to an increase in Total Cost of Ownership (TCO) and a reduction in Return of Investment (ROI) [10]. Further, high energy consumption causes carbon emission that leads to environmental damage. Consequently, it is challenging to reduce power consumption in DCNs. The high computation servers and storage are required to process user requests and respond within a fixed time [10].

In the recent past, different multimedia applications using cloud services are growing each day. Therefore, researchers have proposed several methods to maximize energy efficiency through traffic aggregation and scheduling. We have already discussed the aggregation routing methods in Sect. 10.3. In this method [37, 50, 67] the aggregated traffic is sent onto a few switch ports, whereas idle ports of switches are kept in turn-off mode to save power [65]. Further, several schemes of power saving have been proposed that leverages Energy-Efficient Routing (EER) strategy that includes EAR [50] and ElasticTree [27]. The main idea behind EER is to perform bandwidth scheduling along with flow consolidation to transmit flows only on a subset of links to decrease energy consumption.

One of the main challenges of energy efficiency is the consumption of energy in networks [4]. This problem creates difficulty in the traditional network architecture because of its limited flexibility. SDN treats the network as a logical entity to support enterprises to program, automate, and control data centers. Thus, SDN attempts to solve this problem by adjusting energy consumption proportionate to the amount of traffic.

Now, we will discuss the energy-efficient routing techniques that mainly include energy saving at links and switches. Several methods exist in the state-of-the-art for energy-aware routing in SDN. Some of the methods focus on energy consumption on links [58, 62, 72], whereas others focus on switches [60, 61]. In addition, the queue-based techniques to determine per-port energy requirement have also been considered in [48, 57, 61, 66, 72]. Further, a few experiments show that distributing the traffic flows on underutilized links results in more power saving than turning off the links [23]. But this approach does not apply to all cases.

The controller in the OpenFlow network splits traffic flows onto multiple paths based on the incoming traffic volume. The multipath flow routing using the SDN-based network is more effective than traditional DCN architecture [3]. Besides, bandwidth utilization is improved if traffic splits on multiple paths. A distributed power-saving mathematical model for large-scale SDN-based DCNs is developed, which uses characteristics to optimize energy efficiency [64]. In this model, an EER routing algorithm is used for intradomain traffic flows and power saving using multiple controller architecture. Further, "green cloud computing" considers achieving resource utilization as it efficiently manages the data center's resources by minimizing energy consumption.

We can say that SDN achieves energy-efficient routing in cloud data centers. SDN technology not only minimizes energy consumption but also considers that network performance should not have deteriorated in data centers. We know that we have constrained energy resources and may not be easily attainable in the future [9]. Therefore, it becomes a concern to save the energy consumption of data centers.

10.5 Conclusions

In this chapter, we presented an outline of software-defined networking (SDN) and cloud computing. In particular, we discuss the basic building blocks of SDN and its various applications in different domains. Next, we have given an overview of cloud computing and its various characteristics and service models. Subsequently, we talked about the data center network (DCN), its different challenges that have arisen due to the increase in data center applications, and how SDN plays an important role in overcoming the challenges in data centers.

We then discussed different routing and traffic engineering schemes in DCNs, how they manage network traffic to meet application QoS requirements, and resource management. We also presented an overview of various proposed methods of how resources are managed using SDN in DCNs. Furthermore, energy management issues and SDN-based energy-efficient routing methods have been taken into account to minimize power consumption in DCNs.

References

1. Agarwal, S., Kodialam, M., & Lakshman, T. V. (2013). Traffic engineering in software defined networks. In *Proceedings IEEE INFOCOM* (pp. 2211–2219). Piscataway: IEEE.
2. Akyildiz, I. F., Lee, A., Wang, P., Luo, M., & Chou, W. (2014). A roadmap for traffic engineering in SDN-OpenFlow networks. *Computer Networks, 71*, 1–30.
3. Al-Fares, M., Radhakrishnan, S., Raghavan, B., Huang, N., & Vahdat, A. (2010). Hedera: Dynamic flow scheduling for data center networks. In *Nsdi* (vol. 10, pp. 89–92).
4. Assefa, B. G., & Özkasap, Ö. (2020). RESDN: A novel metric and method for energy efficient routing in software defined networks. *IEEE Transactions on Network and Service Management, 17* (2):736–749.
5. Aujla, G. S., Chaudhary, R., Kumar, N., Kumar, R., & Rodrigues, J. J. P. C. (2018). An ensembled scheme for QoS-aware traffic flow management in software defined networks. In *2018 IEEE International Conference on Communications (ICC)*, (pp. 1–7). Piscataway: IEEE.
6. Aujla, G. S., Jindal, A., Kumar, N., & Singh, M. (2016). SDN-based data center energy management system using res and electric vehicles. In *2016 IEEE Global Communications Conference (GLOBECOM)* (pp. 1–6). Piscataway: IEEE.

7. Aujla, G. S., & Kumar, N. (2018). SDN-based energy management scheme for sustainability of data centers: An analysis on renewable energy sources and electric vehicles participation. *Journal of Parallel and Distributed Computing, 117*, 228–245.
8. Aujla, G. S., Singh, A., & Kumar, N. (2019). AdaptFlow: Adaptive flow forwarding scheme for software-defined industrial networks. *IEEE Internet of Things Journal, 7*(7), 5843–5851.
9. Bahrami, S., Wong, V. W. S., & Huang, J. (2018). Data center demand response in deregulated electricity markets. *IEEE Transactions on Smart Grid, 10*(3), 2820–2832.
10. Beloglazov, A., Abawajy, J., & Buyya, R. (2012). Energy-aware resource allocation heuristics for efficient management of data centers for cloud computing. *Future Generation Computer Systems, 28*(5), 755–768.
11. Bhaumik, S., & Chakraborty, S. (2018). Hierarchical two dimensional queuing: A scalable approach for traffic shaping using software defined networking. In 4th *IEEE Conference on Network Softwarization and Workshops (NetSoft)* (pp. 150–158). Piscataway: IEEE.
12. Bilal, K., Malik, S. U. r. R., Khan, S. U., & Zomaya, A. Y. (2014). Trends and challenges in cloud datacenters. *IEEE Cloud Computing, 1*(1), 10–20 (2014).
13. Bilal, K., Manzano, M., Khan, S. U., Calle, E., Li, K., & Zomaya, A. Y. (2013). On the characterization of the structural robustness of data center networks. *IEEE Transactions on Cloud Computing, 1*(1), 1–1.
14. Black, D. L., Wang, Z., Carlson, M. A., Weiss, W., Davies, E. B., & Blake, S. L. (1998). An Architecture for Differentiated Services. *RFC 2475*. https://doi.org/10.17487/RFC2475, https://rfc-editor.org/rfc/rfc2475.txt
15. Braden, R. T., Clark, D. D. D., & Shenker, S. (1994). Integrated Services in the Internet Architecture: an Overview. *RFC 1633*, https://doi.org/10.17487/RFC1633, https://rfc-editor.org/rfc/rfc1633.txt
    RFC1633: Integrated services in the internet architecture: an overview.
16. Bui, N., Malanchini, I., & Widmer, J. (2015). Anticipatory admission control and resource allocation for media streaming in mobile networks. In *Proceedings of the* 18th *ACM International Conference on Modeling, Analysis and Simulation of Wireless and Mobile Systems* (pp. 255–262).
17. Cao, H., Wu, S., Aujla, G. S., Wang, Q., Yang, L., & Zhu, H. (2019). Dynamic embedding and quality of service-driven adjustment for cloud networks. *IEEE Transactions on Industrial Informatics, 16*(2), 1406–1416.
18. Chao, S.-C., Lin, K. C.-J., & Chen, M.-S. (2016). Flow classification for software-defined data centers using stream mining. *IEEE Transactions on Services Computing, 12*(1), 105–116.
19. Chaudhary, R., Aujla, G. S., Kumar, N., & Zeadally, S. (2018). Lattice-based public key cryptosystem for internet of things environment: Challenges and solutions. *IEEE Internet of Things Journal, 6*(3), 4897–4909.
20. Cui, L., Yu, F. R., & Yan, Q. (2016). When big data meets software-defined networking: SDN for big data and big data for SDN. *IEEE Network, 30*(1), 58–65.
21. Curtis, A. R., Kim, W., & Yalagandula, P. (2011). Mahout: Low-overhead datacenter traffic management using end-host-based elephant detection. In *Proceedings IEEE INFOCOM* (pp. 1629–1637). Piscataway: IEEE.
22. Fortz, B., & Thorup, M. (2000). Internet traffic engineering by optimizing OSPF weights. In *Proceedings IEEE INFOCOM. Conference on Computer Communications. Nineteenth Annual Joint Conference of the IEEE Computer and Communications Societies)* (vol. 2, pp. 519–528). Piscataway: IEEE.
23. Garroppo, R., Nencioni, G., Tavanti, L., & Scutella, M. G. (2013). Does traffic consolidation always lead to network energy savings? *IEEE Communications Letters, 17*(9), 1852–1855.
24. Greenberg, A., Hamilton, J., Maltz, D. A., & Patel, P. (2008). The cost of a cloud: Research problems in data center networks. *ACM SIGCOMM Computer Communication Review, 39*, 68–73.
25. Han, G., Jiang, J., Bao, N., Wan, L., & Guizani, M. (2015). Routing protocols for underwater wireless sensor networks. *IEEE Communications Magazine, 53*(11), 72–78.

26. Han, Y., Seo, S.-s., Li, J., Hyun, J., Yoo, J.-H., & Hong, J. W.-K. (2014). Software defined networking-based traffic engineering for data center networks. In *The* 16th *Asia-Pacific Network Operations and Management Symposium* (pp. 1–6). Piscataway: IEEE.
27. Heller, B., Seetharaman, S., Mahadevan, P., Yiakoumis, Y., Sharma, P., Banerjee, S., & McKeown, N. (2010). ElasticTree: Saving energy in data center networks. In *Nsdi* (vol. 10) (pp. 249–264).
28. Hong, C.-Y., Kandula, S., Mahajan, R., Zhang, M., Gill, V., Nanduri, M., & Wattenhofer, R. (2013). Achieving high utilization with software-driven WAN. In *Proceedings of the ACM SIGCOMM Conference on SIGCOMM* (pp. 15–26).
29. Hopps, C. (2000). *Rfc2992: Analysis of an equal-cost multi-path algorithm.*
30. Jain, S., Kumar, A., Mandal, S., Ong, J., Poutievski, L., Singh, A., Venkata, S., Wanderer, J., Zhou, J., Zhu, M., Zolla, J., Hölzle, U., Stuart, S., Vahdat, A. M. (2013). B4: Experience with a globally-deployed software defined WAN. *ACM SIGCOMM Computer Communication Review, 43*(4), 3–14.
31. Jarschel, M., Zinner, T., Hoßfeld, T., Tran-Gia, P., & Kellerer, W. (2014). Interfaces, attributes, and use cases: A compass for SDN. *IEEE Communications Magazine, 52*(6), 210–217.
32. Jin, H., Pan, D., Liu, J., & Pissinou, N. (2012). OpenFlow-based flow-level bandwidth provisioning for CICQ switches. *IEEE Transactions on Computers, 62*(9), 1799–1812.
33. Kandula, S., Sengupta, S., Greenberg, A., Patel, P., & Chaiken, R. (2009). The nature of data center traffic: Measurements & analysis. In *Proceedings of the* 9th *ACM SIGCOMM Conference on Internet Measurement* (pp. 202–208).
34. Kant, K. (2009). Data center evolution: A tutorial on state of the art, issues, and challenges. *Computer Networks, 53*(17), 2939–2965.
35. Kreutz, D., Ramos, F. M. V., Verissimo, P. E., Rothenberg, C. E., Azodolmolky, S., & Uhlig, S. (2014). Software-defined networking: A comprehensive survey. *Proceedings of the IEEE, 103*(1), 14–76.
36. La, R. J., & Anantharam, V. (2002). Utility-based rate control in the Internet for elastic traffic. *IEEE/ACM Transactions On Networking, 10*(2), 272–286.
37. Li, D., Yu, Y., He, W., Zheng, K., & He, B. (2014). Willow: Saving data center network energy for network-limited flows. *IEEE Transactions on Parallel and Distributed Systems, 26*(9), 2610–2620.
38. Li, L. E., Mao, Z. M., & Rexford, J. (2012). Toward software-defined cellular networks. In *European Workshop on Software Defined Networking* (pp. 7–12). IEEE.
39. Li, Y., & Chen, M. (2015). Software-defined network function virtualization: A survey. *IEEE Access, 3*, 2542–2553.
40. Liu, J., Li, J., Shou, G., Hu, Y., Guo, Z., & Dai, W. (2014). SDN based load balancing mechanism for elephant flow in data center networks. In *International Symposium on Wireless Personal Multimedia Communications* (pp. 486–490). IEEE.
41. Marbach, P. (2002). Priority service and max-min fairness. In *Proceedings. Twenty-First Annual Joint Conference of the IEEE Computer and Communications Societies* (vol. 1, pp. 266–275). IEEE.
42. McKeown, N., Anderson, T., Balakrishnan, H., Parulkar, G., Peterson, L., Rexford, J., Shenker, S., & Turner, J. (2008). OpenFlow: enabling innovation in campus networks. *ACM SIGCOMM Computer Communication Review, 38*(2), 69–74.
43. Michel, O., & Keller, E. (2017). SDN in wide-area networks: A survey. In *Fourth International Conference on Software Defined Systems* (pp. 37–42). IEEE.
44. Mogul, J. C., Tourrilhes, J., Yalagandula, P., Sharma, P., Curtis, A. R., & Banerjee, S. (2010). DevoFlow: Cost-effective flow management for high performance enterprise networks. In *Proceedings of the* 9th *ACM SIGCOMM Workshop on Hot Topics in Networks* (pp. 1–60).
45. Nunes, B. A. A., Mendonca, M., Nguyen, X.-N., Obraczka, K., & Turletti, T. (2014). A survey of software-defined networking: Past, present, and future of programmable networks. *IEEE Communications Surveys & Tutorials, 16*(3), 1617–1634.

46. Oktian, Y. E., Witanto, E. N., Kumi, S., & Lee, S.-G. (2019). ISP network bandwidth management: Using blockchain and SDN. In *International Conference on Information and Communication Technology Convergence* (pp. 1330–1335). IEEE.
47. Rhamdani, F., Suwastika, N. A., & Nugroho, M. A. (2018). Equal-cost multipath routing in data center network based on software defined network. In 6th *International Conference on Information and Communication Technology (ICoICT)* (pp. 222–226). IEEE.
48. Riekstin, A. C., Januário, G. C., Rodrigues, B. B., Nascimento, V. T., Carvalho, T. C. M. B., & Meirosu, C. (2016). Orchestration of energy efficiency capabilities in networks. *Journal of Network and Computer Applications, 59*, 74–87.
49. Rimal, B. P., Choi, E., & Lumb, I. (2009). A taxonomy and survey of cloud computing systems. In *Fifth International Joint Conference on INC, IMS and IDC* (pp. 44–51). IEEE.
50. Shang, Y., Li, D., & Xu, M. (2010). Energy-aware routing in data center network. In *Proceedings of the First ACM SIGCOMM Workshop on Green Networking* (pp. 1–8).
51. Sharkh, M. A., Jammal, M., Shami, A., & Ouda, A. (2013). Resource allocation in a network-based cloud computing environment: design challenges. *IEEE Communications Magazine, 51*(11), 46–52.
52. Shirali-Shahreza, S., & Ganjali, Y. (2015). ReWiFlow: Restricted wildcard OpenFlow rules. *ACM SIGCOMM Computer Communication Review, 45*(5), 29–35.
53. Shu, Z., Wan, J., Lin, J., Wang, S., Li, D., Rho, S., & Yang, C. (2016). Traffic engineering in software-defined networking: Measurement and management. *IEEE Access, 4*, 3246–3256.
54. Staessens, D., Sharma, S., Colle, D., Pickavet, M., & Demeester, P. (2011). Software defined networking: Meeting carrier grade requirements. In 18th *IEEE Workshop on Local & Metropolitan Area Networks (LANMAN)* (pp. 1–6). IEEE.
55. Sun, Y., & Kim, M. S. (2010). Tree-based minimization of TCAM entries for packet classification. In 7th *IEEE Consumer Communications and Networking Conference* (pp. 1–5). IEEE.
56. Swallow, G. (1999). MPLS advantages for traffic engineering. *IEEE Communications Magazine, 37*(12), 54–57.
57. Tajiki, M. M., Salsano, S., Chiaraviglio, L., Shojafar, M., & Akbari, B. (2018). Joint energy efficient and QoS-aware path allocation and VNF placement for service function chaining. *IEEE Transactions on Network and Service Management, 16*(1), 374–388.
58. Thanh, N. H., Nam, P. N., Truong, T.-H., Hung, N. T., Doanh, L. K., & Pries, R. (2012). Enabling experiments for energy-efficient data center networks on OpenFlow-based platform. In *Fourth International Conference on Communications and Electronics* (pp. 239–244). IEEE.
59. Thyagaturu, A. S., Mercian, A., McGarry, M. P., Reisslein, M., & Kellerer, W. (2016). Software defined optical networks (SDONs): A comprehensive survey. *IEEE Communications Surveys & Tutorials, 18*(4), 2738–2786.
60. Vasić, N., Bhurat, P., Novaković, D., Canini, M., Shekhar, S., & Kostić, D. (2011). Identifying and using energy-critical paths. In *Proceedings of the Seventh Conference on emerging Networking Experiments and Technologies* (pp. 1–12).
61. Vu, T. H., Luc, V. C., Quan, N. T., Thanh, N. H., & Nam, P. N. (2015). Energy saving for OpenFlow switch on the NetFPGA platform based on queue engineering. *SpringerPlus, 4*(1), 64.
62. Wang, X., Yao, Y., Wang, X., Lu, K., & Cao, Q. (2012). CARPO: Correlation-aware power optimization in data center networks. In *Proceedings IEEE INFOCOM* (pp. 1125–1133). IEEE.
63. Xia, W., Zhao, P., Wen, Y., & Xie, H. (2016). A survey on data center networking (DCN): Infrastructure and operations. *IEEE Communications Surveys & Tutorials, 19*(1), 640–656.
64. Xie, K., Huang, X., Hao, S., & Ma, M. (2018). Distributed power saving for large-scale software-defined data center networks. *IEEE Access, 6*, 5897–5909.
65. Xu, G., Dai, B., Huang, B., & Yang, J. (2015). Bandwidth-aware energy efficient routing with SDN in data center networks. In *IEEE* 17th *International Conference on High Performance Computing and Communications, IEEE* 7th *International Symposium on Cyberspace Safety and Security, and IEEE* 12th *International Conference on Embedded Software and Systems* (pp. 766–771). IEEE.

66. Xu, G., Dai, B., Huang, B., Yang, J., & Wen, S. (2017). Bandwidth-aware energy efficient flow scheduling with SDN in data center networks. *Future Generation Computer Systems, 68*, 163–174.
67. Yang, Y., Xu, M., & Li, Q. (2014). Towards fast rerouting-based energy efficient routing. *Computer Networks, 70*, 1–15.
68. Yoshioka, K., Hirata, K., & Yamamoto, M. (2017). Routing method with flow entry aggregation for software-defined networking. In *International Conference on Information Networking* (pp. 157–162). IEEE.
69. Yu, M., Rexford, J., Freedman, M. J., & Wang, J. (2010). Scalable flow-based networking with DIFANE. *ACM SIGCOMM Computer Communication Review, 40*(4), 351–362.
70. Zakia, U., & Yedder, H. B. (2017). Dynamic load balancing in SDN-based data center networks. In 8th *IEEE Annual Information Technology, Electronics and Mobile Communication Conference* (pp. 242–247). IEEE.
71. Zhou, L., Bhuyan, L. N., & Ramakrishnan, K. K. (2019). DREAM: Distributed energy-aware traffic management for data center networks. In *Proceedings of the Tenth ACM International Conference on Future Energy Systems* (pp. 273–284).
72. Zhu, H., Liao, X., de Laat, C., & Grosso, P. (2016). Joint flow routing-scheduling for energy efficient software defined data center networks: A prototype of energy-aware network management platform. *Journal of Network and Computer Applications, 63*, 110–124.
73. Zhu, J., Jiang, X., Yu, Y., Jin, G., Chen, H., Li, X., & Qu, L. (2020). An efficient priority-driven congestion control algorithm for data center networks. *China Communications, 17*(6), 37–50.

Chapter 11
QoS-Aware Dynamic Flow Management in Software-Defined Data Center Networks

Ayan Mondal and Sudip Misra

11.1 Introduction

With technological advancement, the Internet of things (IoT) devices are capable of generating a huge amount of data, which are to be stored in data centers or managed by the backbone of the data center networks (DCNs) [4, 9]. Additionally, these IoT devices are heterogeneous in terms of computation and memory capacity [37]. As a result, these devices are capable of handling heterogeneous applications in terms of datarate requirements. To handle these heterogeneous applications, we envision using the fat-tree architecture-based DCN, which enables to reduce the single point failure. On the other hand, we consider that software-defined networking (SDN) is one of the promising technologies which can enable the fat-tree DCN for balanced data traffic in the presence of heterogeneous applications. In SDN, due to having a centralized overview of the network, the network failures also can be reduced [7, 8, 10]. However, in the existing literature, the heterogeneity of the flows while designing the schemes for software-defined DCNs is not considered.

Software-defined DCN is an integrating architecture of fat-tree DCN, which follows a hierarchical architecture, and SDN. We envision that instead of having a single controller for the overall network, each pod has a dedicated controller in addition to the centralized controller. Thereby, it reduces the load on the centralized controller and also helps in the efficient management of the network. In the existing literature, researchers focused on designing schemes, viz. [22, 35] for the data

A. Mondal (✉)
Department of Computer Science, University of Rennes 1, INRIA, CNRS, IRISA, Rennes, France
e-mail: ayan.mondal@irisa.fr

S. Misra
Department of Computer Science and Engineering, Indian Institute of Technology Kharagpur, Kharagpur, India
e-mail: smisra@cse.iitkgp.ac.in

G. S. Aujla et al. (eds.), *Software Defined Internet of Everything*, Technology, Communications and Computing, https://doi.org/10.1007/978-3-030-89328-6_11

management and flow rule placement. However, none of these schemes considers the presence of heterogeneous flows in the network. Additionally, few works, viz. [6, 27], focused on managing the heterogeneous flows in SDN. However, these works are capable of providing a local solution and cannot ensure balanced traffic distribution globally. On the other hand, few data transmission schemes, viz. [25], for DCN are proposed by the researchers. However, none of the schemes considers the heterogeneity among the switches and the presence of SDN architecture. Therefore, we argue that we need a design of an efficient data flow management scheme for software-defined DCN while considering the quality of service (QoS) parameters such as per-flow throughput and delay, and overall network throughput and delay [5, 11].

In this work, we design a QoS-aware flow management scheme, named FASCES, for software-defined DCN to ensure that heterogeneous applications generated by the IoT devices are served efficiently while allocating the network resources dynamically for each application. We use a single-leader-multiple-followers Stackelberg game to design the scheme—FASCES. In FASCES, each controller acts as the leader and decides the flow rule association among the incoming flows and the available switches. On the other hand, the IoT applications are considered to be the followers in FASCES. These followers aim to achieve a high datarate while satisfying a delay bound, which depends on the type of applications. To summarize, the contributions of this chapter are as follows:

1. We design a dynamic flow management scheme, named FASCES, for software-defined DCN in the presence of mobile IoT devices, while ensuring high QoS in terms of throughput and delay.
2. We use a single-leader-multiple-follower Stackelberg game to design the interactions between the IoT applications and the controllers. We also evaluate the existence of the Stackelberg equilibrium for FASCES. Using FASCES, we eventually obtain an optimal distribution of flows in the software-defined DCN and optimal datarate of the IoT applications.
3. We evaluate the performance of FASCES in terms of per-flow throughput and delay, and overall network throughput and delay, while comparing with the existing schemes.

The rest of the chapter is organized as follows. In Sect. 11.2, we briefly present the related works in the area of resource management in SDN as well DCN and identified the lacuna in the existing works. The system model and the proposed FASCES scheme are described in Sects. 11.3 and 11.4, respectively. Thereafter, we analyze the performance of FASCES in Sect. 11.5 while comparing with the existing schemes through simulation. Finally, we conclude the chapter while citing a few future directions in Sect. 11.6.

11.2 Related Works

In this section, we survey the related literature on traffic engineering schemes for DCNs and SDNs in detail. The existing literature related to traffic engineering of SD-DCNs is divided into two categories—resource management in SDNs and DCNs.

Resource Management in SDNs
In the existing literature, Bera et al. [13] studied different aspects of resource allocation in SDN for IoT. Saha et al. [35] proposed a flow-rule aggregation scheme for SDN, while focusing on the problem of over-subscription. The authors used a key-based aggregation policy to reduce the number of flow rules. In another work, Maity et al. [22] proposed a tensor-based flow-rule aggregation scheme in SDN. Sadeh et al. [33] designed a flow-traffic aware rule placement scheme while reducing the usage of TCAM space. On the other hand, an optimal multipath flow management scheme is proposed by Rottenstreich et al. [32] while considering network heterogeneity in terms of network path in SDN. Mondal et al. [28] modeled a data traffic management scheme while considering that the data volume associated with the flows is known a priori. An SDN-based network storage scheme is proposed by Wang et al.[42] in the absence of any physical storage. For reducing the usage of oversubscribed buffer, Li et al. [21] suggested not to store the entire packet at the switch but only the packet header. Hayes et al. [20] studied the traffic-classification in SDN. In another work, Saha et al. [34] proposed a QoS-aware routing scheme for SDN, while maximizing end-to-end delay and considered different types of flows in terms of delay- and loss-sensitivity. Bera et al. [12] studied a mobility-aware SDN and attempted to maximize the overall network performance.

Having a centralized overview of the network in SDN, controller can reduce the packet drop and delay while ensuring efficient data traffic management [1]. Tseng et al. [40] designed a scheme for ensuring path stability in hybrid SDN. In this work, initially, the paths are calculated distributively and locally while reducing the computational complexity. Thereafter, the paths are re-evaluated centrally to ensure high stability. Misra and Bera [23] proposed a task offloading scheme for an SDN-based fog network. The authors minimized the delay in task offloading and computation while selecting the optimal number of fog nodes. Singh et al. [38] proposed a hash-based flow-table to reduce the flow-table lookups. In another work, Aujla et al. [5] proposed a traffic flow management scheme in SDN. Moreover, a traffic engineering scheme is proposed by Moradi et al. [30] for SDN-based ISP networks in the presence of heterogeneous links and switches. A fair resource allocation scheme is designed by Allybokus et al. [3] in multipath SDN. Sanvito et al. [36] also proposed a flow-table reconfiguration scheme, while considering overlapping data flow paths.

Resource Management in DCNs
In existing literature, researchers studied *Fat-tree DCNs* [2, 19]. The different challenges of DCN such as generation, processing, and storage of data are studied

by Chen et al. [18] in the presence of various applications such as social networks, healthcare, smart grid, and managing enterprises. Similarly, Chakraborty et al. designed schemes for provisioning sensor-based services in data center networks while considering economic aspect [14, 16] and resource orchestration [15, 24, 26]. A network selection scheme for multimedia data delivery in ad-hoc networks proposed by Trestian et al. [39]. Moreover, the optimal server positioning scheme is proposed by Paul et al. [31] while minimizing the maintenance cost.

Synthesis
Based on the study of the existing works, we observe that a few schemes are proposed for data traffic management in SDN as well as DCN. However, the researchers have not considered the presence of heterogeneous applications and switches in the existing literature. Additionally, efficient management of heterogeneous data traffic in software-defined DCN while ensuring optimized QoS in terms of high throughput and low delay is one of the important aspects which needs to be addressed.

11.3 System Model

We consider an SDN-enabled Fat-Tree architecture of DCN [25]. A general fat-tree architecture is composed of three layers—edge, aggregation, and core layers, which enables reducing the bottleneck in transmission as well as is capable of handling the single point failures. Additionally, we consider a multi-tier SDN, where there is a dedicated SDN controller for each pod at the aggregation layer. The switches at the aggregation layer are connected with the switches at the core layers. We consider that the switches at the core layer are managed by a single controller. Moreover, in this work, we consider that the IoT devices at the edge layer are mobile and are connected to the switches at the aggregation layer through the access points. The system architecture is depicted in Fig. 11.1. The IoT devices are capable of executing heterogeneous applications having different datarate.

Each application a_n of IoT devices $n \in \cdot N$, where N is the set of IoT devices at the edge layer, denotes a separate flow[1] and has a datarate r_a and connected with a switch s. These switches at the aggregation layers, where the set of switches is denoted by S_2, are connected with the set of switches at the core layer, which is denoted by S_1. We consider that each switch at the aggregation and core layers is heterogeneous in terms of bandwidth and TCAM. Furthermore, in addition to the IoT devices, we also consider the presence of servers at the edge layer.

To achieve a high throughput with an optimal delay, we need to ensure a balanced data traffic in the network. Considering that each switch s has a limited capacity of B_s and there are A_s set of flows associated with switch s, where $\forall s \in S_1 \cup S_2$, the following constraint needs to be satisfied:

[1] For the rest of the chapter, we use a_n to denote flow or application a generated from IoT device n, synonymously.

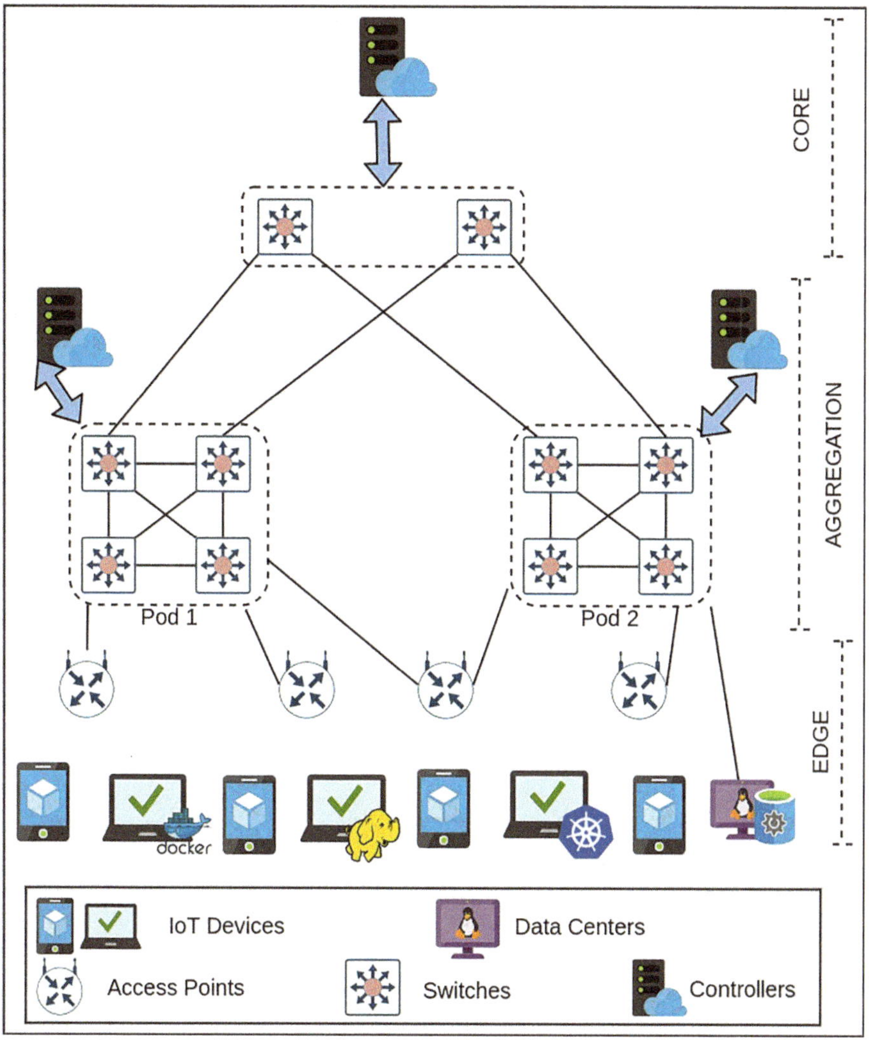

Fig. 11.1 Schematic diagram of software-defined fat-tree DCN

$$B_s \leq \sum_{a_n \in A_s} r_a \tag{11.1}$$

On the other hand, each application a_n has a delay threshold d_a^{th}. Hence, while allocating the flows to the switches, the following constraint also needs to be satisfied.

$$d_a^{th} \leq \sum_{s \in P_a} d_a^s \tag{11.2}$$

where P_a denotes the set of switches associated with the flow a_n; and d_a^s represent the delay at switch s for handling flow a_n. We consider that for handling each flow, the switches follow first-in-first-out (FIFO) scheduling and require a fixed duration Δ. Hence, for processing a single packet, we get:

$$d_a^s = \sum_{s \in P_a} \sum_{a_n \in A_s} \Delta \tag{11.3}$$

Additionally, due to limited TCAM, the maximum number of flows can be handled by each switch s is denoted by M_s and must satisfy the following constraint:

$$M_s \geq |A_s| \tag{11.4}$$

Hence, to serve a high number of flows, the controllers can request to reduce the associated datarate for each application a_n, however, each IoT device n needs to ensure that minimum datarate r_a^{min} is achieved, i.e., $r_a \geq r_a^{min}$.

11.4 FASCES: QoS-Aware Dynamic Flow Management Scheme

We propose a single-leader-multiple-followers Stackelberg game for studying the interaction between the IoT applications and the controllers in software-defined data center networks. The Stackelberg game is a non-cooperative game that deals with the interaction among the leaders and followers. In FASCES, the controller acts as the leader, and the IoT application act as the followers. In this work, the controllers at the aggregation layer deal with the IoT applications directly. However, the controller at the core layer needs to interact with the controllers at the aggregation layer. Hence, the decision of each leader at the aggregation layer is always influenced by the decision of the controllers at the core layer. In the proposed game, the leaders aim to maximize their utility values while maximizing the bandwidth utilization with optimal delay and maximizing the number of applications served. On the other hand, the followers aim to maximize their utility while obtaining a high datarate with less delay. Therefore, the components of FASCES are as follows:

1. Each controller acts as the leader. The utility of each controller at the aggregation layer is influenced by the decision of the controller at the core layer. The decision of the controllers is executed by the switches, hence the switches are not considered active players in the proposed game.

2. Each IoT application acts as the follower and decides the required datarate. The maximum datarate can be achieved by each application depends on the hosted IoT device.
3. The IoT applications run for a finite duration which is not known a priori.

11.4.1 Single-Leader-Multiple-Followers Stackelberg Game: The Justification

The proposed system comprising of the fat-tree DCN and the SDN follows a hierarchical architecture. The different entities, such as the controllers and IoT applications, are non-cooperative and aim to maximize their payoff values. This results in a "oligopolistic market" scenario [17]. On the other hand, Stackelberg game is the most suitable game-theoretic approach to model a hierarchical system with non-cooperative players. Hence, we propose to use the single-leader-multiple-followers Stackelberg game for designing the FASCES scheme.

11.4.2 Game Formulation

To model the game-theoretic interactions in FASCES, we design two utility functions for the controllers and the IoT applications, which are discussed as follows.

Utility Function of Each IoT Application

The utility function $U_a(\cdot)$ of each IoT application signifies the satisfaction of the end-users in data transmission. Each application a_n needs to finalize an optimal datarate r_a^* to ensure that the associated flow rule is active. Considering that each switch $s \in P_a$ handles A_s set of applications, the optimal datarate of flow r_a^* depends on $r_{-\boldsymbol{a}}^*$, where $r_{-\boldsymbol{a}}^* = A_s \setminus \{a_n\}$. This is because the IoT applications are non-cooperative. Therefore, the utility function $U_a(r_a, r_{-\boldsymbol{a}}, A_s, P_a)$ of each IoT application a_n of IoT device n needs to satisfy the following constraints:

1. Each IoT device aims to achieve the maximum datarate r_a^{max}, where $r_a \leq r_a^{max}$. Therefore, the utility function $U_a(r_a, r_{-\boldsymbol{a}}, A_s, P_a)$ is a non-decreasing function. Mathematically,

$$\frac{\partial U_a(r_a, r_{-\boldsymbol{a}}, A_s, P_a)}{\partial r_a} \geq 0 \tag{11.5}$$

2. The payoff value of $U_a(r_a, r_{-\boldsymbol{a}}, A_s, P_a)$ decreases on increasing the datarate beyond the optimal value. Therefore, in the marginal condition, $U_a(r_a, r_{-\boldsymbol{a}}, A_s, P_a)$ is considered to be a non-increasing function. Mathematically,

$$\frac{\partial^2 U_a(r_a, r_{-\boldsymbol{a}}, A_s, P_a)}{\partial (r_a)^2} < 0 \tag{11.6}$$

3. With the increase in the number of applications, i.e., $|A_s|$ managed by each switch s, the probability of flow rule replacement increases. Hence, the payoff of the utility function $U_a(r_a, r_{-\boldsymbol{a}}, A_s, P_a)$ decreases with the increase in $|A_s|$. Mathematically,

$$\frac{\partial U_a(r_a, r_{-\boldsymbol{a}}, A_s, P_a)}{\partial |A_s|} < 0, \quad \forall s \in P_a \tag{11.7}$$

Therefore, motivated by the work of Tushar et al. [41], in FASCES, the utility function $U_a(r_a, r_{-\boldsymbol{a}}, A_s, P_a)$ of IoT application a_n is represented as follows:

$$U_a(r_a, r_{-\boldsymbol{a}}, A_s, P_a) = r_a^{max} r_a - \left(\frac{r_a^{min}}{r_a^{max}}\right) {r_a}^2 - r_a \frac{\sum_{s \in P_a} |A_s|}{|P_a|} \tag{11.8}$$

In FASCES, each IoT application aims to maximize the payoff of $U_a(r_a, r_{-\boldsymbol{a}}, A_s, P_a)$, while deciding an optimal datarate. Mathematically,

$$_{r_a}U_a(r_a, r_{-\boldsymbol{a}}, A_s, P_a) \tag{11.9}$$

while satisfying the constraints mentioned in Eqs. (11.1) and (11.2).

Utility Function of Each Controller

The utility function $B_c(r_a, r_{-\boldsymbol{a}}, A_s)$ of each controller c signifies the utilization of the switch capacity B_s. The controllers aim to maximize the set of applications served as well as maximize the bandwidth allocated to each application or flow. Therefore, the utility function $B_c(r_a, r_{-\boldsymbol{a}}, A_s)$ of each controller c needs to satisfy the following constraints:

1. Each controller tries to allocate high bandwidth possible to ensure high utilization of its capacity. Mathematically,

$$\frac{\partial B_c(r_a, r_{-\boldsymbol{a}}, A_s)}{\partial r_a} \geq 0 \tag{11.10}$$

2. The overall objective of the controllers is to accommodate high number of flows, while satisfying the physical limitations of the switches. Mathematically,

$$\frac{\partial B_c(r_a, r_{-\boldsymbol{a}}, A_s, P_a)}{\partial |A_s|} > 0 \tag{11.11}$$

Therefore, we design the utility function $B_c(r_a, r_{-\boldsymbol{a}}, A_s)$ of each controller c as follows:

$$\sum_{r_a, A_s} B_c(r_a, r_{-a}, A_s) \tag{11.12}$$

Each controller c aims to maximize the payoff of $B_c(r_a, r_{-a}, A_s)$ while satisfying the constraints in Eqs. (11.1) and (11.4).

11.4.3 Existence of Equilibrium

In this section, we evaluate the existence of the Stackelberg equilibrium, defined in Definition 11.4.3, for FASCES in Theorem 11.4.3.

In FASCES, the Stackelberg equilibrium is denoted as a tuple of $\langle r_a^*, A_s^* \rangle$, where r_a^* and A_s^* represent the optimal datarate for application a_n and the optimal set of flows associated with switch $s \in S1 \cup S_2$. The equilibrium condition also needs to satisfy the following constraints:

$$U_a(r_a^*, r_{-a}^*, A_s^*, P_a^*) \geq U_a(r_a, r_{-a}^*, A_s^*, P_a^*) \tag{11.13}$$

$$B_c(r_a^*, r_{-a}^*, A_s^*) \geq B_c(r_a^*, r_{-a}^*, A_s) \tag{11.14}$$

Given an optimal set of flows A_s^* for each switch $s \in P_a$, a Stackelberg equilibrium exists for each IoT application a_n.

Proof The cumulative payoff obtained by the applications A_s associated with switch s is represented as follows:

$$\boldsymbol{U}_s(\cdot) = \sum_{a_n \in A_s} U_a(r_a, r_{-a}, A_s, P_a) \tag{11.15}$$

By considering the Karush–Kuhn–Tucker (KKT) conditions [29] on $\boldsymbol{U}_s(\cdot)$, we get:

$$L_s = \boldsymbol{U}_s(\cdot) + \lambda_1(B_s - \sum_{a_n \in A_s} r_a) + \sum_{a_n \in A_s} \lambda_2^a(d_a^{th} - \sum_{s \in P_a} d_a^s) \tag{11.16}$$

where λ_1 and λ_2^a are the Lagrangian multipliers. By taking the derivative of L_s, we obtain the Hessian matrix $\nabla^2 L_s$ is as follows:

$$\nabla^2 L_s = \begin{bmatrix} -\frac{r_1^{min}}{r_1^{max}} & \cdots & 0 & \cdots & 0 \\ \vdots & \ddots & \vdots & \ddots & \vdots \\ 0 & \cdots & -\frac{r_a^{min}}{r_a^{max}} & \cdots & 0 \\ \vdots & \ddots & \vdots & \ddots & \vdots \\ 0 & \cdots & 0 & \cdots & -\frac{r_{|A_S|}^{min}}{r_{|A_S|}^{max}} \end{bmatrix} \quad (11.17)$$

We observe that the obtained Hessian matrix is a negative diagonal matrix. Hence, we conclude that there exists at least one Stackelberg equilibrium in FASCES. □

11.4.4 Proposed Workflow

To obtain the optimal distribution of flows, FASCES follows the workflow as shown in Fig. 11.2. Initially, each application informs about their minimum datarate requirement to the controllers. On receiving the set of applications to be served, the controllers allocate the flows optimally among the available switches at aggregation and core layers. Thereafter, the controllers inform the path associated with each flow to the IoT devices, and these devices try to find an optimal value of the datarate can be achieved while interacting with the controllers.

11.5 Performance Evaluation

In this section, the performance of FASCES is analyzed through simulation with varying the number of heterogeneous IoT applications. We simulated using the MATLAB simulation platform considering a terrain of 10×10 m^2 [25]. The deployment of switches follows a grid pattern. On the other hand, IoT devices are deployed randomly and follow random waypoint mobility model[25]. We consider that there are 2 pods at the aggregation layer, where each pod is comprised of 4 switches and 2 switches at the core layer. We considered that the IoT applications generate data in chunk having size 800 Mb, as shown in Table 11.1. We consider the datarate requirement distribution of IoT applications, as presented in Table 11.2.

The performance of FASCES is evaluated while comparing two of the existing schemes—data flow management in SDN (FlowMan) [27] and data broadcasting in fat-tree DCN (D2B) [25]. In FlowMan, the authors proposed a Nash bargaining game-based data traffic management scheme for SDN. However, while allocating resources, the authors only considered the flows within one-hop neighbors. In other words, FlowMan is capable of ensuring a local optimum which is not sufficient for

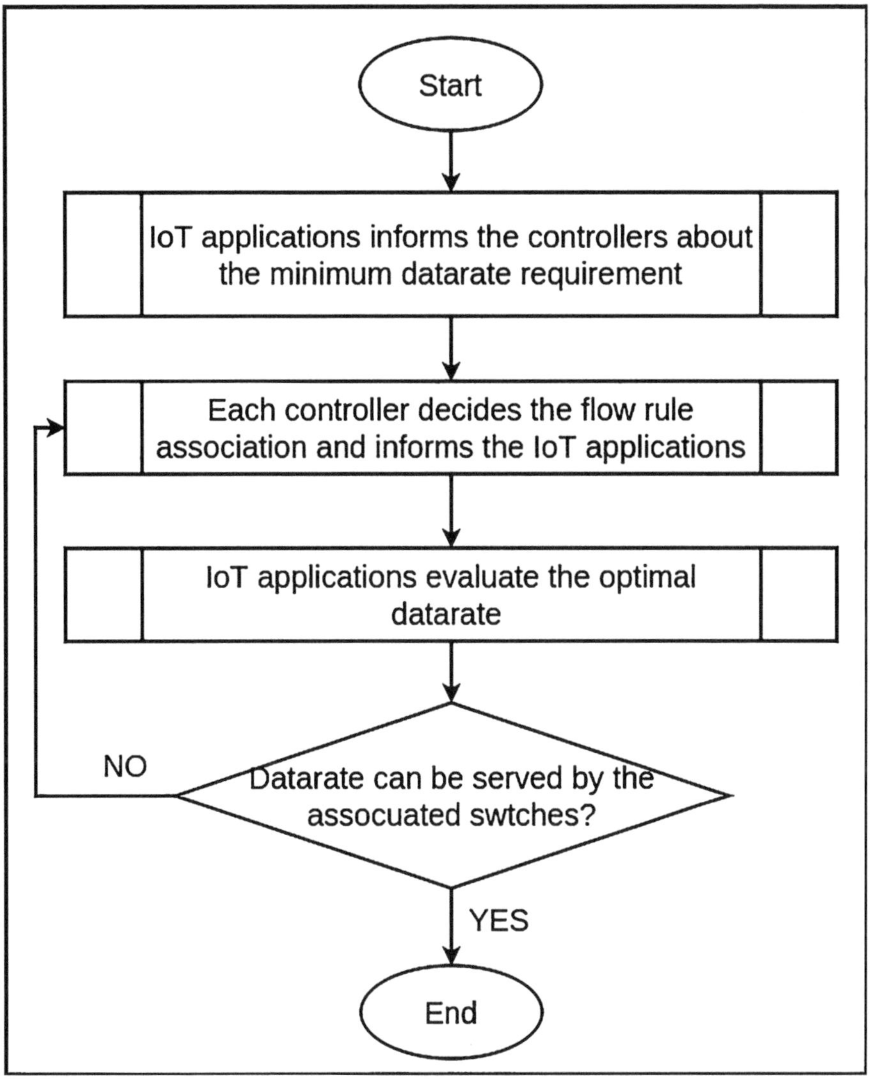

Fig. 11.2 Workflow of FASCES

a network with heterogeneous switches. On the other hand, in D2B, the authors proposed a Stackelberg game-based data broadcasting scheme for DCN. However, in D2B, only a single IoT source is considered. Additionally, the switches are homogeneous and traditional without having any limitation on the set of applications that can be handled by each switch. Using FASCES, we address these lacunae in the existing literature while ensuring balanced data traffic in the network.

Table 11.1 Simulation parameters

Parameter	Value
Number of applications	1000–5000
Maximum datarate of IoT applications	128–5000 Kbps
Velocity of IoT devices	6–10 m/s
Maximum capacity of each switches	5–10 Gbps
Chunk of data generated by each IoT applications	500–2000

Table 11.2 Maximum datarate distribution [25]

Maximum datarate (Kbps)	IoT applications (%)
5000	15
1000	25
1000	25
384	40
128	20

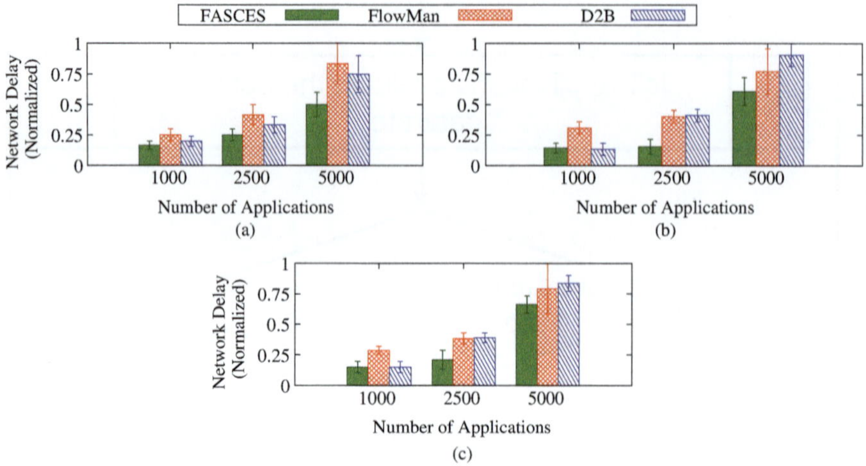

Fig. 11.3 Per-flow throughput. (**a**) Velocity = 6 m/s. (**b**) Velocity = 8 m/s. (**c**) Velocity = 10 m/s

We evaluate the performance of FASCES based on the following parameters—(1) per-flow throughput, (2) per-flow delay, (3) network throughput, (4) network delay, and (5) set of serviced IoT applications.

Figure 11.3 depicts that with the increase in the number of applications, the per-flow throughput increases 15.37–26.91% using FASCES than using FlowMan and D2B. However, with the increase in the velocity of IoT devices, the throughput decreases as the applications need to be associated with new switches very often and a few packets get dropped due to the delay constraint. On the other hand, the delay for each flow also reduces by 27.78–36.67% using FASCES than that of using FlowMan and D2B, as depicted in Fig. 11.4. Additionally, we observe that with the increase in the number of applications, the increase in delay is not significant using FASCES.

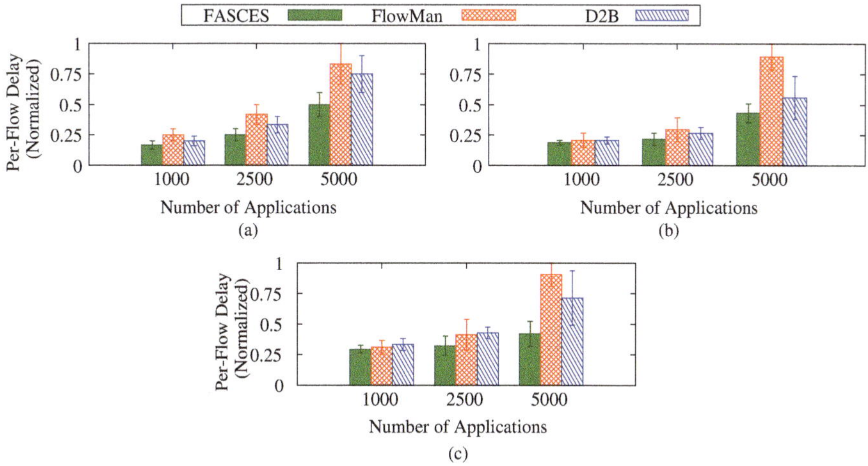

Fig. 11.4 Per-flow delay. (**a**) Velocity = 6 m/s. (**b**) Velocity = 8 m/s. (**c**) Velocity = 10 m/s

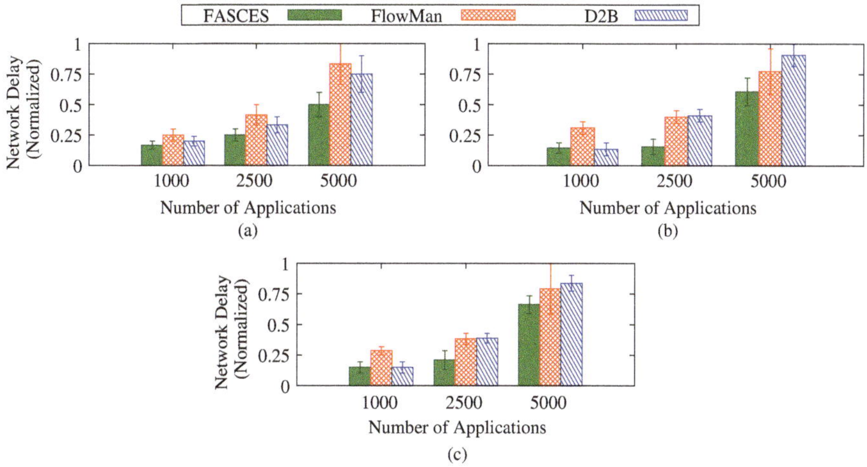

Fig. 11.5 Network delay. (**a**) Velocity = 6 m/s. (**b**) Velocity = 8 m/s. (**c**) Velocity = 10 m/s

Similarly, we observe that the network delay decreases by 16.67–19.45% and the network throughput increases by 16.67–19.45% using FASCES than using FlowMan and D2B, as depicted in Figs. 11.5 and 11.6, respectively. This is because the flows are distributed efficiently among the switches at the aggregation and core layers while ensuring efficient utilization of link capacity and TCAM space.

Furthermore, from Fig. 11.7, we observe that FASCES is capable of serving all the applications with efficient data traffic distribution. However, with the increase in the number of applications, FASCES cannot serve 100% application due to physical

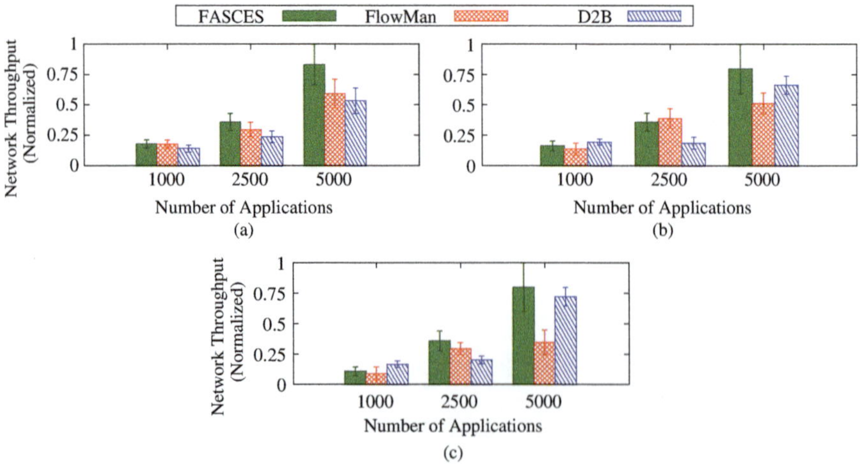

Fig. 11.6 Network throughput. (**a**) Velocity = 6 m/s. (**b**) Velocity = 8 m/s. (**c**) Velocity = 10 m/s

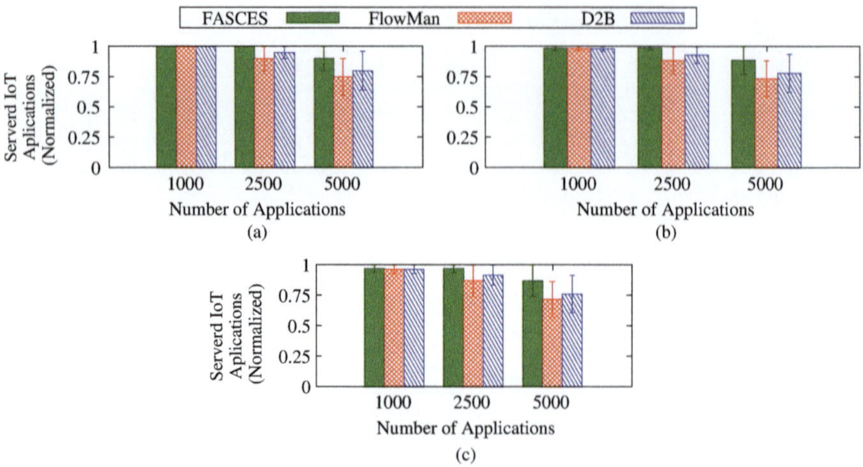

Fig. 11.7 Serviced IoT applications. (**a**) Velocity = 6 m/s. (**b**) Velocity = 8 m/s. (**c**) Velocity = 10 m/s

limitations of the system. Using FASCES, we yield a 4.56–16.67% increase in serviced application than using FlowMan and D2B.

11.6 Conclusion

In this work, we proposed a data traffic management scheme, named FASCES, and modeled the interaction between the controllers and the SDN switches using a single-leader-multiple-followers Stackelberg game. FASCES is capable of ensuring balanced data traffic in the presence of heterogeneous IoT flows and SDN switches. We observed that FASCES reduces the per-flow delay as well as network delay at least by 27.78 and 16.67%, respectively while ensuring an increase in both per-flow throughput and network throughput. Through simulation, we yield that FASCES ensures efficient flow distribution in software-defined DCN.

In future, this work can be extended while designing data traffic management for the recursive architectures of DCN such as B-Cube and DCell. Additionally, we can also explore this work while considering a multi-tier controller structure for each layer in fat-tree DCN. Moreover, this work also can be extended while considering the link and switch failure in software-defined DCN.

References

1. Agarwal, S., Kodialam, M., & Lakshman, T. V. (2013). Traffic engineering in software defined networks, in *Proceedings of IEEE INFOCOM* (pp. 1–9). https://doi.org/10.1109/infcom.2013.6567024
2. Al-Fares, M., Loukissas, A., & Vahdat, A. (2008). A scalable, commodity data center network architecture, in *Proceedings of ACM SIGCOMM Conference on Data Communication* (pp. 63–74). New York: ACM. http://doi.acm.org/10.1145/1402958.1402967
3. Allybokus, Z., Avrachenkov, K., Leguay, J., & Maggi, L. (2018). Multi-path alpha-fair resource allocation at scale in distributed software-defined networks. *IEEE Journal on Selected Areas in Communications, 36*(12), 2655–2666. https://doi.org/10.1109/JSAC.2018.2871293
4. Aujla, G. S., Chaudhary, R., Kaur, K., Garg, S., Kumar, N., & Ranjan, R. (2018). SAFE: SDN-assisted framework for edge–cloud interplay in secure healthcare ecosystem. *IEEE Transactions on Industrial Informatics, 15*(1), 469–480.
5. Aujla, G. S., Chaudhary, R., Kumar, N., Kumar, R., & Rodrigues, J. J. P. C. (2018). An ensembled scheme for QoS-aware traffic flow management in software defined networks, in *2018 IEEE International Conference on Communications (ICC)* (pp. 1–7). Piscataway: IEEE. https://doi.org/10.1109/ICC.2018.8422596
6. Aujla, G. S., Chaudhary, R., Kumar, N., Rodrigues, & J. J., Vinel, A. (2017). Data offloading in 5g-enabled software-defined vehicular networks: A Stackelberg-game-based approach. *IEEE Communications Magazine, 55*(8), 100–108.
7. Aujla, G. S., Garg, S., Batra, S., Kumar, N., You, I., & Sharma, V. (2019). DROpS: A demand response optimization scheme in SDN-enabled smart energy ecosystem. *Information Sciences, 476*, 453–473.
8. Aujla, G. S., Jindal, A., & Kumar, N. (2018). EVaaS: Electric vehicle-as-a-service for energy trading in SDN-enabled smart transportation system. *Computer Networks, 143*, 247–262.
9. Aujla, G. S. S., Kumar, N., Garg, S., Kaur, K., & Ranjan, R. (2019). EDCSuS: Sustainable edge data centers as a service in SDN-enabled vehicular environment. *IEEE Transactions on Sustainable Computing* (pp. 1–14). Early Access. https://doi.org/10.1109/TSUSC.2019.2907110
10. Aujla, G. S., Singh, M., Bose, A., Ku mar, N., Han, G., & Buyya, R. (2020). BlockSDN: Blockchain-as-a-service for software defined networking in smart city applications. *IEEE Network, 34*(2), 83–91.
11. Aujla, G. S., Singh, A., & Kumar, N. (2019). Adaptflow: Adaptive flow forwarding scheme for software-defined industrial networks. *IEEE Internet of Things Journal, 7*(7), 5843–5851.

12. Bera, S., Misra, S., & Obaidat, M. S. (2016). Mobility-aware flow-table implementation in software-defined IoT, in *Proceedings of IEEE GLOBECOM* (pp. 1–6). https://doi.org/10.1109/GLOCOM.2016.7841995
13. Bera, S., Misra, S., & Vasilakos, A. V. (2017). Software-defined networking for internet of things: A survey. *IEEE Internet of Things Journal 4*(6), 1994–2008.
14. Chakraborty, A., Misra, S., & Mondal, A. (2020). QoS-aware dynamic cost management scheme for sensors-as-a-service. *IEEE Transactions on Services Computing*, Early Access, 1–12 (2020). https://doi.org/10.1109/TSC.2020.3011495
15. Chakraborty, A., Misra, S., Mondal, A., & Obaidat, M. S. (2020). Sensorch: QoS-aware resource orchestration for provisioning sensors-as-a-service, in *ICC 2020 - 2020 IEEE International Conference on Communications (ICC)* (pp. 1–6). https://doi.org/10.1109/ICC40277.2020.9148621
16. Chakraborty, A., Mondal, A., & Misra, S. (2018). Cache-enabled sensor-cloud: The economic facet, in *2018 IEEE Wireless Communications and Networking Conference (WCNC)*, pp. 1–6. https://doi.org/10.1109/WCNC.2018.8377069
17. Chakraborty, A., Mondal, A., Roy, A., & Misra, S. (2018). Dynamic trust enforcing pricing scheme for sensors-as-a-service in sensor-cloud infrastructure. *IEEE Transactions on Services Computing, 14*(5), pp. 1345–1356. Early Access. https://doi.org/10.1109/TSC.2018.2873763
18. Chen, M., Mao, S., & Liu, Y. (2014). Big data: A survey. *Mobile Networks and Applications, 19*(2), 171–209. https://doi.org/10.1007/s11036-013-0489-0
19. Guo, Z., & Yang, Y. (2013). Multicast fat-tree data center networks with bounded link oversubscription, in *Proceedings of IEEE INFOCOM* (pp. 350–354). https://doi.org/10.1109/INFCOM.2013.6566793
20. Hayes, M., Ng, B., Pekar, A., & Seah, W. K. G.: (2018). Scalable architecture for SDN traffic classification. *IEEE Systems Journal, 12*, 3203–3214. https://doi.org/10.1109/JSYST.2017.2690259
21. Li, F., Cao, J., Wang, X., Sun, Y., Pan, T., & Liu, X. (2017). Adopting SDN switch buffer: Benefits analysis and mechanism design, in *Proceedings of IEEE International Conference on Distributed Computing Systems (ICDCS)* (pp. 2171–2176). https://doi.org/10.1109/ICDCS.2017.255
22. Maity, I., Mondal, A., Misra, S., & Mandal, C. (2019). Tensor-based rule-space management system in SDN. *IEEE Systems Journal, 13*(4), 3921–3928.
23. Misra, S., & Bera, S. (2020). Soft-VAN: Mobility-aware task offloading in software-defined vehicular network. *IEEE Transactions on Vehicular Technology, 69*(2), 2071–2078.
24. Misra, S., & Chakraborty, A. (2019). QoS-aware dispersed dynamic mapping of virtual sensors in sensor-cloud. *IEEE Transactions on Services Computing* (pp. 1–12). Early Access. https://doi.org/10.1109/TSC.2019.2917447
25. Misra, S., Mondal, A., & Khajjayam, S. (2019). Dynamic big-data broadcast in fat-tree data center networks with mobile IoT devices. *IEEE Systems Journal, 13*(3), 2898–2905. https://doi.org/10.1109/JSYST.2019.2899754
26. Misra, S., Schober, R., & Chakraborty, A. (2020). Race: Qoi-aware strategic resource allocation for provisioning Se-aaS. *IEEE Transactions on Services Computing*, 1–12 (2020). https://doi.org/10.1109/TSC.2020.3001078
27. Mondal, A., & Misra, S. (2020). Flowman: QoS-aware dynamic data flow management in software-defined networks. *IEEE Journal on Selected Areas in Communications, 38*(7), 1366–1373. https://doi.org/10.1109/JSAC.2020.2999682
28. Mondal, A., Misra, S., & Chakraborty, A. (2018). TROD: Throughput-optimal dynamic data traffic management in software-defined networks, in *Proceedings of IEEE Globecom Workshops* (pp. 1–6). https://doi.org/10.1109/GLOCOMW.2018.8644398
29. Mondal, A., Misra, S., & Obaidat, M. S. (2017). Distributed home energy management system with storage in smart grid using game theory. *IEEE Systems Journal, 11*(3), 1857–1866. https://doi.org/10.1109/JSYST.2015.2421941

30. Moradi, M., Zhang, Y., Morley Mao, Z., & Manghirmalani, R. (2018). Dragon: Scalable, flexible, and efficient traffic engineering in software defined ISP networks. *IEEE Journal on Selected Areas in Communications, 36*(12), 2744–2756. https://doi.org/10.1109/JSAC.2018.2871312
31. Paul, D., Zhong, W. D., & Bose, S. K. (2017). Demand response in data centers through energy-efficient scheduling and simple incentivization. *IEEE Systems Journal, 11*(2), 613–624. https://doi.org/10.1109/JSYST.2015.2476357
32. Rottenstreich, O., Kanizo, Y., Kaplan, H., & Rexford, J. (2018). Accurate traffic splitting on SDN switches. *IEEE Journal on Selected Areas in Communications 36*(10), 2190–2201. https://doi.org/10.1109/JSAC.2018.2869949
33. Sadeh, Y., Rottenstreich, O., Barkan, A., Kanizo, Y., & Kaplan, H. (2019). Optimal representations of a traffic distribution in switch memories, in *Proceedings of IEEE INFOCOM* (pp. 2035–2043). https://doi.org/10.1109/infocom.2019.8737645
34. Saha, N., Bera, S., & Misra, S. (2018). Sway: Traffic-aware QoS routing in software-defined IoT. *IEEE Transactions on Emerging Topics in Computing, 9*, 390–401.
35. Saha, N., Misra, S., & Bera, S. (2018). QoS-aware adaptive flow-rule aggregation in software-defined IoT, in *Proceedings of IEEE Global Communications Conference (GLOBECOM)* (pp. 206–212).
36. Sanvito, D., Filippini, I., Capone, A., Paris, S., & Leguay, J. (2018). Adaptive robust traffic engineering in software defined networks, in *Proceedings of IFIP Networking and Workshops* (pp. 1–9). https://doi.org/10.23919/ifipnetworking.2018.8696406
37. Singh, A., Aujla, G. S., Garg, S., Kaddoum, G., & Singh, G. (2019). Deep-learning-based SDN model for Internet of Things: An incremental tensor train approach. *IEEE Internet of Things Journal 7*(7), 6302–6311.
38. Singh, A., Batra, S., Aujla, G. S. S., Kumar, N., & Yang, L. T. (2020). BloomStore: Dynamic bloom filter-based secure rule-space management scheme in SDN. IEEE Transactions on Industrial Informatics pp. 1–11 (2020). https://doi.org/10.1109/TII.2020.2966708
39. Trestian, R., Ormond, O., & Muntean, G. M. (2014). Enhanced power-friendly access network selection strategy for multimedia delivery over heterogeneous wireless networks. *IEEE Transactions on Broadcasting, 60*(1), 85–101.
40. Tseng, S. H., Tang, A., Choudhury, G. L., & Tse, S. (2019). Routing stability in hybrid software-defined networks. *IEEE/ACM Transactions on Networking, 27*(2), 790–804. https://doi.org/10.1109/tnet.2019.2900199
41. Tushar, W., Saad, W., Poor, H. V., & Smith, D. B. (2012). Economics of electric vehicle charging: A game theoretic approach. *IEEE Transactions on Smart Grid, 3*(4), 1767–1778. https://doi.org/10.1109/TSG.2012.2211901
42. Wang, M. H., Chi, P. W., Guo, J. W., & Lei, C. L. (2016). SDN storage: A stream-based storage system over software-defined networks, in *Proceedings of IEEE INFOCOM Workshops* (pp. 598–599). https://doi.org/10.1109/INFCOMW.2016.7562146

Part IV
Security and Trust Applications for Software-Defined Networking

Chapter 12
Trusted Mechanism Using Artificial Neural Networks in Healthcare Software-Defined Networks

Geetanjali Rathee

12.1 Introduction

Software-Defined Networks (SDN) is an architecture which provides the network to behave intelligent and automated programmed and controlled using software applications [4, 5, 7]. The SDN enables the network to behave holistically and consistently regard-less of their technique. It provides centralized controlling using software applications by implementing common SDN layers. In addition, SDN provides abstracting of data plans by supporting the control plane isolation from logical separations running at the network top using three different components such as SDN applications, SDN controller, and SDN networking services consisting of three different layers like infrastructure, control, and application layer. Along with a centralized and automatic controlled system, SDN is vulnerable to many security threats and issues such as (1) incomplete encryption where intruders may sometimes access the information and modify it for their own benefits. Attackers may manipulate the network by compromising some of the communicating nodes by taking data plane and control plane on its own. The most severe threats in SDN are denial-of-service, man-in-middle, and data alteration threats. Therefore, it is needed to propose some security resilient mechanisms for SDN while transferring the information in any IoT based applications such as healthcare systems. The involvement of intelligent analyzing devices (IoT devices) in traditional healthcare schemes improved the overall management and processing of huge amount of information in an efficient and standardized way [2, 3]. The online storage systems such as cloud services may further reduce the storage overhead and computation of large data generated by smart devices. Though, embedding of IoT devices in

G. Rathee (✉)
Department of Computer Science and Engineering, Netaji Subhas University of Technology, New Delhi, India

G. S. Aujla et al. (eds.), *Software Defined Internet of Everything*, Technology, Communications and Computing, https://doi.org/10.1007/978-3-030-89328-6_12

various applications provides several benefits to the modern era, however, the organizations are afraid to fully adopt this technique in current scenarios [9, 11, 15]. The involvement of intruders may not only affect the network performance but also degrade the overall budget of the company. The intruder's aim in network system is to compromise the legal nodes by forging their identity for their own benefits. In case of healthcare systems, intruders may drastically affect the overall performance of network in several ways. During the recording of patient's history, manufacturing of medicines, patient's medical records details can be easily forged by the intruders after compromising the legitimate IoT devices and continuously track their communication step. In order to fully adopt the smart technique by various organizations, it is needed to make a system very secure and efficient. Though researchers have proposed various cryptographic and hypothetical scenarios, however, the amount of time and cost required legitimate the system may further encourage the researchers to propose new security schemes [17–19]. Along with several success stories in various biomedical fields, the Artificial Intelligence (AI) assistance can be considered as a medium to conduct remote automation of activities. The integration of AI technique with smart devices in healthcare applications may further improve the managing and immediate services in the network. The AI-based learning has been widely integrated in medical scenarios for its unprecedented performance still the lack of reliable data sets to analyze malicious activities of smart devices is researched at its early stages.

12.1.1 Contribution

The aim of this paper is to propose a secure and trusted communication mechanism by identifying malicious behavior of nodes in healthcare systems. The proposed mechanism generated an Artificial Neural Network (ANN) to process the inputs from various devices. The generated output of ANN determines the legitimacy of each node by analyzing their certain behavior. Further, a back propagation (BP) algorithm is used that is responsible to generate an error free categorization of devices in healthcare mechanisms. Further, the potential contribution of the paper is discussed as follows:

- A trusted and secure ANN network is proposed where the devices category by ANN is computed by analyzing their previous communication behavior.
- The BP algorithm is further used to improve the secrecy of system by generating an error free computation.
- The proposed mechanism is analyzed over several numerical results against conventional approach.

The remaining structure of the paper is organized as follows. The number of security techniques proposed by several scientists/authors is discussed in section two [6, 14]. The ANN with BP algorithm is used to ensure a secure communication and error free categorization of smart devices in section three. Further, section four illustrates the verification of proposed phenomenon over various security measures such as identification of trusted smart nodes and data alteration against conventional mechanism. Finally, section five discusses the conclusion along with its future directions.

12.2 Related Work

Trusted based schemes can be defined as one of the most promising security schemes in various applications. This section discussed various security techniques proposed by various scientists/researchers. Meng et al. [16] have focused on identification of insider threats by surveying the stakeholders from 12 different healthcare centers. The authors have developed a trusted mechanism using Bayesian rules based on survey outcomes and searched the malicious nodes in healthcare environments. The simulated result demonstrated the effectiveness and feasibility of proposed mechanism for identifying the malicious activity of the nodes.

In addition, the authors claimed the identification of malicious devices much faster as compared to existing schemes. Table 12.1 depicts a latest literature survey

Table 12.1 Literature survey on security schemes in Healthcare SDN

Authors	Techniques	Mechanism
Meng et al. [16]	A trusted mechanism using Bayesian rules	The simulated result demonstrated the effectiveness and feasibility of proposed mechanism for identifying the malicious activity of the nodes.
Jiang et al. [13]	Two blind signature symptom matching schemes	The authors do not rely on any trusted third party that can be realized for ensuring an energy efficient privacy mechanism.
Alduailij et al. [1]	developed the updated version of opponent virtual machine	The authors have illustrated and developed various healthcare applications for monitoring the wellness and medical conditions in the network.
Wu et al. [20]	Fuzzy logic for evaluating and defining the trusts	The authors have offered set of rules for reasoning and analyzing the rules with a certain level of uncertainty.
Hirtan et al [10]	Blockchain based mechanism in healthcare	The proposed mechanism involved public and private chains to protect the confidentiality of data.
Iqbal et al. [12]	Introduction of software-defined networks	The core issue is highlighted by unifying the stakeholders with few outcomes on network based security scheme.
Geng et al. [8]	Security framework by constructing the SDN	The authors have proposed various security schemes to overcome the anti-tamper, anti-forge, anti-replay, and anti-wormhole issues.

on security schemes in various applications. Jiang et al. [13] have proposed two blind signature symptom matching schemes. In addition, the authors have achieved coarse-grained and fine grained matching approaches for realizing the privacy preserving in healthcare networks. The authors do not rely on any trusted third party that can be realized for ensuring an energy efficient privacy mechanism. Further, a comprehensive evaluation scheme is demonstrated to identify the practicality of the proposed phenomenon. Alduailij et al. [1] have validated and developed the updated version of opponent virtual machine that is capable of supporting and diversion various sets of applications. The authors have illustrated and developed various healthcare applications for monitoring the wellness and medical conditions in the network. Wu et al. [20] have introduced a fuzzy logic for evaluating and defining the trusts by providing formal representations of rules. The authors have offered set of rules for reasoning and analyzing the rules with a certain level of uncertainty. Applications of proposed model along with pervasive computing provided a novel scheme to handle the trust management for pervasive and federated networks. Hirtan et al. [10] have pro-posed a blockchain based mechanism in healthcare where the information is analyzed and shared among medical clinics, hospitals, and research institutes defined by patients. The proposed mechanism involved public and private chains to protect the confidentiality of data. The proposed scheme is developed over hyperledger platform to validate the results against amount of time required to detect the malicious activity. Iqbal et al. [12] have illustrated the introduction of software-defined networks along with its deployment models such as de-centralized and centralized mechanism. In addition, the authors have illustrated a secure IoT based SDN mechanisms by deter-mining its overview. Further, the core issue is highlighted by unifying the stakeholders with few outcomes on network based security scheme. The authors have also discussed some future directions of security needs in IoT based SDN. Geng et al. [8] have proposed a security framework by constructing the SDN using network architecture. The authors have proposed various security schemes to overcome the anti-tamper, anti-forge, anti-replay, and anti-wormhole issues. In addition, the authors have focused on various multi-service and anti-replay security modules. Further, an NS2 simulator is analyzed to validate the superiority of proposed scheme over end-to-end delay, packet deliver ratio, and overhead metrics.

12.3 Proposed Approach

An ANN is defined as a computational mechanism inspired by various neurons based on biological neural network concept. The neurons are generally defined as the smallest cells that our brains made of. It is determined as an assortment of billions of neurons considered as the base for modeling an AI in terms of architecture designing and operations performed. The ANN is a mathematical model for performing the non-linear function, data classification and regression approach.

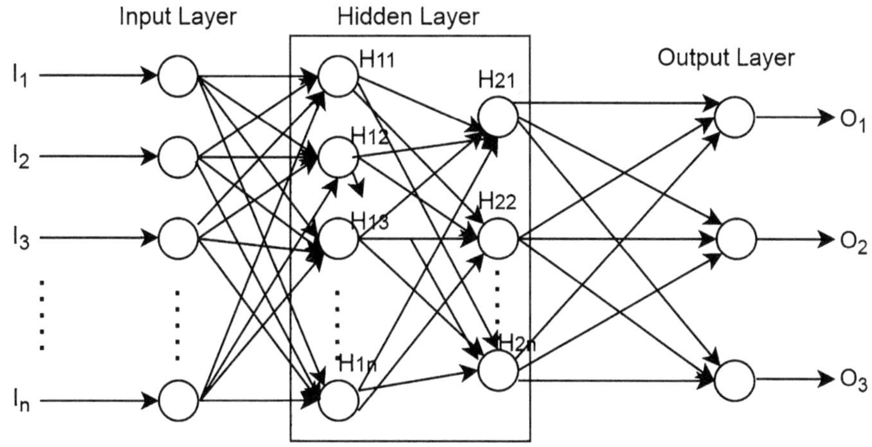

Fig. 12.1 Multi-layered ANN architecture

It is capable in generating the robotic decision modeling through multi-layered perceptron.

The depicted Fig. 12.1 represents a multi-layered ANN architecture having input, middle, and output layer. The network consists of n number of outputs, hm number of hidden layers, and ix number of input nodes as expressed in Eq. (12.1).

$$p_q(t) = \sum_{p=1}^{h_m} W_{qr}^2 F(.) \sum_{a=1} i_x W_{pq}^1 p_r(t)^0 + b_p^1, \text{ where, } 1 <= q <= n \tag{12.1}$$

where w_{pq} and w_{qr} represents the weight connection among input, hidden, and output layer, respectively. F(.) defines the activation function which is considered as sigmoid function.

Further, the number of values in $w_q r$ and $w_p q$ is determined through an appropriate mechanism such as previous history behavior and BP. The BP algorithm is generally used to generate an optimal value with an error-correction rule to accurately categorize the smart devices according to their malicious behaviors.

12.3.1 Back Propagation (BP) Algorithm

BP is gradient descent method to generate derivation values by updating the weights of learning parameters. It is a steepest method where weights between hidden qth layer and pth input layers are according to Eq. (12.2).

$$w_{pq}(t) = w_{pq}(t-1) + \delta w_{pq}(t)$$

$$b_q(t) = b_q(t-1) + \delta b_q(t)$$

In addition, the increments $\delta b_q(t)$ and $\delta w_p q(t)$ are illustrated as:

$$\begin{aligned} \delta w_{pq}(t) &= \eta_w h_q(t) h_p(t) + a_{wpq}(t-1) \\ \delta b_q(t) &= \eta_b p_q(t) + a_b \delta b_q(t-1) \end{aligned} \tag{12.2}$$

where w and b are defined as weights and threshold parameters, respectively. In addition, a_b and a_w are momentum constants which represent the changes in earlier metrics upon movement of directions in metric space. Further, η_b and η_w determine learning values with h_q (t) error signals propagation to the entire network. Since, the output layer activation function is linear; therefore the error signal is computed as Eq. (12.3).

$$h_q(t) = q_r(t) - q_r^{'}(t) \tag{12.3}$$

where q_r (t) is defined as expected output. However, the neurons at hidden layer are represented at Eq. (12.4) as:

$$h_q(t) = F^{'}(h_p(t)) w_{qr}^2(t-1) \tag{12.4}$$

Where, $F^{'}(h_p(t))$ denotes the first derivation function of $F(h_p(t))$ with respect to $h_p(t)$. Further, the number of attributes considered for analyzing the devices category according to their behavior is detailed as:

- Node's activation rate: It is defined as the amount of time a node remains active inside the network to attract its neighboring nodes. The node having malicious nature will be more active with respect to remaining nodes.
- Previous history interaction: The malicious behavior of a node can be further analyzed by checking its previous communication history in the network.

12.4 Performance Evaluation

In order to validate the proposed phenomenon against conventional approach that categorized the behavior of each node using cryptographic scheme, a synthesized dataset is created. The proposed phenomenon is analyzed using BP scheme against various security measures. Table 12.2 represents the analysis of simulated results with their mentioned values.

The security measured used to analyze the proposed phenomenon is discussed as follows:

Table 12.2 Simulation metrics

S. no.	Traditional approach	Blockchain approach
1	Number of nodes	100
2	Network area	500 × 500
3	Node's behavior	Ideal, malicious
4	Algorithm	BP, ANN

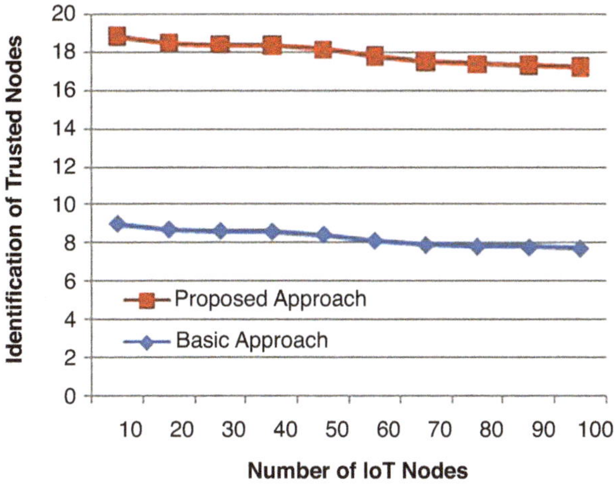

Fig. 12.2 Identification of trusted nodes (%)

- Identification of trusted nodes: It is defined as the number of node's identified as legitimate while analyzing their communication behavior.
- Message alteration: The information tried to be altered by the intruders upon transmitting among devices. It is considered as one of the significant parameters to detect the behavior of a node.

12.4.1 Results and Discussion

Before categorizing the IoT devices, the classification algorithm is analyzed using two statistical measures as accuracy and timely. The accurate analysis of a system within significant time may recognize the efficiency of a mechanism.

Figure 12.2 represents the identification of trusted nodes over proposed and baseline method. The result shows the outperformance of proposed phenomenon because of its accurate decision making and less delay using neural and BP algorithms. Further, the data tried to alter by the proposed phenomenon is represented in Fig. 12.3. The number of compromised nodes may try to forge the data and do some modifications for their own benefits. In case of proposed approach, the

Fig. 12.3 Altered data (%)

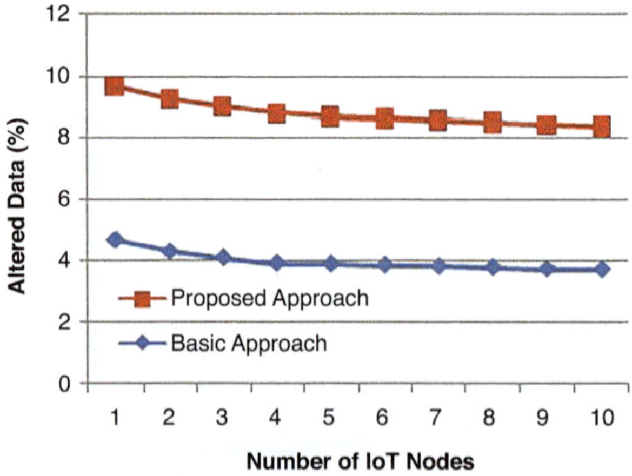

data alteration rate is very less because of error-propagation scheme called BP as compared to existing approach.

12.5 Conclusion

This paper goal is to categorize the smart devices according to their communicational behavior using ANN and BP schemes. The proposed mechanism determines the behavior of a node by analyzing its previous history and active nature in the network. Further, to speed up the analysis process and efficiently categorize the node's behavior according to ANN can be analyzed through back propagation schemes. The simulated simulation results validate the proposed phenomenon as compared to existing scheme. Further, the dynamic pattern of attacks categorized by malicious nodes during the communication process can be further traced using various security measures. The dynamic analysis of patterns behavior can be reported in future research.

References

1. Alduailij, M. A., & Lilien, L. T. (2015). A collaborative healthcare application based on opportunistic resource utilization networks with OVM primitives, in *IEEE 2015 International Conference on Collaboration Technologies and Systems (CTS)* (pp. 426–433).
2. Aujla, G. S., Chaudhary, R., Kumar, N., Kumar, R., & Rodrigues, J. J. (2018). An ensembled scheme for QoS-aware traffic flow management in software defined networks, in *2018 IEEE International Conference on Communications (ICC)* (pp. 1–7). Piscataway: IEEE.
3. Aujla, G. S., & Jindal, A. (2020). A decoupled blockchain approach for edge-envisioned IoT-based healthcare monitoring. *IEEE Journal on Selected Areas in Communications, 39*(2), 491–499.

4. Aujla, G. S., Jindal, A., Kumar, N., & Singh, M. (2016). SDN-based data center energy management system using RES and electric vehicles, in *2016 IEEE Global Communications Conference (GLOBECOM)* (pp. 1–6). Piscataway: IEEE.
5. Aujla, G. S., Kumar, N., Garg, S., Kaur, K., & Ranjan, R. (2019). EDCSuS: Sustainable edge data centers as a service in SDN-enabled vehicular environment. *IEEE Transactions on Sustainable Computing.*
6. Aujla, G. S., Singh, M., Bose, A., Kumar, N., Han, G., & Buyya, R. (2020). BlockSDN: Blockchain-as-a-service for software defined networking in smart city applications. *IEEE Network, 34*(2), 83–91.
7. Aujla, G. S., Singh, A., & Kumar, N. (2019). Adaptflow: Adaptive flow forwarding scheme for software-defined industrial networks. *IEEE Internet of Things Journal, 7*(7), 5843–5851.
8. Geng, R., Wang, R., & Liu, R. (2018). A software defined networking-oriented security scheme for vehicle networks. *IEEE Access, 6*, 58195–58203.
9. Graupe, D. (2013). *Principles of artificial neural networks* (vol. 7). Singapore: World Scientific.
10. Hirtan, L., Krawiec, P., Dobre, P., & Batalla, J. M. (2019). Blockchain-based approach for e-health data access management with privacy protection, in *2019 IEEE 24th International Work-shop on Computer Aided Modeling and Design of Communication Links and Networks (CAMAD)* (pp. 1–7).
11. Holzinger, A., Langs, G., Denk, H., Zatloukal, K., & Müller, H. (2019). Causability and explainability of artificial intelligence in medicine. *Data Mining and Knowledge Discovery, 9*(4), 1312–1321.
12. Iqbal, W., Abbas, H., Daneshmand, M., Rauf, B., & Bangash, Y. A. (2020). An in-depth analysis of IoT security requirements, challenges, and their countermeasures via software-defined security. *IEEE Internet of Things Journal, 7*(10), 10250–10276.
13. Jiang, S., Duan, M., & Wang, L. (2018). Toward privacy-preserving symptoms matching in SDN-based mobile healthcare social networks. *IEEE Internet of Things Journal, 5*(3), 1379–1388.
14. Jindal, A., Aujla, G. S., Kumar, N., & Villari, M. (2019). GUARDIAN: Blockchain-based secure demand response management in smart grid system. *IEEE Transactions on Services Computing, 13*(4), 613–24.
15. Lelcuk, A., Groskop, M., Yehuda, D., Yotam, B. E. N. (2017). U.S. Patent No. 9,825,928. Washington: U.S. Patent and Trademark Office.
16. Meng, W., Choo, K. R. R., Furnell, S., Vasilakos, A. V., & Probst, C. W. (2018). Towards Bayesian-based trust management for insider attacks in healthcare software-defined networks. *IEEE Transactions on Network and Service Management, 15*(2), 761–773.
17. Meneghello, F., Calore, M., Zucchetto, D., Polese, M., & Zanella, A. (2019). IoT: Internet of threats? A survey of practical security vulnerabilities in real IoT devices. *IEEE Internet of Things Journal, 6*(5), pp. 8182–8201.
18. Mitchell, R., Michalski, J., & Carbonell, T. (2013). *An artificial intelligence approach.* Berlin: Springer.
19. Rathee, G., Sharma, A., Kumar, R., Ahmad, F., & Iqbal, R. (2020). A trust management scheme to secure mobile information centric networks. *Computer Communications, 151*, 66–75.
20. Wu, Z., & Weaver, A. C. (2006). Application of fuzzy logic in federated trust management for perva-sive computing, in *IEEE 30th Annual International Computer Software and Applications Conference (COMPSAC'06)* (vol. 2, pp. 215–222).

Chapter 13
Stealthy Verification Mechanism to Defend SDN Against Topology Poisoning

Bakht Zamin Khan, Anwar Ghani, Imran Khan, Muazzam Ali Khan, and Muhammad Bilal

13.1 Introduction

Traditional computer networks control plane and data plane are highly coupled. Each plane has its own and fixed functionalities. Thus the whole architecture is highly decentralized. Due to the decentralize nature it is difficult to extend and modify network functionalities at run time. SDN has recently emerged as a new network paradigm which decouples both control plane and data plane. SDNs offer better network resource utilization, control, and management with minimal operating cost [7, 9, 18, 21, 30]. Due to the simplicity and added features it got attention in both academia and industry [17]. With centralized management SDN simplifies network management and monitoring more easily and precisely. The underlying network switches in the data plane ensure forwarding of packets to the destination switches and provide their current status to the controller.

SDN controllers are responsible for the integrity, accuracy, and credibility of the network topology. Network topology information is not use only for flow table construction, data forwarding but is also essential for the efficient utilization of network resources [5, 15]. Topology discovery mechanism mainly provides

B. Z. Khan · A. Ghani · I. Khan
Department of Computer Science & Software Engineering, International Islamic University Islamabad, Islamabad, Pakistan
e-mail: bakht.phdcs172@iiu.edu.pk; anwar.ghani@iiu.edu.pk; imran.khan@iiu.edu.pk

M. A. Khan
Department of Computer Sciences, Quaid-e-Azam University, Islamabad, Pakistan
e-mail: muazzam.khattak@qau.edu.pk

M. Bilal (✉)
Department of Computer Engineering, Hankuk University of Foreign Studies, Yongin-si, Gyeonggi-do, South Korea
e-mail: mbilal@hufs.ac.kr

G. S. Aujla et al. (eds.), *Software Defined Internet of Everything*, Technology, Communications and Computing, https://doi.org/10.1007/978-3-030-89328-6_13

two key services, i.e., host tracking service and link discovery service [21, 28]. Due to lack of authentication mechanism, SDN controller cannot validate the legitimacy of the received network topology information including hosts, switches, and connecting links between network nodes. If an attacker performs malicious activity and fabricates the network information. Consequently, the SDN controller will get poisoned once it received the falsify topology information. By executing topology poisoning attacks the attacker can easily diverts the intended traffic to a malicious gateway or to black-hole to launch denial of services to the upper layer applications [16, 20]. Maintaining of correct global view of the network topology and identification of malicious activity is the two mainstream activities to be ensure by SDN controllers [15].

Currently, OpenFlow Topology Discovery Protocol is used for SDN topology discovery information by all well-known controllers (Floodlight, OpenDayLight, Maestro, NOX, Rue, etc.) [16, 23]. Current implementation of OFDP is vulnerable to two topology poisoning attacks, i.e., Host Hijacking and Link Fabrication attacks. These two type of attacks are unique to SDN environment, which can cause man-in-the-middle or Denial of Service (DOS) attacks [4, 8, 10, 14, 16, 20]. Host hijacking attack injects fake host-generated packets in the network to blur the controller that the generating host is being migrated to new location. Whereas link fabrication attack announces a new malicious link in the network. By doing this the attacker diverts the traffic to itself and network segments may suffer from denial of service attack [19]. OpenFlow protocol [20] is the default communication protocol widely used for communication between SDN switches and controllers. For smooth operation of SDN and seamless provision of services at application layer, OpenFlow implements TopoGuard [16] which significantly improves and secures the topology discovery mechanism. An OF controller collects the whole network's topology information to form a global and shared view of the network. This global view is used to ensure smooth and efficient networks operations [15]. Most services in the application plane like routing, policy, live migration, virtualization, optimization, etc. are highly dependent on the network topology information constructed and monitored by the SDN controllers. Therefore, the correctness of the topology information has critical importance [16, 20]. Maintaining of correct global view of the network topology and identification of malicious activity is the two mainstream activities to be ensured by SDN controllers [6, 15]. Figure 13.1 shows the structure of SDN architecture.

To overcome the gape, we proposed a light weight Enhanced Stealthy Probing-based Verification (ESPV) mechanism which fulfill the shortcoming of the existing technique and provide a suitable solution for large data center networks as well as resource limited networks. The contribution of our paper is summarized as follows:

- A parametrized framework that considers other network parameters also
- A new topology verification mechanism based on SPV is proposed, which is verified to cope with two loopholes
- Implementation and verification of the proposed scheme
- Through analysis of results for large 7 data center networks

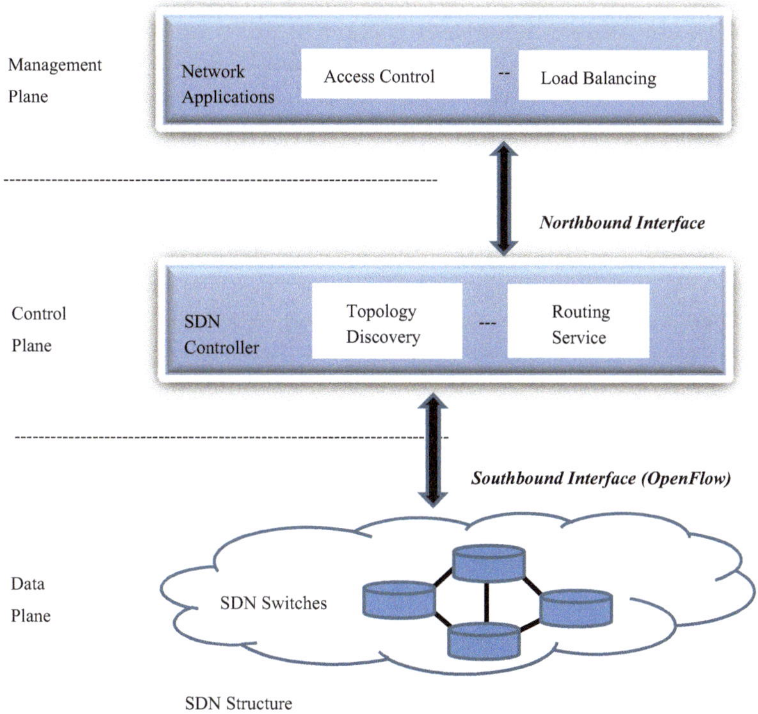

Fig. 13.1 SDN structure

Sect. 13.2 discusses related work on SDN security and feasibility of SDN anti-topology poisoning techniques. Section 13.3 includes the problem statement and Sect. 13.4 illustrates the proposed scheme. Section 13.5 is about implementation and results are discussed in Sect. 13.6. Section 13.7 concludes the paper along with future challenges.

13.2 Related Work

To protect SDN infrastructure from topology poisoning attacks and to verify the legitimacy of topology information several authentication mechanisms are developed. But due to their simplicity, open source authentication method and nature of authentication mechanism that follow a predictable pattern an attacker can evade the controller. In literature various types of techniques have already been proposed to prevent SDN enable networks for two kinds of attacks. In [2], mitigation technique is proposed to avoid link fabrication attack. The proposed method adds a Message Authentication Code (MAC) to each Link Layer Discovery

Protocol (LLDP) to validate each topology discovery packet's integrity. Network flow graphs are used by Dhawan et al. [13] to detect malicious traffic which cross the learned acceptable level of graph patterns. Hong et al. [16] implement a real time low-overhead defensive technique, which add some minor changes to the topology discovery protocol of OpenFlow controller, called TopoGuard. The technique provides effective way and high performance against the two types of attacks. However, TopoGuard has limitation of formal verification for the host that are migrating to new location. If attacker modifies his/her method of attack he/she may easily evade the SDN controller. The proposed scheme in [3] uses an active SPV approach for identifying malicious host and fake link established by a compromised switch. This technique improved the identification mechanism, however, due to its probing mechanism it may face scalability and bandwidth consumption issues in case of large enterprise networks. Furthermore, in resource constraint network like Wireless Sensor Networks (WSNs) the probe packets may consume precious network bandwidth causing delay and loss of network resources.

13.3 Problem Statement

Maintaining of correct global view of the network topology and identification of malicious activity are the two mainstream activities to be ensure by SDN controllers. The SPV technique [3] sufficiently improved the identification mechanism, however, due to its probing mechanism it faces scalability and bandwidth consumption issues in case of large enterprise networks. Furthermore, in resource constraint network including Wireless Sensor Networks (WSNs) the probe packets may consume precious network bandwidth causing delay and loss of precious network resources.

13.4 Proposed Solution

The proposed solution is based on a light weight Enhanced Stealthy Probing-based Verification (ESPV) mechanism which fulfill the shortcoming of the existing technique and provide a suitable solution for large networks as well as resource limited networks. The proposed solution includes to validate a newly added link by using stealthy probing packet generation technique and to trigger the probing packets only in the network segment from where the controller receives the topology change packets. Thus, the probing packets are only initiated when the topology change update messages are being received by the controller.

13.4.1 Methodology

The proposed scheme is implemented and illustrated in Link Verification and malicious node identification as Algorithms 1 and 2 depicted as below.

Algorithm 1 New link verification

Require: Topology_Update_newlink
Ensure: Updates_Topology
1: Check new_triggered_link_update
2: Function topo_update_check(network_segment)
3: **if** $ID \neq 1$ **then**
4: $DST_{i,J} \leftarrow P_{i,j}(d) \leftarrow \frac{p_i G_i G_j \wedge^2}{(4\pi)^2 d^2}$
5: **end if**
6: **if** $0 < DST_{i,j} < DST_{tr}$ **then**
7: push i to StackCNs.Top
8: $StackCNs.Top \leftarrow StackCNs.Top + 1$
9: **end if**

Algorithm 2 Example calculation of CNs for each node in the network

Require: Topology_view
Ensure: triggered_topolog_updates
1: Check new_triggered_link_update
2: StackCNs.top$\leftarrow$ 0
3: **if** $ID \neq 1$ **then**
4: $DST_{i,J} \leftarrow P_{i,j}(d) \leftarrow \frac{p_i G_i G_j \wedge^2}{(4\pi)^2 d^2}$
5: **end if**
6: **if** $0 < DST_{i,j} < DST_{tr}$ **then**
7: push i to StackCNs.Top
8: $StackCNs.Top \leftarrow StackCNs.Top + 1$
9: **end if**

Similar to the proposed work in [2, 11, 20, 21, 27], we assume that an attacker may compromise one or more hosts or network switches connected to each other in SDN. The attacker can send forged information of topology change message by using one of the compromised hosts. By receiving this information SDN's controller modifies communication flows of the network nodes. The intruder can distinguish control traffic and user traffic. He can sniff the control packets, modify them accordingly, and re-transmit them in the network to perform his malicious task. The attacker can also compromise the controller due to its vulnerabilities but we assume that the SDN controller is secure against any attack and the channels between data switches and SDN controller are trusted. Consequently, the attacker attempts to isolate probing traffic from the normal traffic by their patterns. We also assume that the probing packets will be only initiated when the topology change messages are received by the controller. The probing packets will be only flooded after some specified time of interval to reduce the bandwidth consumption in large enterprise and resource constrained wireless sensor networks.

13.5 Implementation

The proposed scheme is implemented by using two widely used open source tools for simulating SDNs. SDN's controller is implemented in OpenDayLigh (ODL) [12] which is an open source platform for simulating and controlling large scale SDNs [24]. Data plane switches, based on OpenFlow protocol, are implemented by using Mininet network emulator [22, 31]. To compare and analyze the proposed and base scheme proposed in [3], implementation parameters like type of SDN controller, network topology, number of data switches, communication links, and virtual machines are kept near similar. The performances of both schemes are compared in terms of processing power, memory usage, and time required to identify any malicious link or network node. To examine the accuracy and performance of proposed scheme, 5000 links were established among all network nodes.

13.5.1 Software-Defined Network Setup

SDN controllers keep centralized global view the network topology to ensure service accessibility to upper layer. SDN controller is implemented by using OpenDayLight (ODL). ODL Carbon release including a virtual machine running Linux Ubuntu Server 16.04 operating system with two Intel(R) E3-1271 v3 CPUs and 6 GB of processing memory. Data switches are programmed to forward packets as per instructions of the controller and send relevant information to the controller back for different operations performed by the controller, like updating and maintaining of topological view. Link Layer Discovery Protocol (LLDP) Ethernet (0x88cc) standard is used as network topology discovery mechanism. LLDP discovers newly added switches and their placement in the SDN topology. To replicate the data plane switches, Mininet 2.2.1 on Linux virtual machine running Ubuntu 16.4 Server with two Intel(R) X eon (R) E3-1271 v3 and 6 GB of RAM were used. Latest version of OpenFlow v1.3 is used for software based open vSwitchs [29] switches.

13.5.2 ESPV Implementation

Most modules of ESPV are implemented in Java. Open source packet generation tool Scapy [26] is used for generation and manipulation of probing packets. The scheme is implemented in both single-threading and multi-threading modes to analyze the performance of link verification mechanism in both methods. Both schemes are simulated for large enterprise data center networks only, having huge network traffic.

13.5.3 Network Topology

The Fate-tree topology [1] is used for simulating the data plane that is a network architecture widely used for implementation of large data center [25]. The topology or size of network depends upon the network switches used at access level to connect end user systems and servers. To examine the performance of proposed scheme, the number of data plane switches is increased gradually from five to 45. In small topology there is one core, two aggregate switches, and two access level switches. The largest topology has eight core switches, 12 aggregate switches, and 25 access switches. The access switches can connect up to 600 servers which comply large size data center requirements.

13.6 Results and Analysis

13.6.1 Performance Analysis

In performance analysis the time required to verify each newly established links is analyzed for both schemes. The scheme proposed in [3] verifies each link of entire topology. While our proposed scheme only verifies the network segment where new link is established. Due to working methodology the proposed scheme significantly reduces the amount of processing required to verify new links, and thus the number of probing packets is significantly reduced. Thus overall performance of the system is improved. The obtained results as shown in Figs. 13.2 and 13.3 show the time required in milliseconds to verify a newly established links between two network nodes in both single-threading and multi-threading modes. While increasing the number of switches from 4 to 45 with a maximum of 106 connecting links among data plane switches, average verification time to verify each link was 86 ms. The proposed scheme significantly reduces the time required to verify a new link and most near to real time behavior. Figure 13.3 shows verification of a group of links by increasing the number of switches from four to 45 data plane switches. In second case the time taken to verify a set of links is 19.1 s. By applying multi-threading it decreases the time interval to 8.2 s.

Now the proposed scheme is applied to the segment from where the topology change messages were triggered and received by the SDN's controller. By dividing the network topology in segments the proposed scheme only takes 9.3 ms in single-threading and 3.3 s in case of multi-threading with same number of network links

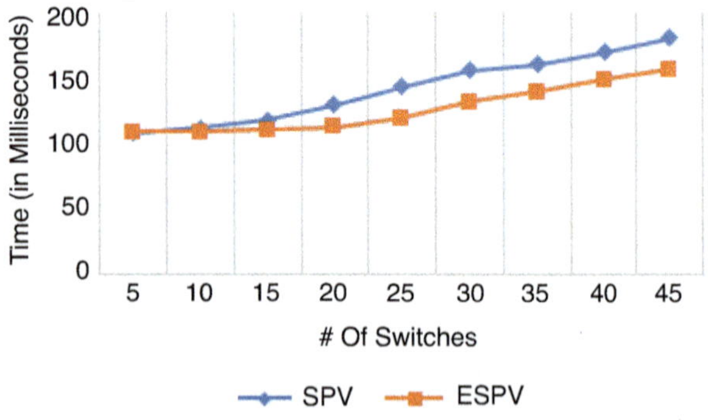

Fig. 13.2 Single-threading

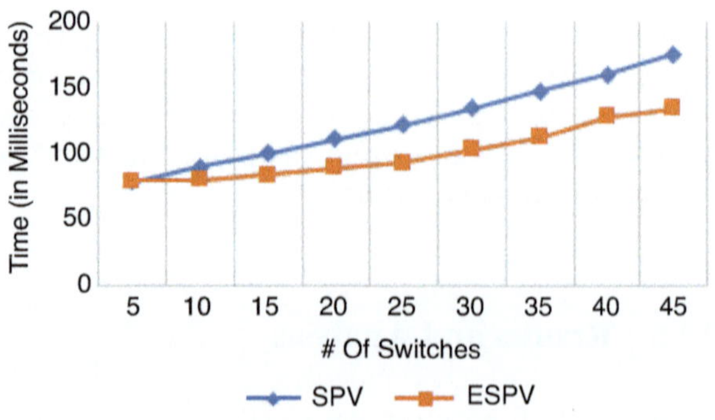

Fig. 13.3 Multi-threading

and switches as shown in Figs. 13.4 and 13.5. The results indicate significant improvement as compared to the scheme proposed in [3]. The proposed scheme produces linear results in case of increasing the number of data switches and their connecting links. By limiting the probing packets to the time based and by propagating the probing packets in specific network segment form where the topology changes messages received by SDN controller. Our proposed scheme significantly reduces the number of probing packets, results to avoid congestion of probing packets in heavy traffic data centers. This phenomenon also consumes less processing power, memory usage and communication bandwidth, make it candidate scheme for resource constraints networks.

13.6.2 Resource Consumption by ESPV Scheme

The second part of ESPV performance is the amount of resources required to execute complete process of verification mechanism. CPU processing, memory consumption, and communication bandwidth parameters are analyzed and compared. Figure 13.6 shows the CPU consumption for the experiments. The proposed scheme only consumes 33% of CPU, reasonably reduce the CPU consumption as compared to the previous scheme and make our proposed solution suitable to be used in

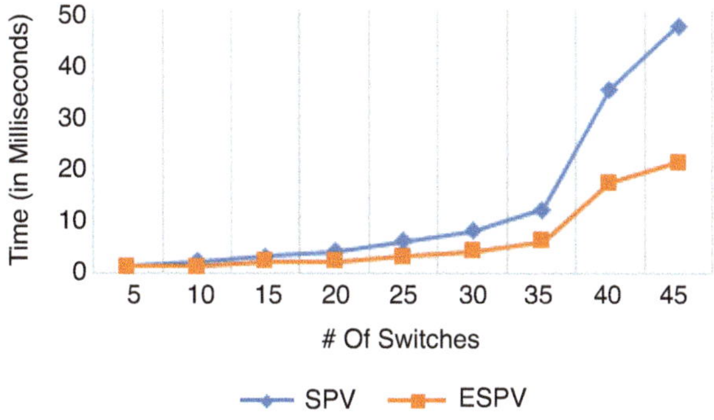

Fig. 13.4 Single-threading

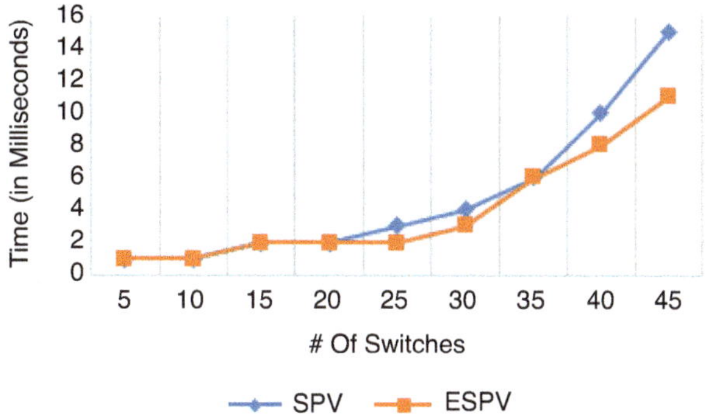

Fig. 13.5 Multi-threading

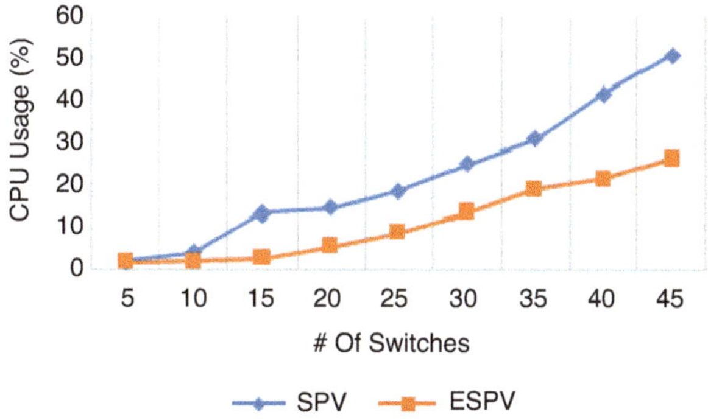

Fig. 13.6 CPU usage

WSNs. Figure 13.7 shows the memory consumption for the experiments. Due to the segmented verification mechanism it consumes 42% less memory of the scheme proposed in [3]. The proposed scheme produced better results in large enterprise networks and data centers due to its verification methodology. The results indicate that proposed scheme is more scalable solution for large data center network. Due to less resource requirements the proposed scheme can be used as a candidate topology verification mechanism in resource limited networks.

Fig. 13.7 Memory usage

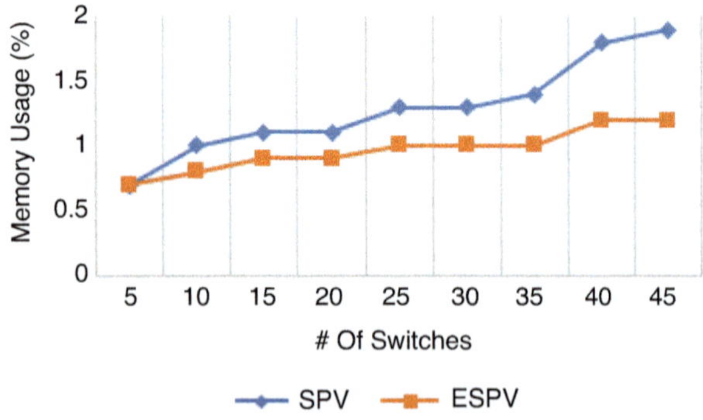

Table 13.1 Details of experimental results with real SDN/cloud topology

Results	SDN/cloud mesh	Mininet fat-tree	Mininet fat-tree [ESPV]
All links verification time (MT)	13.2052 s	10.6406 s	10.1031s
All links verification time (ST)	134.147 s	26.1725 s	25.5471 s
Single links verification time (ST)	100.306 s	94.8919 s	94.3245
CPU consumption (ST)	8.50294%	18.79099%	11.56421%
Memory consumption (ST)	1.81817%	1.81003%	1.23144%

13.6.3 Applicability of ESPV in Real SDN Cloud Topology

The performance of proposed scheme is also measured on real SDN topology. OpenStak [25] cloud is one of the largest communication platforms that provides different virtual network services. OpenStak scenario comprised of 23 communication nodes and each node consists of OpenFlow communication switches and thousands of virtual machines. All 22 computing nodes are connected to each other in a mesh topology with a total of 241 communication links. Table 13.1 illustrates the comparison the proposed scheme results and real topology results. The results of our proposed scheme indicate better results and applicability in large networks.

13.7 Conclusion and Future Work

Recently identified vulnerabilities in Open Flow Discovery Protocol reveal that malicious hosts or switches can poison global view of the network topology and an intruder can launch man-in-the-middle or denial of service attacks. Existing solutions are based on passive approach which work only for known attacks types. The scheme in [3] sufficiently improved the identification mechanism, however, due to its probing mechanism it may face scalability and bandwidth consumption issues in case of large enterprise networks and resource limited networks. The results of our proposed scheme indicate that ESPV is the more scalable and suitable solution to detect and identify the fake links or malicious network hosts in both type networks. It significantly reduces the probing traffic to consume less bandwidth and identify malicious host in real time . In future the proposed scheme will be implemented and evaluated in resource constrained networks.

References

1. Al-Fares, M., Loukissas, A., & Vahdat, A. (2008). A scalable, commodity data center network architecture. *ACM SIGCOMM Computer Communication Review, 38*(4), 63–74.
2. Alharbi, T., Portmann, T., & Pakzad, F. (2015). The (in) security of topology discovery in software defined networks, in *2015 IEEE 40th Conference on Local Computer Networks (LCN)* (pp. 502–505). Piscataway: IEEE.
3. Alimohammadifar, A., Majumdar, S., Madi, T., Jarraya, Y., Pourzandi, M., Wang, L., & Debbabi, M. (2018). Stealthy probing-based verification (SPV): An active approach to defending software defined networks against topology poisoning attacks, in *European Symposium on Research in Computer Security* (pp. 463–484). Berlin: Springer.
4. Aryan, R., Yazidi, R., Engelstad, P. E., & Kure, Ø. (2017). A general formalism for defining and detecting openflow rule anomalies, in *2017 IEEE 42nd Conference on Local Networks (LCN)* (pp. 426–434). Piscataway: IEEE.
5. Aujla, G. S., Chaudhary, R., Kumar, N., Kumar, R., & Rodrigues, J. J. P. C. (2018). An ensembled scheme for QoS-aware traffic flow management in software defined networks, in *2018 IEEE International Conference on Communications (ICC)* (pp. 1–7). Piscataway: IEEE.
6. Aujla, G. S., & Kumar, N. (2018). SDN-based energy management scheme for sustainability of data centers: An analysis on renewable energy sources and electric vehicles participation. *Journal of Parallel and Distributed Computing, 117*, 228–245.
7. Aujla, G. S. S., Kumar, N., Garg, S., Kaur, K., & Ranjan, R. (2019). EDCSuS: Sustainable edge data centers as a service in SDN-enabled vehicular environment. *IEEE Transactions on Sustainable Computing, IEEE*. https://doi.org/10.1109/TSUSC.2019.2907110
8. Aujla, G. Singh, S., M., Bose, A., Kumar, N., Han, G., & Buyya, R. (2020). BlockSDN: Blockchain-as-a-service for software defined networking in smart city applications. *IEEE Network, 34*(2), 83–91.
9. Aujla, G. S., Singh, A., & Kumar, A. (2019). Adaptflow: Adaptive flow forwarding scheme for software-defined industrial networks. *IEEE Internet of Things Journal, 7*(7), 5843–5851.
10. Aujla, G. S., Singh, A., Singh, M., Sharma, S., Kumar, N., & Choo, K.-K. R. (2020). Blocked: Blockchain-based secure data processing framework in edge envisioned v2x environment. *IEEE Transactions on Vehicular Technology, 69*(6), 5850–5863.
11. Azzouni, A., Trang, N. T. M., Boutaba, R., & Pujolle, G. (2017). Limitations of openflow topology discovery protocol, in *2017 16th Annual Mediterranean Ad Hoc Networking Workshop (Med-Hoc-Net)* (pp. 1–3). Piscataway: IEEE.
12. Badotra, S., & Singh, J. (2017). Open daylight as a controller for software defined networking. *International Journal of Advanced Research in Computer Science, 8*(5), 1105–1111.
13. Dhawan, M., Poddar, R., Mahajan, K., & Mann, K. (2015). Sphinx: Detecting security attacks in software-defined networks, in *Network and Distributed System Security (NDSS)* (vol. 15, pp. 8–11).
14. Fei, Y., Zhu, H., Wu, X., Fang, H., & Qin, S. (2018). Comparative modelling and verification of pthreads and dthreads. *Journal of Software: Evolution and Process, 30*(3), e1919.
15. Gude, N., Koponen, T., Pettit, T., Pfaff, B., Casado, M., McKeown, N., & Shenker, S. (2008). NOX: Towards an operating system for networks. *ACM SIGCOMM Computer Communication Review, 38*(3), 105–110.

16. Hong, S., Xu, L., Wang, L., & Gu, G. (2015). Poisoning network visibility in software-defined networks: New attacks and countermeasures, in *Network and Distributed System Security (NDSS)* (vol. 15, pp. 8–11).
17. Huang, X., Shi, P., Liu, Y., & Xu, Y. (2020). Towards trusted and efficient SDN topology discovery: A lightweight topology verification scheme. *Computer Networks, 170*, 107119.
18. Jarraya, Y., Madi, Y., & Debbabi, M. (2014). A survey and a layered taxonomy of software-defined networking. *IEEE Communications Surveys & Tutorials, 16*(4), 1955–1980.
19. Jindal, A., Aujla, G. S., Kumar, N., & Villari, M. (2019). Guardian: Blockchain-based secure demand response management in smart grid system. *IEEE Transactions on Services Computing, 13*(4), 613–624.
20. Khan, S., Gani, A., Wahab, A. W. A., Guizani, M., & Khan, M. K. (2016). Topology discovery in software defined networks: Threats, taxonomy, and state-of-the-art. *IEEE Communications Surveys & Tutorials, 19*(1), 303–324.
21. Kreutz, D., Ramos, F. M. V., Verissimo, P. E., Rothenberg, C. E., Azodolmolky, S., & Uhlig, S. (2014). Software-defined networking: A comprehensive survey. *Proceedings of the IEEE, 103*(1), 14–76.
22. Lantz, B., Heller, B., & McKeown, N. (2010). A network in a laptop: Rapid prototyping for software-defined networks, in *Proceedings of the 9th ACM SIGCOMM Workshop on Hot Topics in Networks* (pp. 1–6).
23. McKeown, N., Anderson, T., Balakrishnan, H., Parulkar, G., Peterson, L., Rexford, J., Shenker, S., & Turner, J. (2008). Openflow: Enabling innovation in campus networks. *ACM SIGCOMM Computer Communication Review, 38*(2), 69–74.
24. Medved, J., Varga, R., Tkacik, A., & Gray, K. (2014). Opendaylight: Towards a model-driven SDN controller architecture, in *Proceeding of IEEE International Symposium on a World of Wireless, Mobile and Multimedia Networks 2014* (pp. 1–6). Piscataway: IEEE.
25. Open source software for creating private and public clouds (2017). http://www.openstack.org/
26. Scapy: Packet manipulation program (2017). http://www.secdev.org/projects/scapy/
27. Thanh Bui, T. (2015). 'Analysis of topology poisoning attacks in software-defined networking', KTH, School of Information and Communication Technology (ICT). *Dissertation*.
28. ur Rasool, R., Wang, H., Ashraf, U., Ahmed, K., Anwar, Z., & Rafique, W. (2020). A survey of link flooding attacks in software defined network ecosystems. *Journal of Network and Computer Applications, 172*, 102803.
29. Open vSwitch (2016). Production quality, multilayer open virtual switch. http://openvswitch.org/
30. Xia, W., Wen, Y., Foh, C. H., Niyato, D., & Xie, H. (2014). A survey on software-defined networking. *IEEE Communications Surveys & Tutorials, 17*(1), 27–51.
31. Xia, W., Zhao, P., Wen, Y., & Xie, H. (2016). A survey on data center networking (DCN): Infrastructure and operations. *IEEE Communications Surveys & Tutorials, 19*(1), 640–656.

Chapter 14
Implementation of Protection Protocols for Security Threats in SDN

Amanpreet Singh Dhanoa

14.1 Introduction

Traditional computer networks used dedicated devices like routers, switches, etc. to manage the network payload. Software-defined network (SDN) generates and maintains the traffic using application programs virtually. Using virtualization, helps the organizations to create a single physical network from the different virtual networks and create single virtual networks where it connects the devices on the different physical networks [1, 2]. According to the researchers, due to centralization, lots of attacks are detected during the transmission of packets. Some packets are identified but some not. With software-defined networking, administrators work through central locations to configure network services and allocate virtual resources to make real-time changes. This allows network administrators to optimize the flow of data on the network and prioritize applications that require the highest availability [3–6].

SDN provides visibility to the whole network, and it defines more clarity on security threats. Using smart devices connected to the Internet, SDN has obvious advantages over traditional networks. An administrator creates different spaces for high-security devices, and if these devices are infected, then immediately isolated all the infected devices to secure the other devices as well as the network in which the devices are connected [7–9]. The structure of SDN architecture using OpenFlow is shown in Fig. 14.1.

A. S. Dhanoa (✉)
Department of Computer Science and Engineering, Chandigarh University, Mohali, India
e-mail: amanpreete7280@cumail.in

G. S. Aujla et al. (eds.), *Software Defined Internet of Everything*, Technology, Communications and Computing, https://doi.org/10.1007/978-3-030-89328-6_14

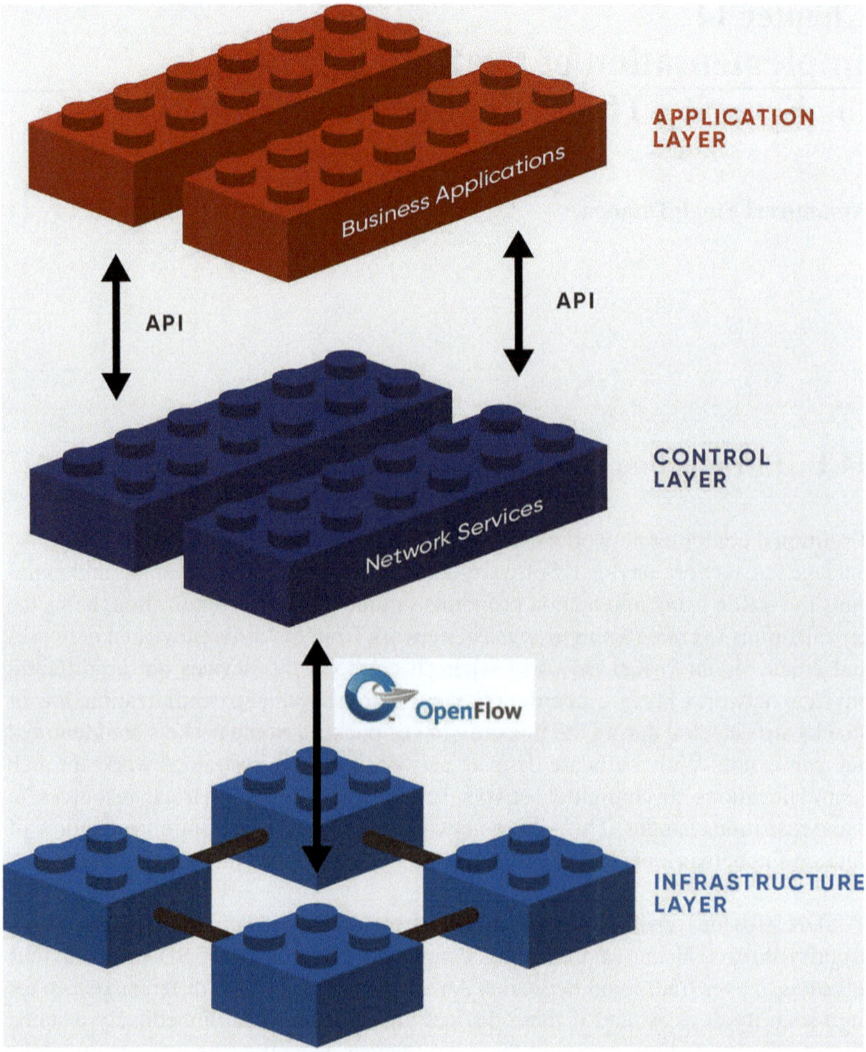

Fig. 14.1 SDN structure using OpenFlow [10]

14.2 Security Issues

The main aim of this chapter is to protect the data plane and control plane from hidden attacks like [11–13]:

Forwarding device attacks: Network traffic can be divided using routers or switches, causing intruders to launch denial of service (DoS) attacks, resulting in network outage or failure [14].

Threats to the control plane: Using a central controller, the problem that occurs in the network will cause to the failure of the central controller. The solution to resolve this issue is to use a horizontal or hierarchical controller distribution.

Communication channel vulnerabilities: OpenFlow protocol of SDN southbound APIs uses TLS to ensure the security of data control channel communication, but it is disabled by management and vulnerable to attacks from intermediaries, so it is not secure for the channel safety.

Fake traffic: This is generated by the non-malicious attackers or faulty devices or DoS attacks to consume resources on the forwarding device or controller. Authenticity: It is the entity of the network that is the entities that claim. The authenticity of forwarding equipment on SDWN networks is like traditional networks; it can hamper network performance.

Confidentiality: It prevents the information from being leaked to unauthorized users. If data is not guaranteed, then malicious users can use the information or data on the network.

Availability: It provides the access to authorized users to access the data, equipment, and services at any time.

Open programmable API: The openness of the API makes vulnerabilities more transparent to attackers.

The main concern of security is to maintain integrity, availability, and confidentiality, but some attacks affected these three parameters as shown in Table 14.1.

Attacks in SDN Control Layer DDoS Attacks: When the unauthorized users generate the request to access the data, then it is difficult to control that as a central data controller. These attacks increase the risk of network failure. The control layer uses the methods to maintain the flow control including attacking controller, northbound API, southbound API, westbound API, or eastbound API. For example, different applications use different flow rules it creates conflicts for the control plane may lead to DDoS attacks. In SDN operations, the data plane is not able to handle the arrival of new packets. So, the data plane sends the request to the control plane to maintain the flow. If the new flow does not match with the flow table, then two options are used to handling such kinds of problems: either the complete packet or a portion of the packet is transmitted to the controller to resolve the query. The

Table 14.1 Comparison of drone delivery cases

Threats	Effected SDN layer	Availability	Confidentiality	Integrity
Distributed DoS attack	Control and data	×		
DoS attack	Control and data	×		
Man-in-the middle	Control, data, link between control and data		×	×
Eavesdropping Application	Control and data		×	

more packets consume the more bandwidth while sending the complete packet to the controller [15].

Infrastructure Layer DDoS Attacks The two mechanisms create DDoS attacks at the infrastructure layer: attack the switch or attack the southbound API. For example, in the first mechanism, only the header information of the packet is passed to the controller, and the data of the packet is stored in the memory of the node itself until the table of flow does not return the value. In such a case, it makes it easy for an authorized person to change the flow rule by generating the known and unknown flows on the node by performing a DoS attack. The attackers sending lots of unknown packets create the bottleneck and overload the switch memory. It makes it difficult to maintain the normal data flows and violating the flow rules. It makes the SDN flow very complex and causes packet flooding: lots of packets are to be sent to the controller by the attackers, it is meant for invisible packets and uses all the resources of the control plane. This flow creates an unpredictable state for the controller. Control message operation: It provides communication between the control plane and the data plane. Attackers modify the control messages like change table flooding, switch spoofing, and malformed control messages [16].

OpenFlow enables the network controller to determine the path of network packets through the switch network. The controller is different from the switch. Compared with the use of access control lists (ACLs) and routing protocols, this separation of control and forwarding allows for more complex traffic management. In addition, OpenFlow allows the use of a single open protocol to remotely manage switches from different vendors, usually each vendor has its own interface and proprietary scripting language. OpenFlow can remotely manage the message forwarding table of Layer 3 switches, add, modify, and delete message matching rules and actions. In this way, the controller can make routing decisions periodically or temporarily. It converts the decisions into rules and actions. The controller implements the rules in the flow table of the switch and compares it with the actual forwarding packets. The data packets that do not match with the switch table can be forwarded to the controller. The controller takes a decision to modify the existing flow table rules on one or more switches or implement new rules to prevent structured traffic flow between the switch and the controller [17, 18].

14.3 Related Work

The protection of SDN layers is more important, but due to a centralized network, it is more difficult to provide direct security to the layers of SDN and the devices attached to the network. To protect the layers of SDN, several mechanisms are developed, but some attacks are not identified due to the simplicity of these mechanisms [19, 20]. These mechanisms are useful for the authentication to validate each topology discovery packet's integrity. The new scheme proposed in this chapter is the FRESCO framework. The FRESCO framework is providing a new

development environment for security applications. It works with the protocol of SDN like OpenFlow that is used to separate the data plane and the control plane. It provides a set of 7 new intelligent security actions, for example, block, deny, allow, redirect, quarantine. The FRESCO works with the application layer of SDN where all the protocols are generated. It divides the application layer into two segments: development environment and resource controller. The development environment takes a link from the sender and checks the validation of that link using the programmable code for that and sends it to the receiver if that link or request is valid. The FRESCO also generates a database to encrypt the value or data using a key. It creates the modules for the database in the form of pairs. The development environment creates a FRESCO script that defines the interface between the modules that connect each module with another module [21, 22].
FRESCO Script:

Syntax

- Case(tag((#(on(load)((#(on(return)((
- **Class:** type of module
- **Load:** module data
- **Return:** module value
- **Variable:** define some parameters
- **Occurrence:** Activate module
- **Activity:** perform activity

Input_Resource((1)(1)((
((((Class:(Resource(
((((Occurrence:(ADD(
((((Load:(destination_output
((((Return:(Resource_output
((((Variable:(23(
((((activity:(=(

14.4 Problem Statement

Security is the main concern in today's technical era to the identification of malicious activity during the communication between two same and different modules. The FRESCO is one of the most useful techniques to protect the protocols of SDN through which the identification is to be done. Other methods are taking more bandwidth to implement the security.

14.5 Proposed Solution

The proposed solution is based on the previous mechanism which does not fulfill the security requirements, and FRESCO creates the security application in the application layer to protect the protocols. It works on the TCP protocol of the application layer. It encrypts the port number of protocols using the unique key.

14.6 Implementation

The proposed solution is implemented by using the FRESCO technique for protocols as shown in Figs. 14.2 and 14.3. It works with the OpenFlow protocol, which is itself an OpenFlow application. It provides the implementation and composition methods to detect modules. It operates on NOX version 0.1.5 using OpenFlow 1.1.0 protocol. The NOX source code provides the FRESCO SEK, which is implemented as an extension that reaches 1160 lines of C++ code.

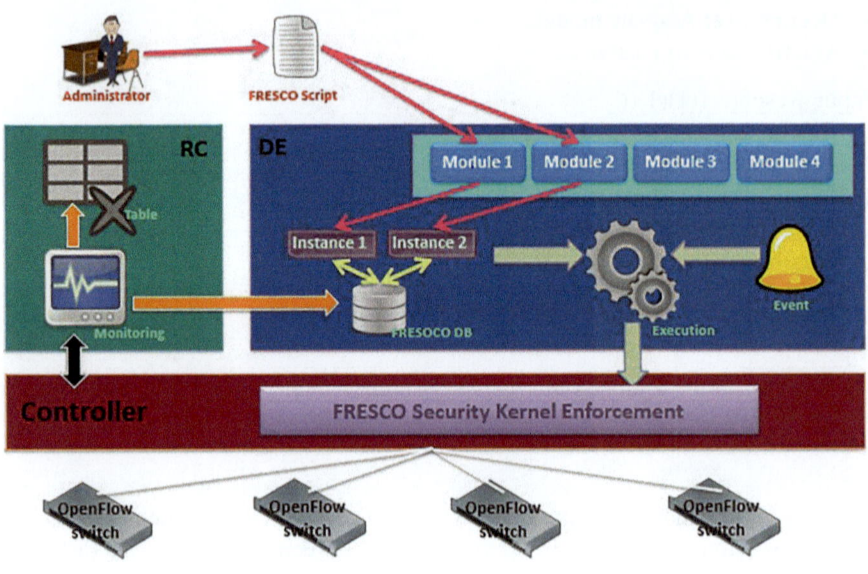

Fig. 14.2 Operational scenario [23]

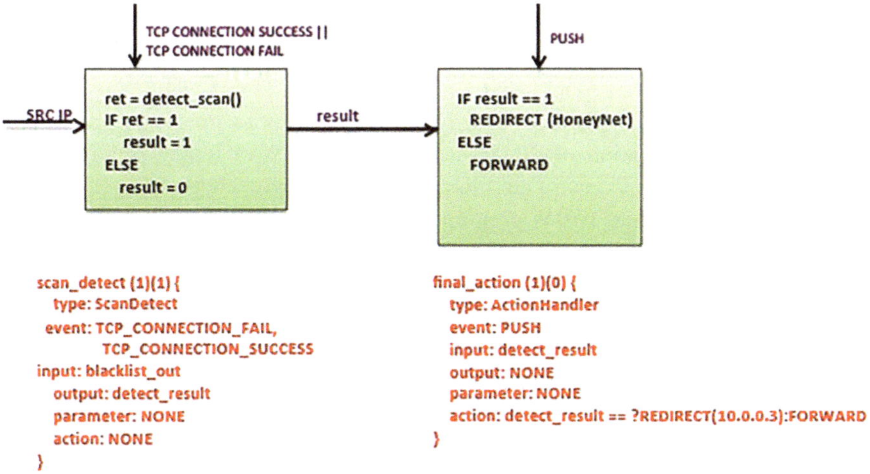

Fig. 14.3 Threshold-based scan detection [23]

```
def Unit_start(input_dis, parameter_list):
FRES_FDataBase = input_dis['FR_FDataBase']
FRES_action = in_dis['FRES_action']
FRES_input = input_dis['FRES_input']
FRES_dir_dis =
FRES_dir_dis['output'] = [0]
FREs_dir_dis['Event'] = Default
if variable_name[0] == FRES_load[0]:
return = 1
else:
return = 0
FRES_dir_dis['output']. ADD(output)
return FRES_dirt_dis
```

14.7 Conclusion

Securing the protocols of SDN using OpenFlow application of FRESCO, creating and implementing mechanism of security for protocols is an important challenge. We present an application of NOX such as the FRESCO framework for such kinds of complex problems. The FRESCO architecture works with the NOX OpenFlow controller and writes program using FRESCO scripting language. We use the FRESCO enforcement kernel to produce possible flow rules that secure the network packets as threats are detected. Over these mechanisms, FRESCO produces minimum overhead, and the scripting language takes a few lines of code to fast the process of detecting threats in a minimal span of time and also disable the fake links. It provides a powerful framework for the security of protocols.

References

1. Aujla, G. S., Singh, A., & Kumar, N. (2019, November 4). Adaptflow: Adaptive flow forwarding scheme for software-defined industrial networks. *IEEE Internet of Things Journal, 7*(7), 5843–5851.
2. Aujla, G. S., Garg, S., Batra, S., Kumar, N., You, I., & Sharma, V. (2019). DROpS: A demand response optimization scheme in SDN-enabled smart energy ecosystem. *Information Sciences, 476*, 453–473.
3. Hussein, A., Chadad, L., Adalian, N., Chehab, A., Elhajj, I. H., & Kayssi, A. (2020). Software-Defined Networking (SDN): The security review. *Journal of Cyber Security Technology, 4*(1), 1–66.
4. Coughlin, M. (2014). *A Survey of SDN Security Research*. University of Colorado Boulder.
5. Jose, T., & Kurian, J. (2015, December). Survey on SDN security mechanisms. *International Journal of Computer Applications, 132*(14), 0975-8887.
6. Iqbal, M., Iqbal, F., Mohsin, F., Rizwan, M., & Ahmad, F. (2019). Security issues in Software Defined Networking (SDN): Risks, challenges and potential solutions. *(IJACSA) International Journal of Advanced Computer Science and Applications, 10*(10), 298–303.
7. Shin, S., Porras, P., Yegneswaran, V., Fong, M., Gu, G., & Tyson, M. (2013, February). FRESCO: Modular composable security services for software-defined networks. In *20th Annual Network & Distributed System Security Symposium*.
8. Mehdi, S. A., Khalid, J., & Khayam, S. A. (2011). Revisiting traffic anomaly detection using software defined networking. In *Proceedings of Recent Advances in Intrusion Detection.*
9. Robertson, S., Alexander, S., Micallef, J., Pucci, J., Tanis, J., & Macera, A. (2015). CINDAM: Customized information networks for deception and attack mitigation. In *IEEE International Conference on Self-Adaptive and Self-Organizing Systems Workshops (SASOW), London, United Kingdom* (pp. 114–119).
10. https://opennetworking.org/sdn-definition/
11. Lara, A., & Ramamurthy, B. (2016). Opensec: Policy-based security using software-defined networking. *IEEE Transactions on Network and Service Management, 13*(1), 30–42.
12. Sahay, R., Blanc, G., Zhang, Z., Toumi, K., & Debar, H. (2017). Adaptive policy-driven attack mitigation in SDN. In *Proceedings of the 1st International Workshop on Security and Dependability of Multi-Domain Infrastructures, Belgrade, Serbia* (p. 1).
13. Karmakar, K. K., Varadharajan, V., & Tupakula, U. (2017). Mitigating attacks in software defined network (SDN). In *Fourth International Conference On Software Defined Systems (SDS), Valencia, Spain* (pp. 112–117).

14. Aujla, G. S., Singh, M., Bose, A., Kumar, N., Han, G., & Buyya, R. (2020). BlockSDN: Blockchain-as-a-service for software defined networking in smart city applications. *IEEE Network, 34*(2), 83–91.
15. Chen, X., Yu, S. (2016). CIPA: A collaborative intrusion prevention architecture for programmable network and SDN. *Computers & Security, 58*, 1–19.
16. Feamster, N., Rexford, J., & Zegura, E. (2013, December). The road to SDN. *ACM Queue, 11*(12), 20–40.
17. Kreutz, D., Ramos, F. M., & Verissimo, P. (2013). Towards secure and dependable software-defined networks. In *Proceedings of the Second ACM SIGCOMM Workshop on Hot Topics in Software Defined Networking HotSDN '13* (p. 55).
18. Braga, R., Mota, E., & Passito, P. (2010). Lightweight DDoS Flooding attack detection using NOX/OpenFlow. In *IEEE Local Computer Network Conference* (pp. 408-415). IEEE.
19. Aujla, G. S., Chaudhary, R., Kaur, K., Garg, S., Kumar, N., & Ranjan, R. (2018). SAFE: SDN-assisted framework for edge–cloud interplay in secure healthcare ecosystem. *IEEE Transactions on Industrial Informatics, 15*(1), 469–480.
20. Singh, M., Aujla, G. S., Singh, A., Kumar, N., & Garg, S. (2020). Deep-learning-based blockchain framework for secure software-defined industrial networks. *IEEE Transactions on Industrial Informatics, 17*(1), 606–616.
21. Scott-Hayward, S., O'Callaghan, G., & Sezer, S. (2013). SDN security: A survey. In *2013 IEEE SDN For Future Networks and Services (SDN4FNS)* (pp. 1–7). IEEE.
22. Kreutz, D., Ramos, F. M. V., & Verissimo, P. (2013, August). Towards secure and dependable software defined networks. In *Proceedings of the Second ACM SIGCOMM Workshop on Hot Topics in Software Defined Networking* (pp. 55–60).
23. https://www.ndss-symposium.org/wp-content/uploads/2017/09/Presentation07_2.pdf

Part V
Application Use Cases of Software-Defined Networking

Chapter 15
SDVN-Based Smart Data Dissemination Model for High-Speed Road Networks

Deepanshu Garg, Neeraj Garg, Rasmeet Singh Bali, and Shubham Rawat

15.1 Introduction

One of the major concerns in road travel is the increasing number of road accidents happening all over the world on the roads. This has been a matter of growing concern among the governments, researchers and automotive industry. There have been increasing researches being conducted in the community to address this concern [1]. There are a number of reasons that can be attributed to accidents. One primary reason is the development of high-speed road networks that are being constructed to reduce the travel time and vehicles capable of moving at high speeds along these road networks of expressways. Since the drivers are expected to maintain a minimum speed while they are driving on the expressway, a sudden reduction in speed of any vehicle may result in a large number of high-speed crashes due to reduced reaction time available with drivers of the following vehicles. Therefore, whenever a vehicle breaks due to some unavoidable reasons, there must be a mechanism to let the following vehicles know about the breaks applied by the car that is ahead. The development of such an early warning system for high-speed vehicles that move on expressways can help in eliminating accidents caused by sudden breaking of any vehicle [2–4].

Although Vehicular Ad Hoc Networks (VANETs) have been around for a long time and one of their primary objectives is to avoid road accidents, manage traffic in an efficient way and ensure road safety from multiple aspects, they have not been able to achieve the above goal [5]. In fact, this problem has increased due to increased vehicle speeds, and it is expected that an effective solution has to be developed especially for automated vehicles of future [6–9]. The Software-Defined Vehicular Network (SDVN) is expected to play a key role in identifying the probable

D. Garg · N. Garg · R. S. Bali (✉) · S. Rawat
Department of Computer Science and Engineering, Chandigarh University, Mohali, India

G. S. Aujla et al. (eds.), *Software Defined Internet of Everything*, Technology, Communications and Computing, https://doi.org/10.1007/978-3-030-89328-6_15

collision and generating alert messages to the target vehicles. SDVN is based on Software-Defined Network with primary focus around VANET. SDVN offers a range of key benefits as compared to traditional networks in terms of efficient network utilization, routing and better control and management of complete network from a centralized control plane. The traditional network has data and forwarding plane and control plane on each of the network node that is generally a router. This leads to unpredictability in deciding about the availability of a vehicular node due to its high mobility. Also, in case a node goes out of the network, it has to be rectified with a lot of effort. Similarly, in case some patch has to be applied on all the networks, this will lead to another pain point for the network administrator or maintainer. These availability issues and other control-related issues can be addressed easily with SDVNs in most of the situations.

The SDVN can be split into different components viz. control plane, data plane and management plane. The responsibility of control plane is to control the network nodes at two levels. One is centralized controller that will control at the level of Road Side Units (RSUs) and another controller that is the Road Side Software-Defined Controller can control the network nodes at the level of vehicles. This allows the flexibility that is required in case of fast moving vehicles moving in and out of the VANET. The main responsibility of the data plane is to forward the data to the node suggested by SDN controller. The responsibility of the management plane is to monitor, configure and maintain the state of the network nodes. The high-level architecture of SDVN is shown in Fig. 15.1.

In this chapter, we propose a prototype that leverages the recent advances in the field of SDVN so as to control network of RSUs. This model will continuously monitor every vehicle on the expressway and identify any vehicle that shows a varying behaviour such as running at a lower speed than the recommended speed or if a vehicle has stopped at a point suddenly. After identification of such vehicles, the proposed system will generate warning alerts to the vehicles coming towards them through SDVN-based infrastructure. This chapter has also listed down the considerations for high availability and robustness of the system. This chapter is structured into four main sections. Section 15.2 uncovers the literature review. In Sect. 15.3 of this book chapter, the proposed model is presented. This section has a couple of sub-sections with the first sub-section focusing on different components of the proposed system. Section 15.3 details out the various steps in the working model, and the various types of services which can be provided by the proposed model are covered in this section. Section 15.4 presents a case study about implementation of the proposed system. Finally, towards the end, the conclusion along with future directions is presented.

15.2 Literature Review

Bhatia et al. [10] have proposed a hierarchical Software-Defined Network (SDN)-based architectural framework for VANET with support of network virtualization, which would make the deployment of a particular VANET flexible in time, space

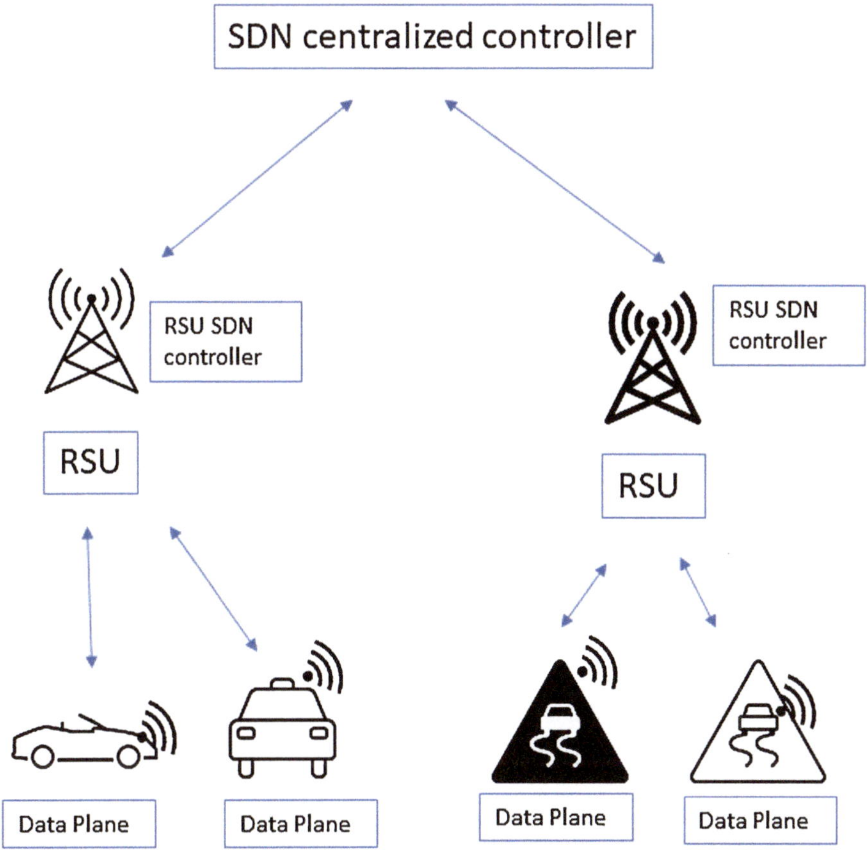

Fig. 15.1 Software-Defined Vehicular Network

and the type of services offered by it. They have introduced the concept of virtual private VANETs to support multi-tenancy in VANET. This will allow a VANET service provider to deploy its services over a single physical infrastructure, probably owned by a third party, quickly in a cost effective manner and in isolation to the other services running on the same infrastructure but owned by different service providers. Liuwei et al. [11] proposed a novel accurate and SDN-based fine-grained traffic measurement approach to obtain comprehensive traffic in vehicular communication network. Zhang et al. [12] have proposed a novel SDFC-VeNET architecture which leverages the concept of SDN and fog computing. Based on the SDFC-VeNET architecture, mobility management and resource allocation are analysed. Truong et al. [13] have proposed combining VANET architecture with fog computing as a prospective solution to address many difficulties in deployment and management because of poor connectivity, less scalability, less flexibility and less intelligence.

Sony et al. [14] proposed a Smart Ranking-based Data Offloading (SRDO) algorithm for selecting an RSU and to improve the Quality of Service. In SRDO algorithm, Q-Learning is utilized for RSU selection. This algorithm is modelled in Software-Defined Network controller to deal with the problem of choosing the RSU in an intelligent way for data offloading. An architecture by Rehman and Kapoor [15] is proposed to introduce hierarchical distributions of controllers where the main controller on the top tier is distributed over the region and at bottom tier some RSUs are selected as local controllers, which keep localized global view of network, where the main controller on the top tier is responsible for operating local controllers at the bottom tier and local controllers are responsible for transferring data among RSUs and vehicles. Bhatia et al. [16] have combined the flexibility, scalability and adaptability leveraged by the SDVN architecture along with the machine learning algorithms to model the traffic flow efficiently.

A VANET monitoring system to provide a resilient and efficient routing for better services for communication between vehicles is proposed by Kalokhe et al. [17]. It also provided the secure channels between the SDN controller and the network devices for reliable communication. Zhao et al. [18] proposed a new software-defined routing method, namely, Novel Adaptive Routing and Switching Scheme (NARSS), deployed in the controller. This adaptive method can dynamically select routing schemes for a specific traffic scenario. The authors presented a method for collecting road network information to describe traffic condition and extracted the feature data used to generate the routing scheme switching model. The authors also trained the feature data through an artificial neural network with high training speed and accuracy. Finally, the model was used as a basis for establishing the NARSS and deploying it in the controller.

A three-level routing hierarchy in improved Software-Defined VANET architecture based on Mobile Edge Computing (MEC) to improve routing performance and enrich the data transmission mode for the VANET is proposed by Xuefeng et al. [19]. Moreover, it can be applied to almost all VANET protocols, enabling protocol-independent forwarding. Besides, this improved architecture can coordinate different edge devices to timely adjust the service delivery strategy under the predictive correction from controllers, providing high-bandwidth and low-delay transmission for Internet of Vehicles. Additionally, MEC technology is introduced to perform local control, leveraging the storage and computing capabilities of edge devices to reduce the processing pressure of the controller. Sanagavarapu et al. [20] have attempted to address the efficiency of SDN's performance in multi-cloud environments by proposing SDPredictNet, a Recurrent Neural Network framework deployed on the SDN Controller that can predict the traffic in the network and update flow tables of the higher layer switches to perform routing based on the perceived bottlenecks in the network. Kaihan et al. [21] have proposed a traffic state prediction method based on Hidden Markov Model and then choose different routing methods according to different traffic states.

Wahid et al. [22] in their paper have proposed the use of Global Positioning System-based floating car data to develop a server communication reduction policy which can help in providing a solution to the perpetual problem of traffic congestion in large urban areas. Khatri et al. [23] have presented and explored the possibility of using machine learning algorithms to address safety, communication and traffic-related issues in VANET systems. Maad et al. [24] in their review paper have discussed the techniques for traffic monitoring along with the concept of early incidents detection so as to address some key issues such as collision, drone navigation, and less computational and communicational overheads.

Kumar et al. [25] proposed the collision-free drone-based movement strategies for road traffic monitoring using Software-Defined Networking. Nama et al. [26] have compiled the review of various tools and techniques to collect traffic data along with machine learning algorithms being applied in the field of modern ways of addressing traffic dilemmas. Emna et al. [27] have introduced the smart city concept and presented a general overview of research that focuses on leveraging the benefits of Software-Defined Networking in specific application domains of smart cities. Paranjothi et al. [28] presented a general architecture of vehicular communication in urban and highway environment as well as a state-of-the-art survey of recent congestion detection and control techniques.

15.3 The Proposed System Model

Although there are numerous data transmission techniques that have been proposed for vehicular networks, the proposed system emphasizes on providing an efficient communication network along with low network overhead for a high-speed vehicular network by using various services of SDN controller. The proposed system utilizes the inherent properties of SDN for taking the decisions smartly according to the requirement of the user. In this section, system model for data transmission system within the vehicles is discussed along with the various services provided by the centralized controller. In the network model shown in Fig. 15.2, there are a certain number of vehicles, RSUs, RSU controllers and base stations that are connected together for achieving fast data transmission for the underlying vehicular nodes. All the RSU controllers are connected to the centralized controller of SDN for the smart data transmission as shown in Fig. 15.2.

The whole system model follows a centralized strategy for data transmission with a centralized SDN controller acting as the logical hub of the system. Each RSU in the network is responsible for transmitting information received through RSU controller to the high-speed vehicles that are within its communication range, and these RSUs are connected together for creating a captive network that continuously monitor vehicles moving on the expressway. These controllers are also used for load balancing on the centralized controller so as to provide fast responses to the high-speed vehicles. It also uses the capabilities of SDN to provide multiple services to end users and can also support on-demand services. The system model assumes that each vehicle is equipped with a Global Positioning System (GPS) technology so as to accurately determine the location of the vehicles. Each vehicle sends its current information to the base stations such as position, direction, source and

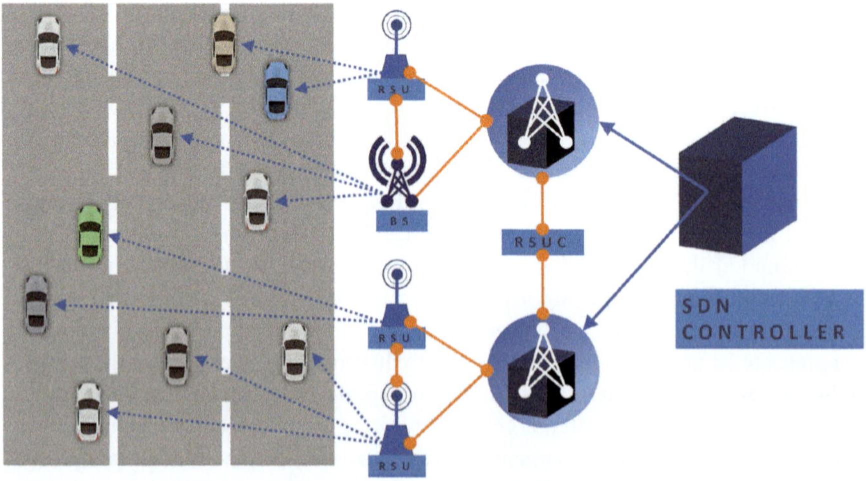

Fig. 15.2 Network Architecture for Data Dissemination in Proposed Model

destination address along with its current speed with the help of RSUs. Figure 15.2 illustrates the network architecture of the proposed model along with the positioning of components and their relationships. All the tasks performed by each component of the network are also depicted in Fig. 15.2 and are discussed below in detail.

SDN Controller This is the core part of the proposed network model as each task of the system would be performed under the supervision of the SDN controller. Traffic data management, routing, network utilization, data transmission and all other decisions regarding smart transmission would be taken by the SDN controller only. It also helps in creating global view of vehicular network and collects all information of the participating entity that can also be integrated with standard applications. This controller also defines the policies, such as privacy laws, computing resources and required memory.

Road Side Unit SDN Controller (RSUC) RSUCs are the controllers with SDN controller capabilities that are managed by the centralized SDN controller. They are required to perform routing within their respective coverage areas based on information received from the base station that can cover multiple RSUs under their network coverage. RSUC communicates with RSUs and base stations via wireless communication and the OpenFlow protocol.

Road Side Unit/Base Station RSUs and base stations are network points with capabilities such as collecting data and transmitting data to RSUC. These are managed through the Road Side Unit SDN controller.

Vehicles Here, we will assume that each vehicular node is equipped with an on-board unit to transmit data in both vehicle-to-vehicle and vehicle-to-infrastructure communication modes, and these are the components of data plane. For establishing communication with network entities, different technologies have been employed in the proposed system.

Communication Devices Vehicle-to-infrastructure communications and vehicle-to-vehicle communications are primarily implemented through IEEE 802.11.x-based wireless communication protocols. For implementing vehicle-to-base station-based communication, a number of different remote wireless communication mechanisms such as cellular-based 4G, cellular-based 5G, wimax or LTE can be utilized. Communications between SDN, RSUC, RSU and BS controllers come into the category of broadband and high-speed connections.

15.3.1 Working Model

The whole process of network model would take place in the following steps that are shown in Fig. 15.3.

Step I When an autonomous vehicle enters on an expressway, it will need to confirm its identity through the SDN controller. The controller keeps a record of all registered vehicles and verifies the identity of that vehicle and its registration status. If the vehicle is not registered, then firstly it would transmit a *beacon* message to RSU through RSU controller that contains its identity for registering that vehicle by providing a unique key. After this, the vehicle will be validated and register in SDN database. Once the registration has been done, that vehicle does not need register again for that expressway as well as other expressways in the same geographical

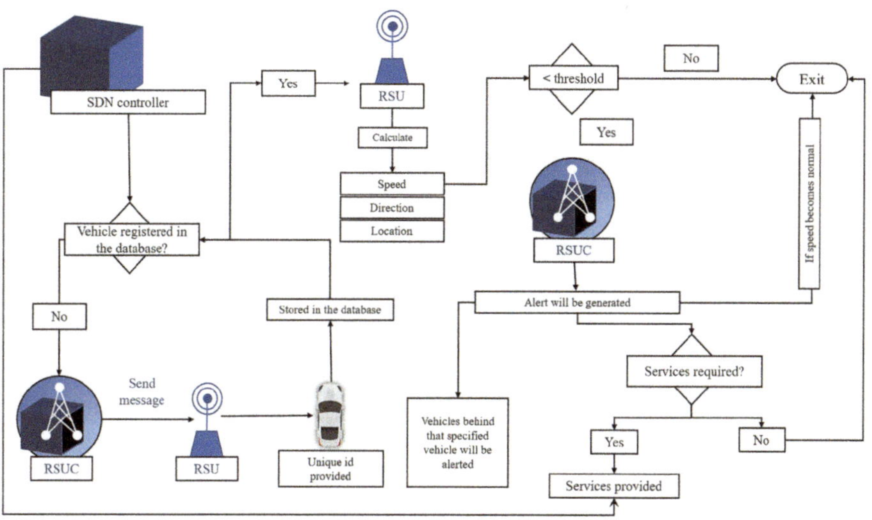

Fig. 15.3 Schematic Description of Data Dissemination in High Speed Road Networks

area as all these expressways maintain a centralized repository containing vehicle details. All previously registered vehicles would be provided a unique key that will be used for validating their identity, while they are moving on the expressway.

Step II In the second phase, each vehicle that is the part of the network will be monitored by the RSUs. Parameters like vehicle speed, location and the direction of movement of every vehicle will be periodically transmitted through *beacon* messages and updated in the SDN storage for further use.

Step III The system maintains threshold values for validating the condition of vehicles. There would be minimal or maximal threshold values that will be used for validating the parameters of vehicles. A minimal threshold value would be set depending on the road conditions on the express highway, and whenever the speed of any vehicle falls below this threshold value, then an alert message will be generated by the closest RSU controller for that particular vehicle. To ward against false alerts, a timer would be activated for re-verification of the vehicle speed. After time out, if the vehicle speed comes back to its normal value, then no further action would be initiated. However, if the speed is still below threshold, then the RSU would broadcast an alert message that would then be forwarded to all nearby vehicles. This process would be cascaded to all the RSUs so that every vehicle travelling in that direction can reduce its speed as they approach the halted vehicle.

Step IV Simultaneously, a *Service Required* message will be generated by the RSU controller that will be received by the closest service vehicle in the vicinity of halted vehicle for providing emergency assistance. This *Service Required* message will also contain a message type to indicate the type of emergency for which service is required. The service vehicle will then initiate appropriate action at its end. For example, in the case of an accident, if the vehicle reports to RSU controller through RSU, then an SDN controller would take the decision and provide assistance to the vehicle. Furthermore, backward cascading message about accident will be sent to all vehicles for avoiding further accidents.

By following the above steps based on SDVN-based infrastructure, the system will be able to provide timely alerts to vehicles travelling through the expressway. By using SDN-based infrastructure, this process is also able to reduce the overhead on the centralized controller as well as to provide various services to the vehicles which are discussed in the next section. Another facet that can be accounted for is the economic overheads. As most of these expressways already use some payment mechanisms at the time of entry, an additional small cost can be added to account for the cost of required infrastructure required for setting up the SDVN-based captive network. The system will also provide a number of useful services to the users. Some of the prominent services that can be offered by this system are as follows.

15.3.2 Types of Services Provided by the Centralized Controller

There are various types of services provided by the SDN controller, and these services are achieved by the vehicles through RSUs or RSU controllers. Depending on the speed or request sent by the vehicle, decision would be taken by the centralized controller, and then service would be provided. The various types of services which we are considering are discussed below

Emergency Services In the case of an accident, emergency service would be provided to the vehicles for providing assistance through the SDN controller.

Repair Services Repair service would be provided in the case if there would be any technical problem occurring in some vehicle while travelling on the express highway. In such a case, RSU controller would supervise the arrival of service vehicle.

Fuel Services If a car is about to run out of fuel on the way, then it can send a request to the centralized controller through RSUs, and the vehicle can be guided to the closest fuel pump or charging point available on the expressway.

Network Services If the cellular network of passengers travelling inside the car is not working, then the users can use high-speed network services by connecting to the SDN servers. The passengers can also extend their available network bandwidth in case they want higher bandwidth for some application that they are using.

Entertainment Services If two cars want to share data between each other, they can easily do so by using the fast network provided by SDN. These services can also be used for sharing any type of infotainment such as downloading multimedia applications. For availing such applications, cars can be connected using the SDN network of SDVN system.

15.4 Case Study: On-Demand Network Service for High-Speed Vehicles

To handle the network requirements of vehicles on the expressway, the proposed system will provide support for utilizing the available bandwidth of SDVN. This on-demand network service can be used to enhance the bandwidth for the vehicles. By utilizing this service, vehicles can avail an additional network bandwidth for running various applications. By leveraging the control plane in SDN, the system can effectively collect and maintain individual vehicle status and also service their network requirements in a logically centralized way, which would not be achievable in a conventional network. In actual fact, control plane stores each vehicle's information provided by the RSU controllers for taking the decisions like switching and routing.

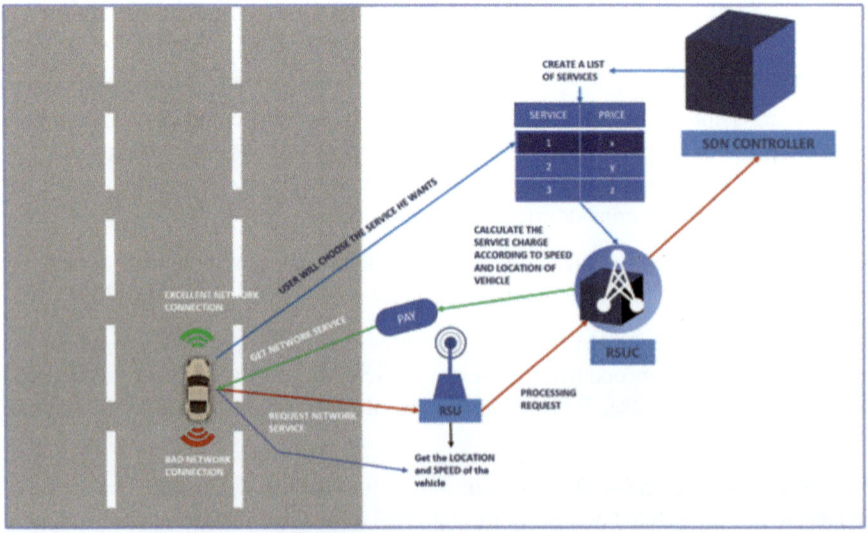

Fig. 15.4 High bandwidth request scenario

To service the on-demand network requirements, we assume that some of the vehicles travelling on expressway are trying to download or upload a large amount of data as shown in Fig. 15.4. But because of relatively slow network connection and high speed of vehicles, they are unable to do so.

These vehicles will send a request packet to RSU to extend the network bandwidth for availing high-speed network. This request is forwarded to SDN by using the RSU controller. After analysing the speed and location of the vehicle, the SDN will provide a list of customized services that are available to the vehicle through the RSU controller. The user can choose the network services according to his/her need, for example, uploading and downloading speed required, time period for which data requires. After choosing the services, this request will be further processed by an SDN controller. The SDN controller will send this data along with parameters of network service, which are the speed of data and amount of data to the RSU controller. The RSU controller will then send a message specifying the details for the same by taking the location and speed of the vehicle into account. The RSU will also calculate the total usage and other accounting information of the user once the service has been availed.

15.5 Conclusion

With the aim to propose smart data communication model for high-speed vehicles, this proposed system presents a data communication scheme based on the SDN paradigm by combining the RSU and the RSU controllers in order to efficiently support the vehicle requirements on express highways. The whole model works under the supervision of a centralized SDN controller to improve the efficiency and reliability of vehicular networks for high-speed vehicles. Along with providing a safe journey through the expressway, the proposed system also provides a number of other services for vehicles moving in high speed through the network. By effectively utilizing this system, the vehicle users can avail both comfortable and safe road transport that are the two primary objectives of the Intelligent Transportation System. We are now working on integrating each of these services and evaluating them so as to provide benefit of the proposed architecture while also developing a holistic service framework for high-speed roads and future automated vehicles.

References

1. Bali, R. S., Kumar, N., & Rodrigues, J. J. P. C. (2014). An intelligent clustering algorithm for VANETs. In *2014 International Conference on Connected Vehicles and Expo (ICCVE)* (pp. 974–979). IEEE.
2. Jindal, A., Aujla, G. S., Kumar, N., Chaudhary, R., Obaidat, M. S., & You, I. (2018). SeDaTiVe: SDN-enabled deep learning architecture for network traffic control in vehicular cyber-physical systems. *IEEE Network, 32*(6), 66–73.
3. Aujla, G. S., Jindal, A., & Kumar, N. (2018). EVaaS: Electric vehicle-as-a-service for energy trading in SDN-enabled smart transportation system. *Computer Networks, 143*, 247–262.
4. Gulati, A., Aujla, G. S., Chaudhary, R., Kumar, N., & Obaidat, M. S. (2018, May). Deep learning-based content centric data dissemination scheme for internet of vehicles. In *2018 IEEE International Conference on Communications (ICC)* (pp. 1–6). IEEE.
5. Bali, R. S., & Kumar, N. (2016). Learning automata-assisted predictive clustering approach for vehicular cyber-physical system. *Computers & Electrical Engineering, 52*, 82–97.
6. Gulati, A., Aujla, G. S., Chaudhary, R., Kumar, N., Obaidat, M., & Benslimane, A. (2019). Dilse: Lattice-based secure and dependable data dissemination scheme for social internet of vehicles. *IEEE Transactions on Dependable and Secure Computing*. https://doi.org/10.1109/TDSC.2019.2953841
7. Garg, S., Singh, A., Aujla, G. S., Kaur, S., Batra, S., & Kumar, N. (2020). A probabilistic data structures-based anomaly detection scheme for software-defined Internet of vehicles. *IEEE Transactions on Intelligent Transportation Systems, 22*(6), 3557–3566.
8. Aujla, G. S., Singh, A., Singh, M., Sharma, S., Kumar, N., & Choo, K. K. R. (2020). BloCkEd: Blockchain-based secure data processing framework in edge envisioned V2X environment. *IEEE Transactions on Vehicular Technology, 69*(6), 5850–5863.
9. Singh, A., Aujla, G. S., & Bali, R. S. (2020). Intent-based network for data dissemination in software-defined vehicular edge computing. *IEEE Transactions on Intelligent Transportation Systems, 22*(8), 5310–5318.
10. Bhatia, A., Haribabu, K., Gupta, K., & Sahu, A. (2018). Realization of flexible and scalable VANETs through SDN and virtualization. In *2018 International Conference on Information Networking (ICOIN)* (pp. 280–282). IEEE.
11. Huo, L., et al. (2019). A SDN-based fine-grained measurement and modeling approach to vehicular communication network traffic. *International Journal of Communication Systems*. https://doi.org/10.1002/dac.4092
12. Zhang, Y., Zhang, H., Long, K., Zheng, Q., & Xie, X. (2018). Software-defined and fog-computing-based next generation vehicular networks. *IEEE Communications Magazine, 56*(9), 34–41.

13. Truong, N. B., Lee, G. M., & Ghamri-Doudane, Y. (2015). Software defined networking-based vehicular adhoc network with fog computing. In *2015 IFIP/IEEE International Symposium on Integrated Network Management (IM)* (pp. 1202–1207). IEEE.
14. Guntuka, S., Shakshuki, E. M., Yasar, A., & Gharrad, H. (2020). Vehicular data offloading by road-side units using intelligent software defined network. *Procedia Computer Science, 177*, 151–161.
15. Rehman, S., & Kapoor, N. (2019). A review on delay efficient architecture for Software Defined Vehicular Networks (SDVN). In *2019 6th International Conference on Computing for Sustainable Global Development (INDIACom)* (pp. 1094–1100). IEEE.
16. Bhatia, J., Dave, R., Bhayani, H., Tanwar, S., & Nayyar, A. (2020). SDN-based real-time urban traffic analysis in VANET environment. *Computer Communications, 149*, 162–175.
17. Kalokhe, K. N., Park, Y., & Chang, S.-Y. (2018). Resilient SDN-based communication in vehicular network. In *International Conference on Wireless Algorithms, Systems, and Applications* (pp. 865–873). Cham: Springer.
18. Zhao, L., Zhao, W., Al-Dubai, A., & Min, G. (2019). A novel adaptive routing and switching scheme for software-defined vehicular networks. In *ICC 2019-2019 IEEE International Conference on Communications (ICC)* (pp. 1–6). IEEE.
19. Ji, X., Xu, W., Zhang, C., & Liu, B. (2020). A three-level routing hierarchy in improved SDN-MEC-VANET architecture. In *2020 IEEE Wireless Communications and Networking Conference (WCNC)* (pp. 1–7). IEEE.
20. Sanagavarapu, S., & Sridhar, S. (2021). SDPredictNet-a topology based SDN neural routing framework with traffic prediction analysis. In *2021 IEEE 11th Annual Computing and Communication Workshop and Conference (CCWC)* (pp. 0264–0272). IEEE.
21. Gao, K., Ding, X., Xu, J., Yang, F., & Zhao, C. (2020). HMM-based traffic state prediction and adaptive routing method in VANETs. In *International Conference on Collaborative Computing: Networking, Applications and Worksharing* (pp. 236–252). Cham: Springer.
22. Wahid, A., Rao, A. C. S., & Goel, D. (2019). Server communication reduction for GPS-based floating car data traffic congestion detection method. In *Integrated Intelligent Computing, Communication and Security* (pp. 415–425). Singapore: Springer.
23. Khatri, S., Vachhani, H., Shah, S., Bhatia, J., Chaturvedi, M., Tanwar, S., & Kumar, N. (2021). Machine learning models and techniques for VANET based traffic management: Implementation issues and challenges. *Peer-to-Peer Networking and Applications, 14*(3), 1778–1805.
24. Hamdi, M. M., Audah, L., Rashid, S. A., & Al Shareeda, M. (2020). Techniques of early incident detection and traffic monitoring centre in VANETs: A review. *Journal of Communications, 15*(12), 896–904.
25. Kumar, A., Krishnamurthi, R., Nayyar, A., Luhach, A. K., Khan, M. S., & Singh, A. (2021). A novel Software-Defined Drone Network (SDDN)-based collision avoidance strategies for on-road traffic monitoring and management. *Vehicular Communications, 28*, 100313.
26. Nama, M., Nath, A., Bechra, N., Bhatia, J., Tanwar, S., Chaturvedi, M., & Sadoun, B. (2021). Machine learning-based traffic scheduling techniques for intelligent transportation system: Opportunities and challenges. *International Journal of Communication Systems, 34*(9), e4814.
27. Rbii, E., & Jemili, I. (2020). Leveraging SDN for smart city applications support. In *International Workshop on Distributed Computing for Emerging Smart Networks* (pp. 95–119). Cham: Springer.
28. Paranjothi, A., Khan, M. S., & Zeadally, S. (2020). A survey on congestion detection and control in connected vehicles. *Ad Hoc Networks, 108*, 102277.

Chapter 16
Advanced Deep Learning for Image Processing in Industrial Internet of Things Under Software-Defined Network

Zhihan Lv, Liang Qiao, Jingyi Wu, and Haibin Lv

16.1 Introduction

The rapid development of Internet of Things (IoT) has built a new digital world. According to statistics, the global expenditure of IoT in 2019 has exceeded 745 billion US dollars and is expected to reach 1 trillion US dollars in 2022 [1, 2]. As the basis and premise of modernization, the highly developed industrial society is an important symbol of modernization. Hence, industrial Internet of Things (IIoT) has become one of the main development directions of IoT technology in the future. IIoT contains hundreds of millions of industrial devices. Even the smallest devices can be connected, monitored, and tracked, share the status data of each device, and communicate with other devices. Then, all the acquired data can be collected and analyzed to improve the efficiency of business process [3]. At present, IIoT has gradually become a research hotspot in manufacturing, general industry, transportation, and other fields. It can help better understanding the operation of the production line and predicting the maintenance time of industrial equipment in time, thus reducing the unexpected downtime [4–6]. Traditional network architecture cannot meet the management requirements of many sensor nodes in IIoT, and the software-defined network (SDN), a new network architecture, provides a new possibility.

The popularity of IIoT devices marks the progress of technology. IIoT based on network sensor and cloud resource technology provides two-way movement of local and remote assets between enterprises and business partners [7]. IoT hardware

---

Z. Lv (✉) · L. Qiao · J. Wu
College of Computer Science and Technology, Qingdao University, Qingdao, China

H. Lv
North China Sea Offshore Engineering Survey Institute, Ministry of Natural Resources NorthSea Bureau, Qingdao, China

G. S. Aujla et al. (eds.), *Software Defined Internet of Everything*, Technology, Communications and Computing, https://doi.org/10.1007/978-3-030-89328-6_16

and software can generate valuable operation data in the "industry 4.0" era, which can be applied to mechanical pistons, bearings, and other small parts, as well as drilling, mining, and other large-scale systems. In IIoT, the support of machine vision technology is indispensable in all aspects of transportation, supply chain management, material processing, and security [8]. As a key intelligent technology, image processing in machine vision can complete higher level conscious decision-making through higher level image recognition. Image processing is not only necessary for inspection but also helps locating and training industrial robot system. Based on the application of image processing system, more and newer efficient business models will appear in the "industry 4.0" era [9, 10].

Image processing is an important branch of machine vision. In recent years, it has been widely used in the field of security camera in IIoT to improve the image quality and avoid the influence of defects in the image on the image analysis. As an important branch of machine learning, deep learning can obtain multilevel image feature information directly from the original image through unsupervised learning. In the field of image processing of IIoT, deep learning also contributes to solving some abstract concepts and selecting useful feature information.

The dynamic computing framework is constructed based on SDN. The connection between hosts is realized through virtual switch, and all hosts are controlled by the controller of SDN. The dynamic computing framework sends instructions through the controller to realize the interaction between the control layer and the data layer of SDN. To sum up, image processing technology based on deep learning has been widely concerned, but there are rare studies on applying advanced deep learning in IIoT devices and industrial defect identification. Thus, the image processing technology in IIoT is analyzed, and the application of deep learning technology in image processing is explored, thereby improving the effectiveness of data acquisition and fault diagnosis in IIoT.

16.2 Related Works

In addition to realizing intelligent life, IoT has also made remarkable development in the industrial field. In the primary stage of modern industrial development, some scholars have explored the challenges faced by the development of IIoT. Zhang et al. [11] held that a large amount of data generated by IIoT is very valuable for understanding the operation status of basic equipment and proposed a method of analyzing the operation status of equipment based on sensor data. In addition, they established a prediction model of the operation status of equipment, designed a deep neural network (DNN) model of the operation data prediction of equipment, and improved the prediction accuracy through system feature engineering and optimal super parameter search [11]. Sisinni et al. [12] found that compared with the wireless industry, IIoT usually requires less throughput for applications, and the capacity of each node can be greatly released. Meantime, there are higher requirements for the performance of IIoT hardware, such as latency, energy efficiency, and reliability [12].

To effectively supervise the equipment and process level of IIoT, the application of image processing technology in IIoT has gradually become a research hotspot. Liu et al. [6] studied the image compression and transmission technology in the image processing system of IIoT. Based on the deep perception and local image characteristics and taking the image quality evaluation index (QAM) as the standard, they proposed an automatic perceptual evaluation of the image quality received by the terminal display device in the environment of IoT. This is of great significance for the prediction of equipment failure of IIoT [6]. He et al. [13] applied deep learning to single image processing and proposed a new DNN to infer image accuracy by effectively fusing the middle-level information of fixed focus data set [13]. Artusi et al. [14] proposed a line monitoring based on deep learning and image processing technology for industrial pantograph slide and a method of surface defect detection and recognition. Moreover, they analyzed the method of equipment wear edge based on image processing technology and verified that the image processing had a higher evaluation accuracy in equipment wear prediction [14].

In summary, the image processing technology based on deep learning has been applied in many fields, such as medicine, industry, and manufacturing. However, regarding IoT, especially in the environment of IIoT, the studies on image processing technology are still in its infancy. The application of deep learning technology in image processing of IIoT is explored here, and then the equipment and processing technology in industrial production is effectively supervised, providing convenient and powerful guarantee for industrial production.

16.3 Method

16.3.1 Architecture and Key Technologies of IIoT

IoT is changing the way people interact with things around them, and it has a great influence on the industry. The emergence of IIoT connects all industrial assets, including machines, control systems, information systems, and business processes [15]. A large amount of data is collected to provide analysis solutions to achieve the best industrial operation. Hence, IIoT affects the entire industrial value chain, which is the inevitable requirement of intelligent manufacturing [16]. IIoT can be regarded as a subset of "Industrial Internet," and sensing is its basic feature, thus realizing industrial operation in industrial management.

IIoT covers the field of industrial communication technology for machine-to-machine (M2M) and automation applications, enabling people to better understand the industrial production process and achieve efficient and sustainable production. Through the application of the new generation technology concept, the industrial Internet realizes the intelligent cooperation of massive industrial entities, thus changing the future industrial infrastructure of industrial production form [17–19]. Based on IIoT, different types of industrial entities and even the entire industrial

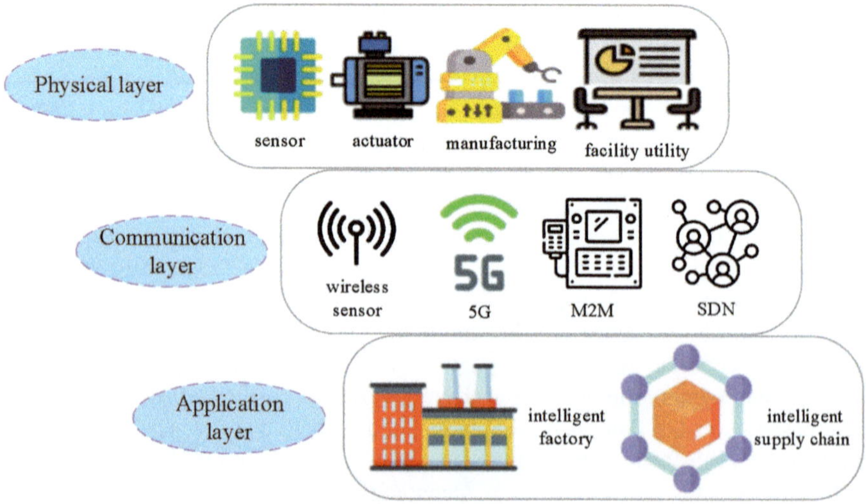

Fig. 16.1 System architecture of IIoT

network are managed and controlled, and the industrial role and social resources are effectively integrated to further realize the intelligent development of industrial entities.

The architecture of IIoT can be divided into three layers: physical layer, communication layer, and application layer, as shown in Fig. 16.1. The physical layer is composed of widely deployed physical devices, including sensors, actuators, manufacturing equipment, facility utilities, and other industrial manufacturing and automation-related objects. The communication layer is the integration of many communication networks, including wireless sensor and actuator networks (WSANs), 5G, M2M, SDN, and others, which is responsible for supporting the connectivity between sensors and actuators. The application layer of IIoT is composed of a series of industrial applications, such as intelligent factory and intelligent supply chain [20]. It realizes real-time monitoring and effective management of industrial production through sensors and actuators.

IIoT is a special field based on IoT, so the key technologies of IIoT mainly include identification and tracking technology, communication technology, network technology, and service management technology. Among them, the identification and tracking technology includes intelligent sensors, radio frequency identification (RFID) technology, and barcode. RFID system has the ability of identification and tracking. It can be combined with wireless sensor network (WSN) to further promote the realization of industrial services and create the IoT application more suitable for industrial environment. For communication technology, gateway can promote the communication between various devices on the network and can also be used to deal with the complex nodes involved in the communication on the network [21]. As the devices in the industrial Internet usually have different communication and computing capabilities, the cross-layer protocol of wireless network needs to be modified properly before it is applied to the industrial internet, thus better using the Internet for information exchange and data communication.

16.3.2 Dynamic Computing Framework of SDN-Based IIoT

The core of SDN is to divide the whole network into control layer and data forwarding layer. The controller in the control layer realizes all decision-making and scheduling, while the switch and other network devices in the data forwarding layer are only responsible for executing the tasks determined by the control layer.

In the actual deployment, SDN software switch is implemented through Open Switch, and its advantages mainly include open forwarding function, standard interface support, and programmable expansion and control. The control of network is realized by SDN controller Open Day Light. Its advantages include supporting many south interface protocols and north interface protocols, which can meet the requirements of network control in dynamic computing framework. The Docker container is utilized to achieve the separation of computing functions, and applications and dependent environments can be packaged into an independent image.

The transmission configuration generated by the dynamic configuration module is used to coordinate the transmission and calculation process of various devices and is the core of the whole dynamic transmission framework. The computing configuration generated by the dynamic computing module is used to coordinate the transmission and calculation process of various devices and is the core of the whole dynamic transmission framework, which is realized by Json here. There are three important parts in the computing configuration: the transfer module, the computing function, and the transfer statistics. In the dynamic computing framework, not only the server in the cloud can run the algorithm to perform the calculation, but the edge host and the middle SDN switch as the starting point can also do this. Thus, the users can choose one of the data acquisition modules, the virtual switch module, and the result processing modules to perform the calculation. Because the execution modules selected in the computing configuration are different, the devices used are also different, and the performance of each transfer will be different. This requires the dynamic computing framework to automatically count the important data in the process of transmission and calculation, so that users can view it in time after the transfer process.

16.3.3 Application of Software-Defined Internet of Everything

Smart city cloud platform requires the network of data center to provide tenants with relatively independent network of autonomous opening, configuration, and management. In Metropolitan Area Network (MAN) and Backbone Network, edge control equipment is the core control unit of user and service access. Based on the SDN technology, the functions of edge access control equipment can be promoted to MAN controller, and the flexible and fast deployment of services can be realized by virtualization. The SDN network controller supports the autonomous discovery and registration of various remote devices and supports maintaining the connection between the remote node and the master node. The edge access control device only needs to realize the physical resource configuration of user access, which greatly reduces the burden of the edge control device and improves the utilization rate of network devices. SDN's forwarding and control are separated, and the controllers are deployed in a centralized way. It can collect the traffic demand among various services in the data center, carry out unified calculation and scheduling, realize the flexible allocation of bandwidth, optimize the network to the greatest extent, and improve the utilization of network resources. For the present stage, SDN technology needs to evolve to access devices closer to the bottom layer for the present stage. It needs to be compatible with more devices and protocols, so that more and more front-end detectors can be managed and controlled by network/software.

Home is not only the smallest part of each city but also the smallest node of smart city. Smart home is the carrier of home, which is the application and embodiment of the concept and technology of smart city at the home level. Based on the SDN technology, the network strategy is adjusted on the Internet, so many smart home users have realized the project research and application of differentiated IPTV service. SDN technology is used to realize the policy scheduling of IP Internet network elements. The ability of network equipment carrying high-definition IPTV service is deeply mined and flexibly used. The access control template of broadband remote access server (BRAS) carrying high-definition IPTV service is designed and deployed. When the service is triggered, the remote service management platform (control layer) directly sends the differentiated change of authorization (COA) policy to the BRAS device (forwarding layer) reference policy control template. According to the user differentiated control policy, BRAS realizes the service access control and differentiated dynamic QoS scheduling on the network to achieve the service control and forwarding, thereby realizing the differentiated IPTV product service function.

SDN-based IIoT heralds a new wave of modernization. In many industries, customers and internal stakeholders demand to achieve more progress in productivity, management, security, and flexibility. Figure 16.2 shows a typical factory where services and workloads are more information technology centric (such as factory data center) and gradually become operational technology centric (such as factory machines) as they move down the hierarchy. Software-defined resource allocation and management is gaining momentum in the fog paradigm because it enables factory operators to better adapt to future needs. From the perspective of network, this will be transformed into using SDN to realize the virtual network function of the whole factory. The wireless combination of "5G + wireless cluster + Wi-Fi" is used to solve the problems of wide coverage, discrete distribution, and extensive connection of mass industrial equipment. The remote monitoring and diagnosis, remote guidance, and operation safety behavior analysis of mass

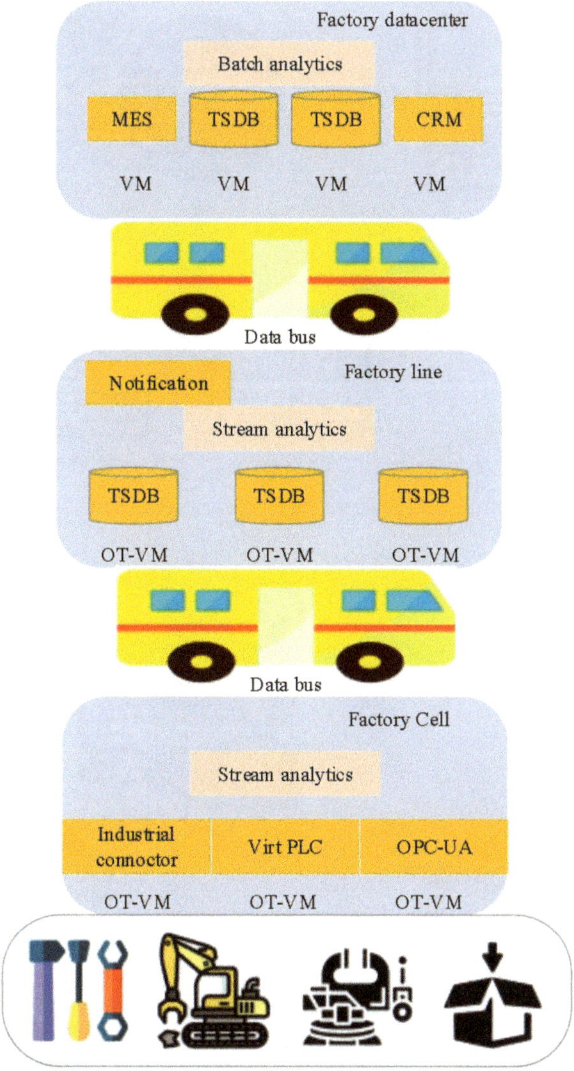

Fig. 16.2 The architecture of smart factory based on software-defined IoT

industrial equipment are realized, making the scene of unmanned railway water transportation and unmanned driving lifting realized.

16.3.4 Image Processing System for Machine Vision of IIoT

Modern industry has realized automation, which takes the place of traditional manpower as the main productivity, and puts forward higher requirements for

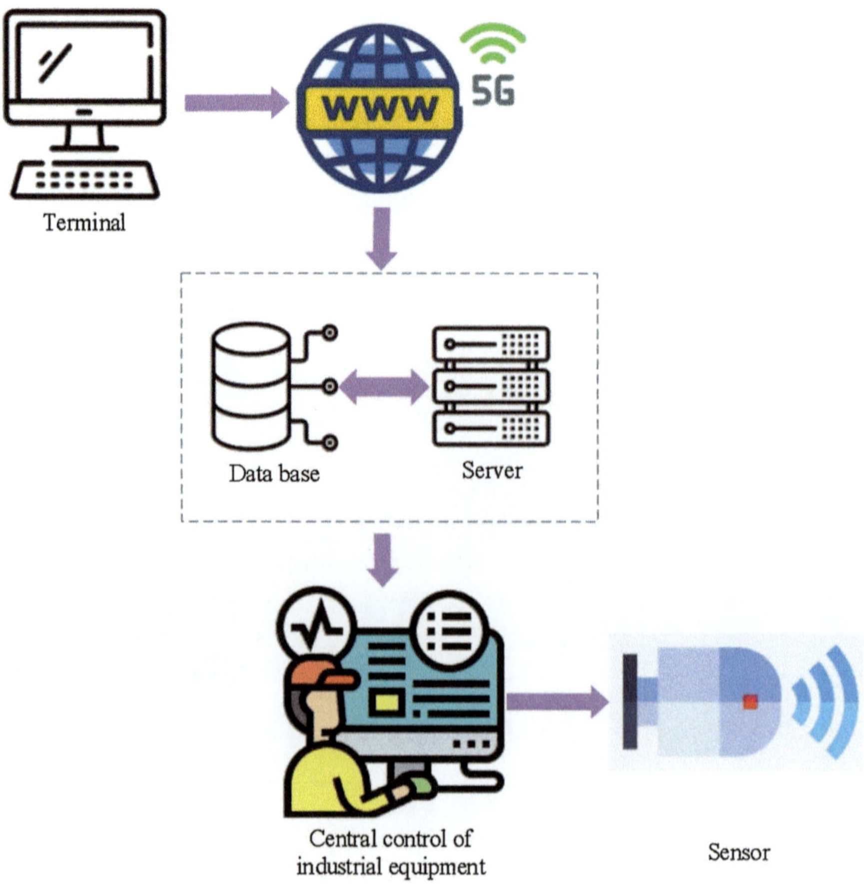

Fig. 16.3 Structure of IIoT monitoring system

production management personnel [22]. The powerful monitoring system of IIoT can realize the supervision of the real-time status of production equipment and its operation and statistics of real-time data information. Figure 16.3 shows the system structure.

As the automation is improved, industrial imaging and machine vision applications are becoming more popular in industrial control to support the power to improve productivity. To increase efficiency, the image processing system must run at a high speed to ensure that the average time between each new unit's start of production is as short as possible. At present, there are high-resolution cameras on the market, which can record high-definition data and support the imaging of the very fine details of the components on the production line [23]. The performance of image processing is very important due to the high-speed transmission of higher quality image using Gigabit Ethernet.

It is very important to reflect the flexibility of industrial image processing through key algorithms in IIoT hardware, especially for different types of industrial equipment components, which need to focus on some visual features [24]. For example, the inspection of the circuit board after the assembly of the equipment is usually to detect the fracture and extrusion in the weld bead, indicating the possible failure source. Through image edge detection and pattern matching to identify areas of interest, each module will perform a complete inspection task according to its own set of image processing programs. Image processing is not only necessary for inspection but also helps to locate and train robot system. The machine vision detection system uses Charge Coupled Device (CCD) or Complementary Metal Oxide Semiconductor (CMOS) camera to convert the detected object into image signal, which is transmitted to the special image processing system [25]. According to the pixel distribution, brightness, color, and other information, it is transformed into digital signal. The image processing system performs various operations on these signals to extract the characteristics of the object, such as quantity, size, position, and volume [26]. Then, according to the preset permissibility and other conditions output results, the automatic recognition function is realized. In the process of mass repetitive industrial production, the detection method using machine vision image processing can greatly improve the efficiency and automation of production. Figure 16.4 shows the application of image processing technology based on machine vision in IIoT.

The image processing system based on deep learning realizes the remote monitoring of the basic process of industrial production, avoids the trouble of manually visiting the production site, and brings users a good experience. Additionally,

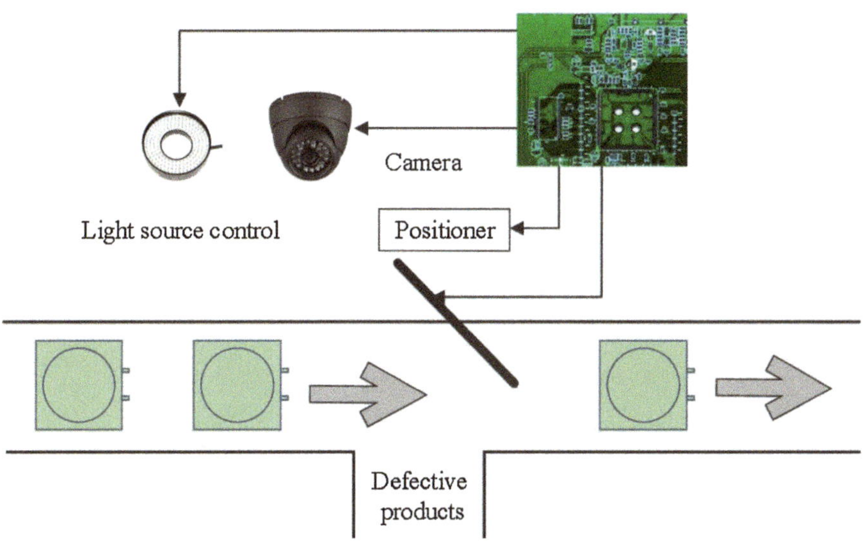

Fig. 16.4 Application of image processing technology based on machine vision in IIoT

through the early warning function and automatic processing function, the safety of equipment operation is greatly improved, and the work efficiency is improved. The device can communicate with the server and transmit the real-time status of the device to the server through the network, and the server can store the device status data in the database. In the later stage, the historical information can be obtained by accessing the data.

16.3.5 Image Processing Technology Based on Deep Learning

As an important branch of machine learning, deep learning can obtain multilevel image feature information directly by unsupervised learning from the original image and simulating the representation of multilevel features in the data by using multiple nonlinear transformation framework [27, 28]. At present, the spatial adaptive image smoothing algorithm based on unsupervised learning has been successfully applied to a series of visual applications, such as image detail enhancement, texture removal, and image processing, and Fig. 16.5 shows the specific process.

The image smoothing aims to reduce the interference of unimportant image details and keep the subject of image structure clear and complete. The energy function can be expressed as

$$\mathcal{E} = \mathcal{E}_d + \lambda_f g \mathcal{E}_f + \Lambda_e g \mathcal{E}_e, \tag{16.1}$$

where $\mathcal{E}_d$ represents data item, $\mathcal{E}_f$ is smooth item, $\mathcal{E}_e$ indicates the edge reserved item of the picture, and λ_f λ_g are balanced weights. The data item measures the color difference between the output and the input images. The data item in the red, green, and blue color space (RGB) can be expressed as

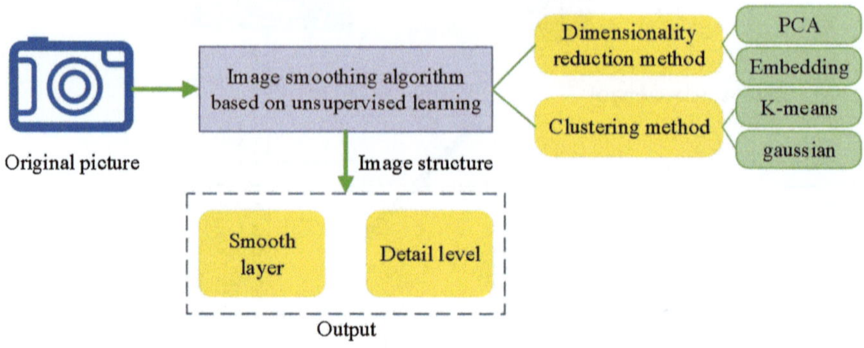

Fig. 16.5 Image smoothing process based on unsupervised learning

$$\mathcal{E}_d = \frac{1}{N}\sum_{1=1}^{N}\left|\left|T_i - I_i\right|\right|_2^2, \tag{16.2}$$

where N is the total number of pixels, and i is the pixel index.

In the process of image smoothing, some important edges may be weakened, so it is necessary to retain the important pixel information. The edge response form of an image can be defined as the sum of the local gradient amplitude of the image:

$$E_i(I) = \sum_{j\in N(i)}\left|\sum_{c}(I_{i,c} - I_{j,c})\right|, \tag{16.3}$$

where $N(i)$ represents the neighborhood of point i, and c is the color channel of input image I. Supposing that the binary mapping is B, $B_i = 1$ is the important edge point and $B_i = 0$ is the unimportant edge point. The edge reservation term can be expressed as

$$\mathcal{E}_e = \frac{1}{N_e}\sum_{i=1}^{N} B_i\, g\left|\left|E_i(T) - E_i(I)\right|\right|_2^2 \tag{16.4}$$

$$N_e = \sum_{i=1}^{N} B_i, \tag{16.5}$$

where N_e represents the sum of important edge points, and I and T suggest the input image and the output smooth image, respectively.

To delete the unimportant details of the image, the smoothing item can control the degree of smoothing by punishing the color difference between adjacent pixels:

$$\mathcal{E}_f = \frac{1}{N_e}\sum_{i=1}^{N}\sum_{j\in N_h(i)} w_{i,j}\, g\left|T_i - T_j\right|^{pi}, \tag{16.6}$$

where $N_h(i)$ is the adjacent pixel of the current pixel, $w_{i,j}$ indicates the weight of the pixel pair, and $\left|T_i - T_j\right|^{pi}$ represents the norm of the smoothing item. $w_{i,j}$ is the Gaussian weight on the color domain and the spatial domain, expressed as the following equations:

$$w_{i,j}^r = exp(-\frac{\sum_c (I_{i,c} - I_{j,c})^2}{2\sigma_r^2}), \tag{16.7}$$

$$w_{i,j}^s = exp(-\frac{(x_i - x_j)^2 + (y_i - y_j)^2}{2\sigma_s^2}), \tag{16.8}$$

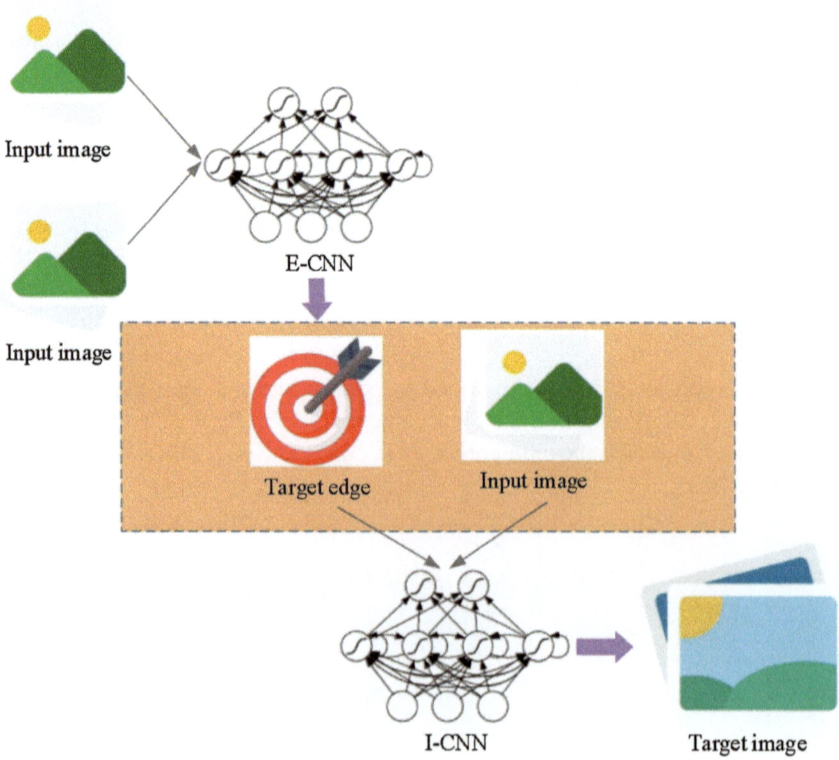

Fig. 16.6 System flowchart of edge learning algorithm

where σ_r represents the standard deviation calculated by the Gaussian kernel in the color domain, and σ_s means the standard deviation calculated by the Gaussian kernel in the space domain.

When dealing with edge-sensitive tasks, the edge detection image E^t of the target image I^t is learned by convolution neural network (CNN). The implementation process of CNN network includes the following: first, convert images into matrices. Each matrix is loaded with corresponding pixel values of different colors and then input into the computer. Second, make data regularization, convolution operation, activation, pooling, full connection, and other calculation operations after completing the preliminary image processing operations, and each convolution is equivalent to a process of extracting basic image graphics. Figure 16.6 shows the system flow of edge learning algorithm, edge learning convolution neural network (E-CNN).

- Step 1: filtering. Edge detection algorithm is mainly based on the first and second derivatives of image intensity, and the filter is used to improve the performance of edge detector. Therefore, there is a trade-off between edge enhancement and noise reduction.

- Step 2: enhancement. The basis of edge enhancement is to determine the change value of the neighborhood intensity of each point in the image. The enhancement algorithm can highlight the points with significant changes in the neighborhood intensity value.
- Step 3: detection. There are many points with relatively large gradient amplitude in the image, and these points are not all edges in a specific application field, so some methods are used to determine which points are edge points.
- Step 4: positioning. If an application scenario needs to determine the edge position, the position of the edge can be estimated on the sub-pixel resolution, and the orientation of the edge can also be estimated.

E-CNN can be approximated by the following function f:

$$E^t = f(I^S, E^S). \tag{16.9}$$

To solve the problem of color attenuation in deep network training, the calibration process of output image color information can be expressed as follows:

$$S_c \; arg \; min_{S_c} \left|\left| I_c^S - S_c.I_c^t \right|\right|_2^2 \tag{16.10}$$

$$I_c^t \leftarrow S_c g I_c^t. \tag{16.11}$$

The subnet is trained by minimizing the mean square error (MSE) of the predicted image and the target image. The difference between the loss function of the image edge prediction and the gradient of the minimized MSE is expressed as Eqs. 16.12 and 16.13, respectively,

$$l_E(\theta) = \left|\left| E^t - E^{t*} \right|\right|_2^2 \tag{16.12}$$

$$l_I(\theta) = \alpha \left|\left| I^t - I^{t*} \right|\right|_2^2 + \beta \left|\left| \nabla_x I^t - \nabla_x I^{t*} \right|\right|_1 + \left|\left| \nabla_y I^t - \nabla_y I^{t*} \right|\right|_1, \tag{16.13}$$

where * represents the true value. In the joint training stage, the whole network is trained by minimizing the loss.

$$l(\theta) = l_I(\theta) + \gamma l_E(\theta) \tag{16.14}$$

When two subnetworks are trained, the learning rate is set to 0.01 in the initial iteration, and then the value is reduced to 0.001 to tune the network.

The loss function is optimized by gradient descent algorithm in DNN. The whole network is trained by unsupervised learning. DNN implicitly learns the optimization process, and it only needs a feed-forward propagation to predict the smooth image, without redundant optimization steps.

16.4 Simulation Experiment

Natural images in Pascal Visual Object Classes (VOC) data set are selected for training [29]. To evaluate the performance of deep learning algorithm, the images in Pascal VOC data set are randomly selected to filter to produce true value labels. The synthetic damaged image is taken as the input and the clear natural image as the target image. DNN is used to simulate several image smoothing methods: the weighted least square (WLS) method, relative total variation (RTV), and L0 smoothing [30]. Peak signal-to-noise ratio (PSNR) and structural similarity (SSIM) are selected as evaluation indexes.

On this basis, the image processing method based on deep learning is applied to the detection of welding defects in industrial production, further proving the effectiveness of this method in practical application. The data set adopts the welding seam ray image in the open database GDX-ray and cuts it into 32×32 image blocks with different defects as the learning sample data set. Texture is an important feature in image pattern recognition. In this experiment, feature vectors are obtained when the distance is 1, 2, and 3, and the Haralick feature is composed of the feature vectors in series. To more accurately compare the performance of various features, three data sets (RUS, ROS, and SMOTE) are selected to compare the Haralick features, directional gradient histogram (HOG) features, deep learning features based on stack sparse automatic encoder (SSAE), and deep CNN (DCNN) proposed. As the texture is formed by the gray distribution in the spatial position repeatedly, there will be a certain gray relationship between two pixels in the image space, that is, the spatial correlation characteristics of the gray level in the image. HOG feature is a feature descriptor for object detection in computer vision and image processing. It constructs the feature by calculating and counting the histogram of the gradient direction of the local region of the image, and the image and shape of the local object can be well described by the density distribution of gradient or edge.

The evaluation indexes are the classification accuracy and average accuracy (ACCR) of crack (CR), lack of penetration (LOP), no fusion (ND), pore (PO), and slag inclusion (SI) [31]. To further analyze the feasibility of applying migration learning in the recognition and classification of weld image defect, experiments are conducted on three data sets to explore the performance of the classic AlexNet and Visual Geometry Group-16 (VGG16) models in weld image classification. The model in this section inherits the lower level of these classical models, changes the top three layers, and changes the classification category to 5. Enlarge the input image block to the appropriate size to adapt to the input of the model. For example, for the AlexNet model, the size of the image block is enlarged to $225 \times 225 \times 3$ (the original image is a gray image, and the values on the three channels are set to be the same). The experiment runs on a single GPU, with 4 cycles and 1380 iterations. Figure 16.7 shows the effect sketch of weld image defect.

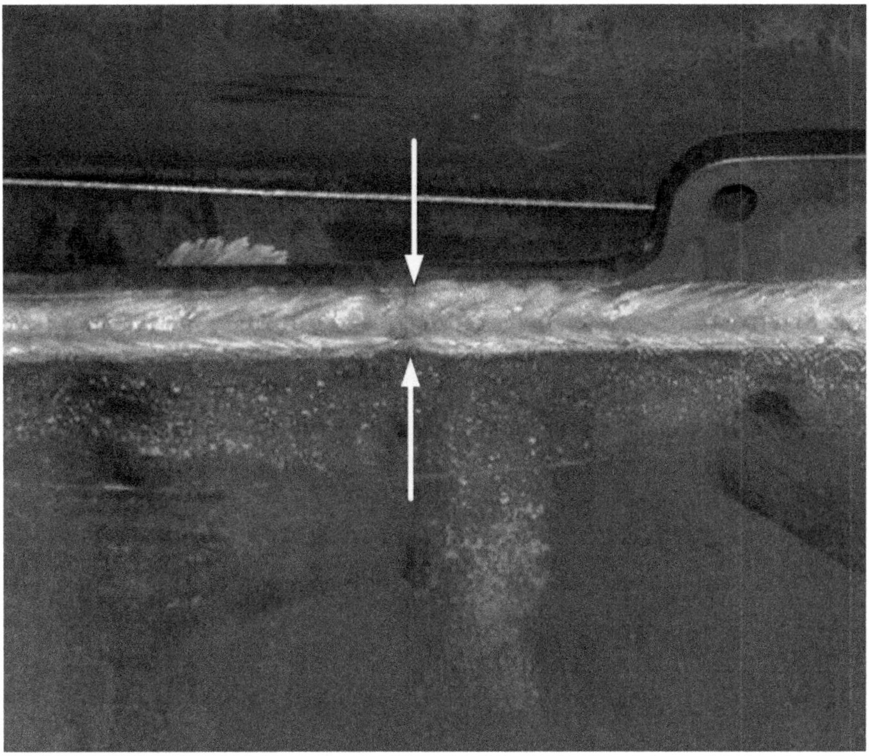

Fig. 16.7 Effect sketch of weld image defect

16.5 Results and Discussion

16.5.1 Comparison of Image Smoothing Technologies Based on Deep Learning

To evaluate the performance of the algorithm based on DNN, a network is trained separately for each filter's single parameter value, and the performance of the network is evaluated with PSNR and SSIM as the indicators. Figures 16.8 and 16.9 show the results. The RTV algorithm based on deep learning achieves better results in PSNR and SSIM indicators, which indicates that the visual results of RTV algorithm under the framework of deep learning can present better visual effect whether in a single parameter value mode or a random parameter value mode.

For each image filter, the deep learning algorithm only needs a joint training network. Even if different filters are used for different image processing and their implementation details are different, the deep learning algorithm can still learn all the filters, thus verifying the stability of the algorithm.

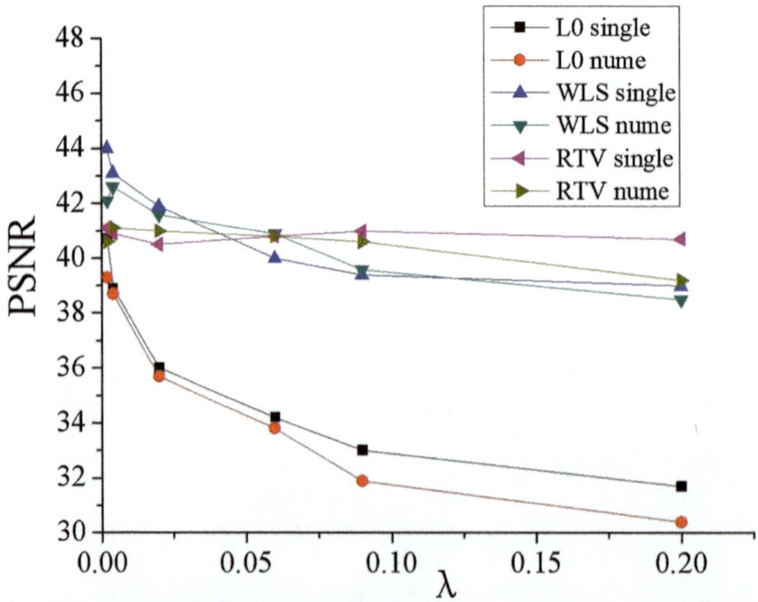

Fig. 16.8 PSNR of different algorithms based on deep learning

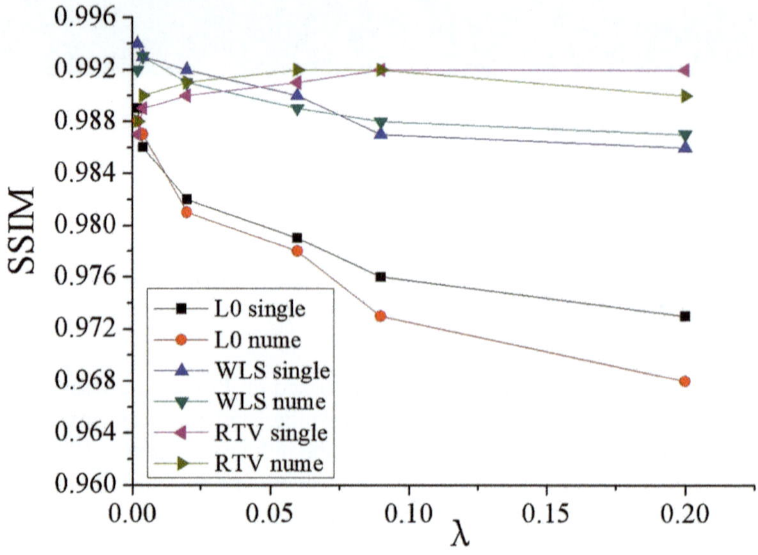

Fig. 16.9 SSIM of different algorithms based on deep learning

Figures 16.10 and 16.11 show the visual results based on the deep learning algorithm. The analysis suggests that the single network trained by the algorithm for continuous random parameters can predict high-quality images with different

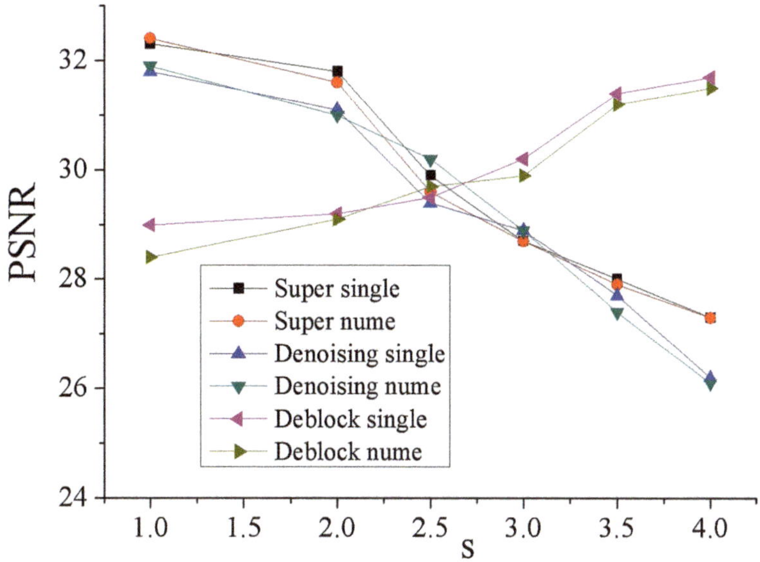

Fig. 16.10 PSNR of image restoration trained by a single parameter value and a random parameter value

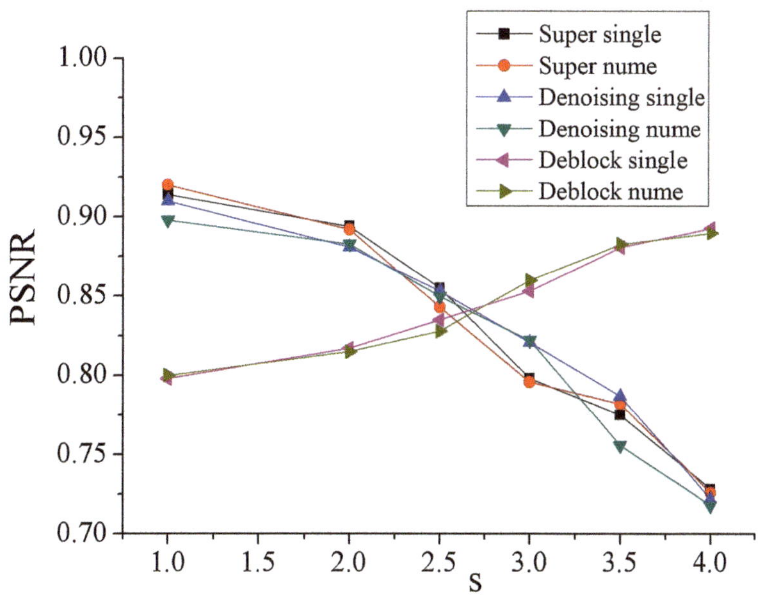

Fig. 16.11 SSIM of image restoration trained by a single parameter value and a random parameter value

smoothing intensities. Taking the clear image as the true value and the damaged image as the input, the effectiveness of the deep learning algorithm in more extensive image processing tasks is further verified.

16.5.2 Classification Effect of Image Processing on Welding Defects in Industrial Production

Figures 16.12, 16.13, and 16.14 show the comparison of the classification accuracy of the three data sets by the five features. Under any balanced data method, the deep features extracted by SSAE, DCNN1, and DCNN2 deep networks show better classification ability. It is proved that the image processing method based on deep learning proposed is effective for the classification of welding defects in industrial production.

In SMOTE data set, DCNN2 has the strongest classification ability, and the classification accuracy can reach 97.5%. However, in RUS data set, the classification accuracy of DCNN1 and DCNN2 is 75–80%, which is significantly lower than that of SSAE (87.9%). This shows that the depth of DCNN will affect the classification ability of features, but DCNN proposed is not suitable for small sample training data set. In addition, the classification accuracy of different types of defects is obviously different. Among them, the pore defect type is easier to identify than other defect types, and the crack defect image is most difficult to identify. Furthermore, the

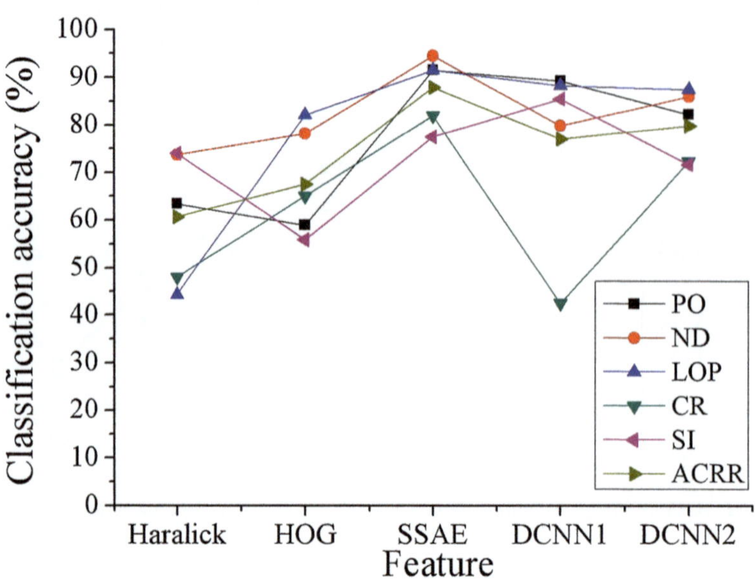

Fig. 16.12 Comparison of classification accuracy on RUS data set

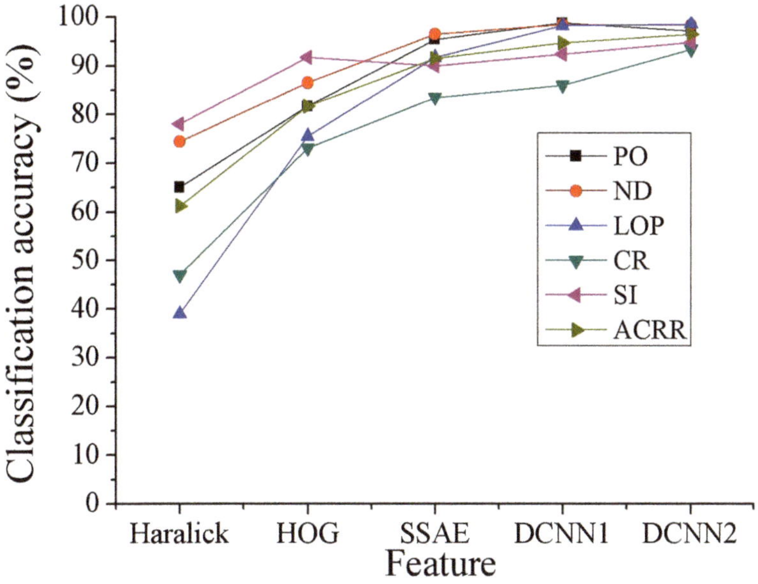

Fig. 16.13 Comparison of classification accuracy on ROS data set

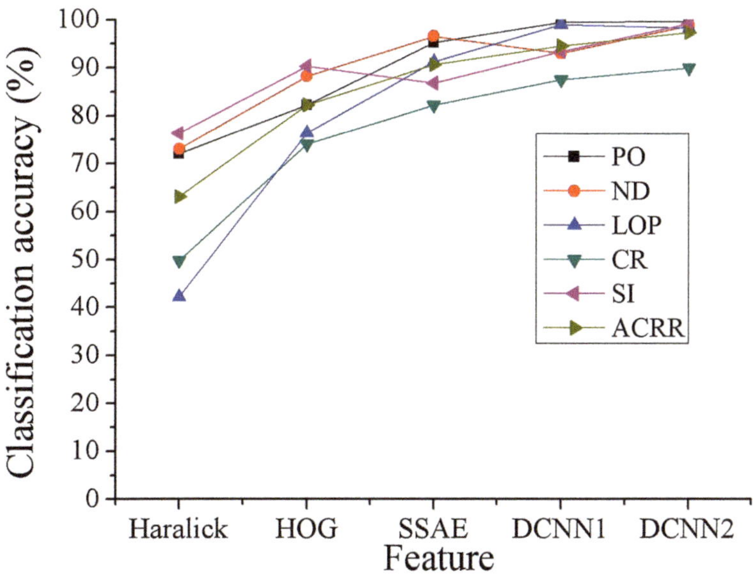

Fig. 16.14 Comparison of classification accuracy on SMOTE data set

classification accuracy of AlexNet, VGG16 model, and DCNN model proposed is compared on three kinds of data sets, and Figs. 16.15, 16.16, and 16.17 show the results.

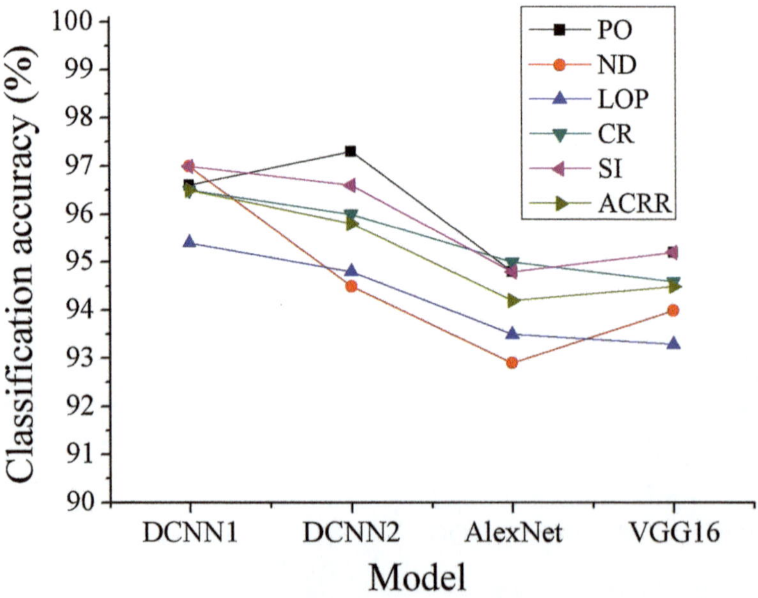

Fig. 16.15 Classification accuracy of different models in RUS data set

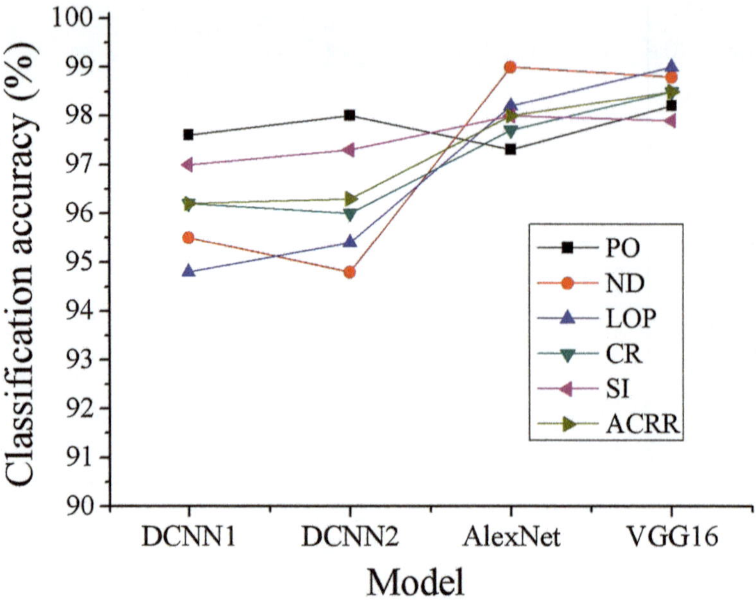

Fig. 16.16 Classification accuracy of different models in ROS data set

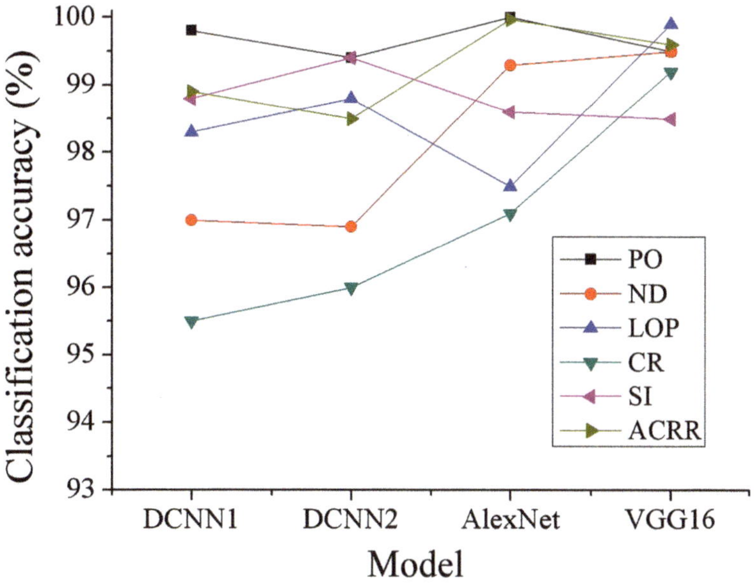

Fig. 16.17 Classification accuracy of different models in SMOTE data set

Figures 16.15, 16.16, and 16.17 suggest that the migration learning of AlexNet and VGG16 models is also feasible in the classification of industrial weld defect. The DCNN model proposed is simple in structure and fast in operation, while the AlexNet and VGG16 models have better training effect on large data sets. In general, migration learning can also achieve good results in small sample data sets. DCNN trained by image pre-training can be applied to weld image defect classification, which is suitable for processing modern industrial images.

Transfer learning uses the existing knowledge to solve the existing problems. Both VGG16 and AlexNet models are trained on big data sets, so it is not necessary to use too many weld image block samples for training in theory. In the case of missing tag data, the model pre-trained by natural image can be used to learn the features of weld image. It shows that although the contents of weld image and natural image are different, there are similar expressions in the underlying features.

16.6 Conclusions

With the rise of modern industry, IIoT further optimizes the industrial production process and production efficiency and uses intelligent terminals with perception ability and mobile communication networks and other technologies in all aspects of industrial production and management. Image processing based on machine vision can achieve higher level consciousness decision-making through higher level image recognition, which is very essential for the inspection in the industrial production process. The image processing technology based on deep learning is explored to improve the level of industrial management through image processing technology in the process of production and processing. The present work designs and implements a dynamic computing framework based on SDN. Open Switch is used as virtual switch and Open Day Light as controller, and an SDN environment that can be deployed in the factory environment is built.

The image processing system of IIoT machine vision application is explored, the application of deep learning in IIoT image processing is analyzed, and the deep learning algorithm is evaluated by simulation experiment. It is found that the deep learning algorithm is effective in a wider range of image processing tasks, and the image processing method based on the deep learning proposed is effective in the image classification of industrial production welding defects. However, in the process of image processing based on deep learning, the parameters of each layer of convolution network need to be extracted and learned separately. Learning such a large number of convolution parameters is also a great burden for network training. Therefore, a method needs to be explored in the future, which can reduce the network storage space and speed up the network training by learning the convolution parameters of one layer alone. This study provides a convenient and powerful guarantee for the effective supervision of equipment and processing technology in industrial production. However, only the industrial welding process is focused on, and the other common processing methods such as cutting and smelting need further deep analysis, which will be the main research direction in the future.

References

1. Zeng, L., Li, E., Zhou, Z., et al. (2019). Boomerang: On-demand cooperative deep neural network inference for edge intelligence on the industrial Internet of Things. *IEEE Network, 33*(5), 96–103.
2. Molanes, R. F., Amarasinghe, K., Rodriguez-Andina, J., et al. (2018). Deep learning and reconfigurable platforms in the Internet of Things: Challenges and opportunities in algorithms and hardware. *IEEE Industrial Electronics Magazine, 12*(2), 36–49.
3. Liu, M., Yu, F. R., Teng, Y., et al. (2019). Performance optimization for blockchain-enabled industrial Internet of Things (IIoT) systems: A deep reinforcement learning approach. *IEEE Transactions on Industrial Informatics, 15*(6), 3559–3570.
4. He, Y., Guo, J., & Zheng, X. (2018). From surveillance to digital twin: Challenges and recent advances of signal processing for industrial Internet of Things. *IEEE Signal Processing Magazine, 35*(5), 120–129.
5. Li, L., Ota, K., & Dong, M. (2018). Deep learning for smart industry: Efficient manufacture inspection system with fog computing. *IEEE Transactions on Industrial Informatics, 14*(10), 4665–4673.
6. Liu, X., Sun, C., Kang, K., et al. (2016). Joint 3-D image quality assessment metric by using image view and depth information over the networking in IoT. *IEEE Systems Journal, 10*(3), 1203–1213.
7. Lyu, L., Bezdek, J. C., He, X., et al. (2019). Fog-embedded deep learning for the Internet of Things. *IEEE Transactions on Industrial Informatics, 15*(7), 4206–4215.

8. Yan, Q., Huang, W., Luo, X., et al. (2018). A multi-level DDoS mitigation framework for the industrial Internet of Things. *IEEE Communications Magazine, 56*(2), 30–36.
9. Li, P., Chen, Z., Yang, L. T., et al. (2018). An incremental deep convolutional computation model for feature learning on industrial big data. *IEEE Transactions on Industrial Informatics, 15*(3), 1341–1349.
10. Yan, H., Wan, J., Zhang, C., et al. (2018). Industrial big data analytics for prediction of remaining useful life based on deep learning. *IEEE Access, 6*, 17190–17197.
11. Zhang, W., Guo, W., Liu, X., et al. (2018). LSTM-based analysis of industrial IoT equipment. *IEEE Access, 6*, 23551–23560.
12. Sisinni, E., Saifullah, A., Han, S., et al. (2018). Industrial Internet of Things: Challenges, opportunities, and directions. *IEEE Transactions on Industrial Informatics, 14*(11), 4724–4734.
13. He, L., Wang, G., & Hu, Z. (2018). Learning depth from single images with deep neural network embedding focal length. *IEEE Transactions on Image Processing, 27*(9), 4676–4689.
14. Artusi, A., Banterle, F., Carra, F., et al. (2019). Efficient evaluation of image quality via deep-learning approximation of perceptual metrics. *IEEE Transactions on Image Processing, 29*, 1843–1855.
15. Gonzalez-Manzano, L., Fuentes, J. M. D., & Ribagorda, A. (2019). Leveraging user-related Internet of Things for continuous authentication: A survey. *ACM Computing Surveys (CSUR), 52*(3), 1–38.
16. Zhang, S., Yao, L., Sun, A., et al. (2019). Deep learning based recommender system: A survey and new perspectives. *ACM Computing Surveys (CSUR), 52*(1), 1–38.
17. Ioannidou, A., Chatzilari, E., Nikolopoulos, S., et al. (2017). Deep learning advances in computer vision with 3d data: A survey. *ACM Computing Surveys (CSUR), 50*(2), 1–38.
18. Sha, L. T., Xiao, F., Huang, H. P., et al. (2019). Catching escapers: A detection method for advanced persistent escapers in industry Internet of Things based on Identity-based Broadcast Encryption (IBBE). *ACM Transactions on Embedded Computing Systems (TECS), 18*(3), 1–25.
19. Li, B., Qin, Y., Yuan, B., et al. (2019). Neural network classifiers using a hardware-based approximate activation function with a hybrid stochastic multiplier. *ACM Journal on Emerging Technologies in Computing Systems (JETC), 15*(1), 1–21.
20. Ludwig, T., Boden, A., & Pipek, V. (2017). 3D printers as sociable technologies: Taking appropriation infrastructures to the Internet of Things. *ACM Transactions on Computer-Human Interaction (TOCHI), 24*(2), 1–28.
21. Stolpe, M. (2016). The Internet of Things: Opportunities and challenges for distributed data analysis. *ACM SIGKDD Explorations Newsletter, 18*(1), 15–34.
22. Yan, Q., Huang, W., Luo, X., et al. (2018). A multi-level DDoS mitigation framework for the industrial Internet of Things. *IEEE Communications Magazine, 56*(2), 30–36.
23. Koroniotis, N., Moustafa, N., & Sitnikova, E. (2019). Forensics and deep learning mechanisms for botnets in Internet of Things: A survey of challenges and solutions. *IEEE Access, 7*, 61764–61785.
24. Aazam, M., Harras, K. A., & Zeadally S. (2019). Fog computing for 5G tactile industrial Internet of Things: QoE-aware resource allocation model. *IEEE Transactions on Industrial Informatics, 15*(5), 3085–3092.
25. Azmoodeh, A., Dehghantanha, A., & Choo, K. K. R. (2018). Robust malware detection for internet of (battlefield) things devices using deep eigenspace learning. *IEEE Transactions on Sustainable Computing, 4*(1), 88–95.
26. Zhu, Q., Chen, Z., & Soh, Y. C. (2018). A novel semisupervised deep learning method for human activity recognition. *IEEE Transactions on Industrial Informatics, 15*(7), 3821–3830.
27. Choo, K. K. R., Gritzalis, S., & Park, J. H. (2018). Cryptographic solutions for industrial Internet-of-Things: Research challenges and opportunities. *IEEE Transactions on Industrial Informatics, 14*(8), 3567–3569.
28. Jindal, A., Aujla, G. S., Kumar, N., et al. (2018). SeDaTiVe: SDN-enabled deep learning architecture for network traffic control in vehicular cyber-physical systems. *IEEE Network, 32*(6), 66–73.

29. Li, G., Wu, J., Li, J., et al. (2018). Service popularity-based smart resources partitioning for fog computing-enabled industrial Internet of Things. *IEEE Transactions on Industrial Informatics, 14*(10), 4702–4711.
30. Yang, X., Wang, H., Liu, K., et al. (2019). Minimax and WLS designs of digital FIR filters using SOCP for aliasing errors reduction in BI-DAC. *IEEE Access, 7*, 11722–11735.
31. Zhu, J., Ge, Z., & Song, Z. (2017). Distributed parallel PCA for modeling and monitoring of large-scale plant-wide processes with big data. *IEEE Transactions on Industrial Informatics, 13*(4), 1877–1885.

Index

G. S. Aujla et al. (eds.), *Software Defined Internet of Everything*, Technology, Communications and Computing, https://doi.org/10.1007/978-3-030-89328-6